Collins

GCSE 9-1 Biology in a week

Tom Adams

Revision Planner

DAY 1

Page	Time	Topic	Date	Time	Completed
4	30 minutes	Structure of cells			
6	35 minutes	Organisation and differentiation			
8	40 minutes	Microscopy and microorganisms			
10	40 minutes	Cell division			
12	45 minutes	Metabolism – respiration			
14	40 minutes	Metabolism – enzymes			

DAY 2

Page	Time	Topic	Date	Time	Completed
16	35 minutes	Cell transport			
18	25 minutes	Plant tissues, organs and systems			
20	40 minutes	Transport in plants			
22	40 minutes	Transport in humans 1			
24	40 minutes	Transport in humans 2			
26	40 minutes	Photosynthesis			

DAY 3

Page	Time	Topic	Date	Time	Completed
28	45 minutes	Non-communicable diseases			
30	30 minutes	Communicable diseases			
32	40 minutes	Human defences			
34	40 minutes	Fighting disease			
36	45 minutes	Discovery and development of drugs			
38	30 minutes	Plant diseases			

DAY 4

Page	Time	Topic	Date	Time	Completed
40	25 minutes	Homeostasis and negative feedback			
42	35 minutes	The nervous system			
44	40 minutes	Nervous control			
46	45 minutes	The eye			
48	35 minutes	The endocrine system			
50	45 minutes	Water and nitrogen balance			

Note: You will see the following logos throughout this book:

HT – indicates content that is Higher Tier only.

WS – indicates 'Working Scientifically' content, which covers practical skills and data-related concepts.

Structure of cells

Prokaryotes

Prokaryotes are simple cells such as bacteria and archaebacteria. Archaebacteria are ancient bacteria with different types of cell wall and genetic code to other bacteria. They include microorganisms that can use methane or sulfur as a means of nutrition (methanogens and thermoacidophiles).

Prokaryotic cells:

- are smaller than eukaryotic cells
- have cytoplasm, a cell membrane and a cell wall
- have genetic material in the form of a DNA loop, together with rings of DNA called plasmids
- do not have a nucleus.

The chart below shows the relative sizes of prokaryotic and eukaryotic cells, together with the magnification range of different viewing devices.

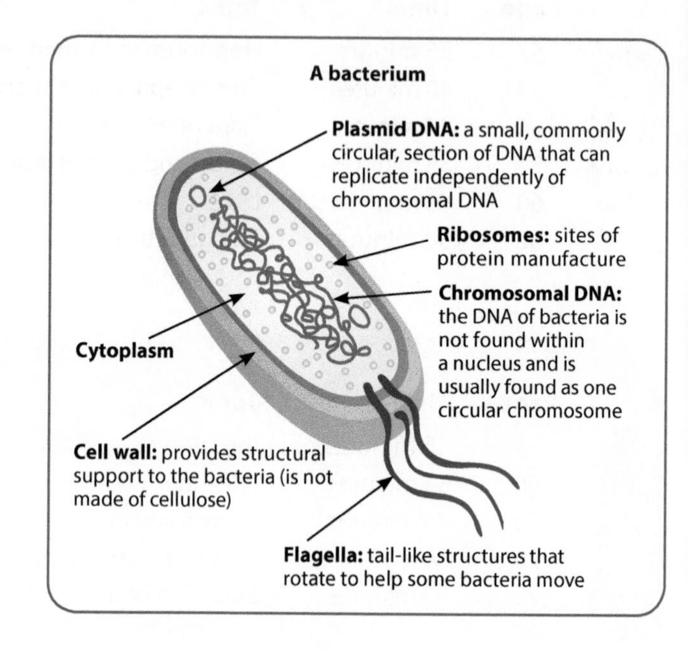

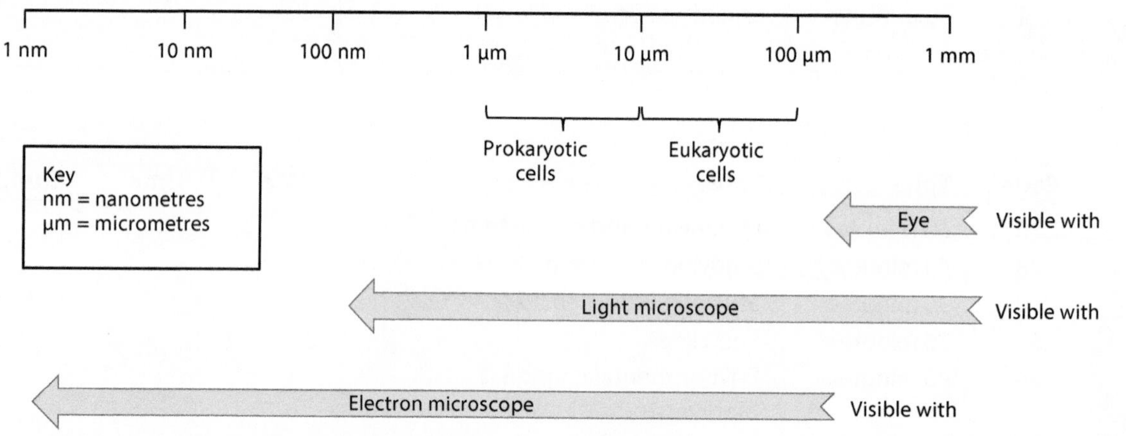

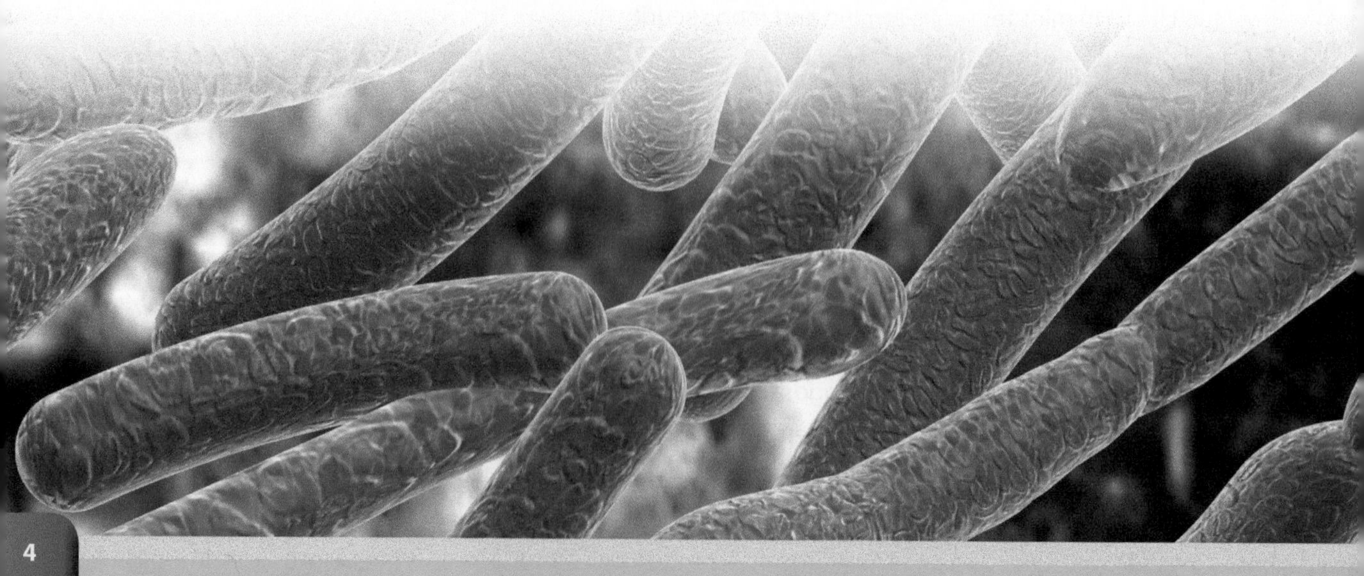

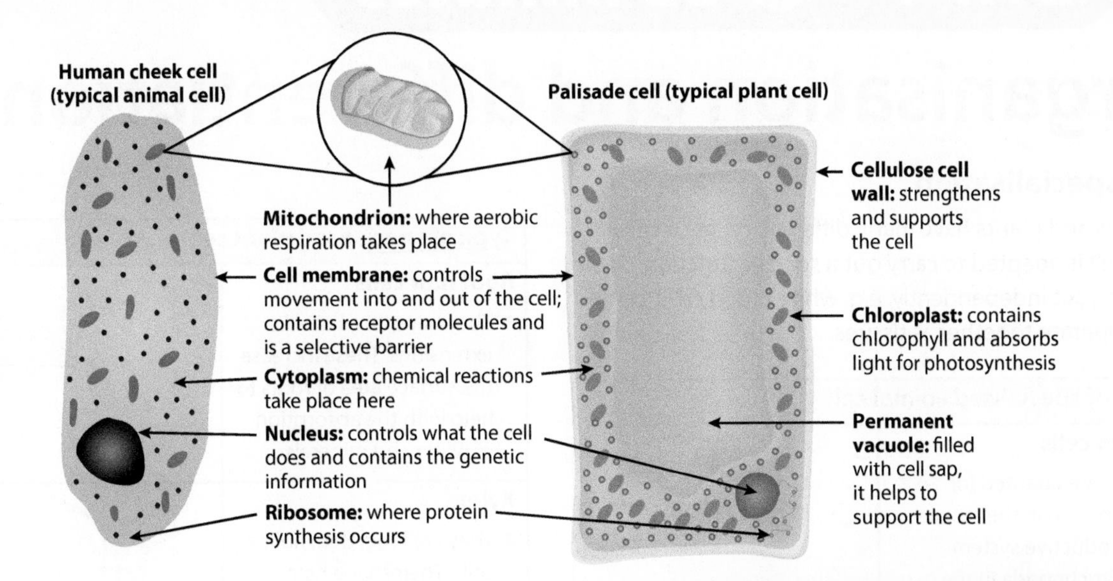

Eukaryotes

Eukaryotic cells:

- are more complex than prokaryotic cells (they have a cell membrane, cytoplasm and genetic material enclosed in a nucleus)
- are found in animals, plants, fungi (e.g. toadstools, yeasts, moulds) and protists (e.g. amoeba)
- contain membrane-bound structures called **organelles**, where specific functions are carried out.

Pictured above are the main organelles in plant and animal cells.

Plant cells tend to be more regular in shape than animal cells. They have additional structures: cell wall, sap vacuole and sometimes chloroplasts.

SUMMARY

- **Cells are the basis of life. All processes of life take place within them.**
- **There are many different kinds of specialised cell but they all have several common features.**
- **The two main categories of cells are prokaryotic (prokaryotes) and eukaryotic (eukaryotes).**

QUESTIONS

QUICK TEST

1. Prokaryotic cells are more complex than eukaryotic cells. True or false?
2. All cells have a nucleus. True or false?
3. Which organelle carries out the function of protein synthesis?

EXAM PRACTICE

1. a) Bacteria contain circular sections of DNA within their cells.
   Name these structures. **[1 mark]**

   b) Plant cells differ to animal cells in which one of the following ways? Tick **one** box. **[1 mark]**

   They contain a nucleus. ☐
   They do not have a cell membrane. ☐
   They possess a cell wall. ☐
   They contain ribosomes. ☐

   c) Muscle cells contain many mitochondria. Explain why this is. **[2 marks]**

   d) The average magnified length of a bacterium is 10mm.

   Using a magnification of 500, calculate its **actual** length in **μm**
   (1 μm is 1×10^3 mm).
   Show your working. **[2 marks]**

Organisation and differentiation

Cell specialisation

Animals and plants have many different types of cells. Each cell is adapted to carry out a specific function. Some cells can act independently, e.g. white blood cells, but most operate together as tissues.

Type of specialised animal cell	
Sperm cells ● They are adapted for swimming in the female reproductive system – mitochondria in the neck release energy for swimming. ● They are adapted for carrying out fertilisation with an egg cell – the acrosome contains enzymes for digestion of the ovum's outer protective cells at fertilisation.	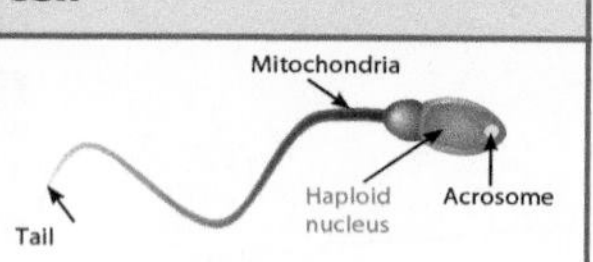
Egg cells (ova) ● They are very large in order to carry food reserves for the developing embryo. ● After fertilisation, the cell membrane changes and locks out other sperm.	Cell membrane Cytoplasm Mitochondria Haploid nucleus
Ciliated epithelial cells ● They line the respiratory passages and help protect the lungs against dust and microorganisms.	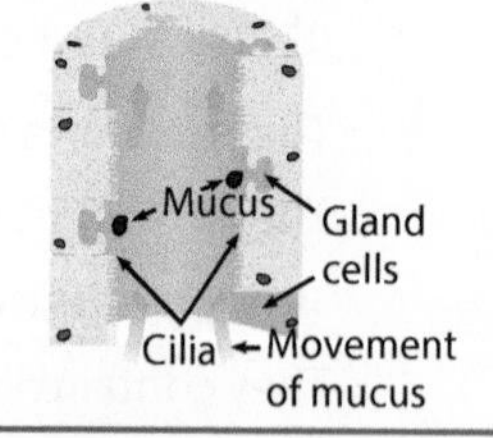
Nerve cells ● They have long, slender extensions called axons that carry nerve impulses.	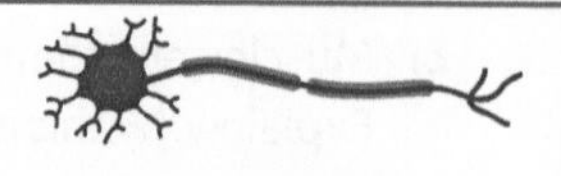
Muscle cells ● They are able to contract (shorten) to bring about the movement of limbs.	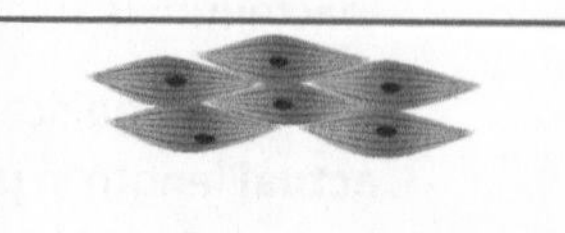

Type of specialised plant cell	
Root hair cells ● They have tiny, hair-like extensions. These increase the surface area of roots to help with the absorption of water and minerals.	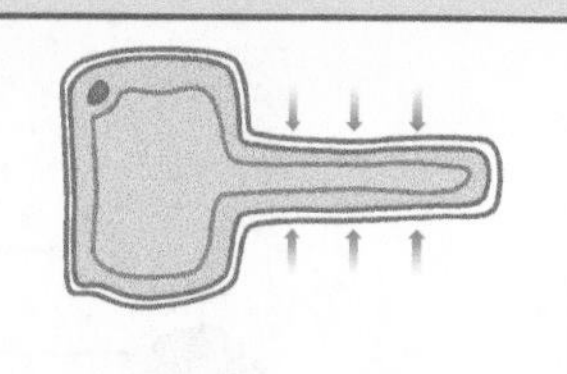
Xylem ● They are long, thin, hollow cells. Their shape helps with the transport of water through the stem, roots and leaves.	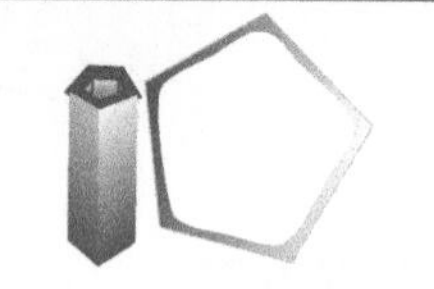
Phloem ● They are long, thin cells with pores in the end walls. Their structure helps the cell sap move from one phloem cell to the next.	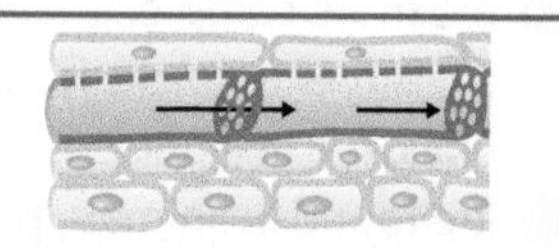

Principles of organisation

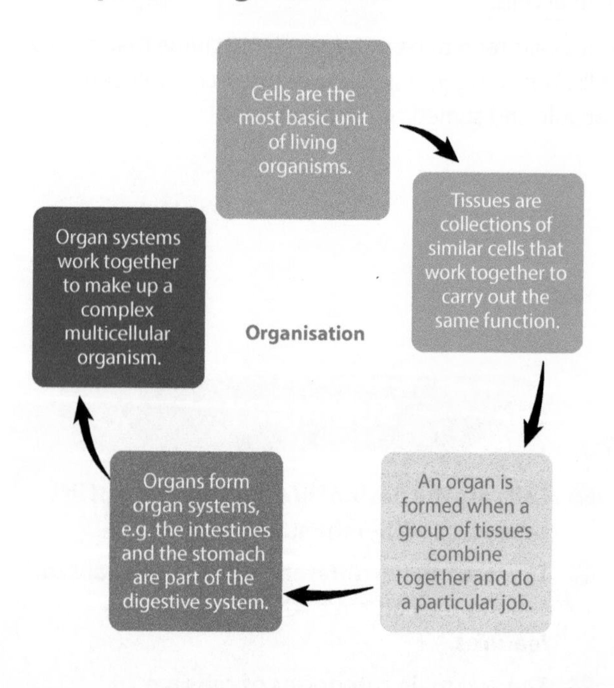

Cell differentiation and stem cells

Stem cells are found in animals and plants. They are unspecialised or **undifferentiated**, which means that they have the potential to become almost any kind of cell. Once the cell is fully specialised, it will possess sub-cellular organelles specific to the function of that cell. Embryonic stem cells, compared with those found in adults' bone marrow, are more flexible in terms of what they can become.

Animal cells are mainly restricted to repair and replacement in later life. Plants retain their ability to **differentiate** (specialise) throughout life.

Using stem cells in humans

Therapeutic cloning treats conditions such as diabetes. Healthy pancreas cells can be cloned in order to achieve this. Embryonic stem cells (that can specialise into any type of cell) are produced with the same genes as the patient. If these are introduced into the body, they are not usually rejected.

Treating paralysis is possible using stem cells that are capable of differentiating into new nerve cells. The brain uses these new cells to transmit nervous impulses to the patient's muscles.

There are benefits and objections to using stem cells.

Benefits	Risks and objections
• Stem cells left over from in vitro fertilisation (IVF) treatment (that would otherwise be destroyed) can be used to treat serious conditions. • Stem cells are useful in studying how cell division goes wrong, e.g. cancer. • In the future, stem cells could be used to grow new organs for transplantation.	• Some people believe that an embryo at any age is a human being and so should not be used to grow cells or be experimented on. • One potential risk of using stem cells is transferring viral infections. • If stem cells are used in an operation, they might act as a reservoir of cancer cells that spread to other parts of the body.

Using stem cells in plants

Plants have regions of rapid cell division called **meristems**. These growth regions contain stem cells that can be used to produce clones cheaply and quickly.

Meristems can be used for:

- growing and preserving rare varieties to protect them from extinction
- producing large numbers of disease-resistant crop plants.

SUMMARY

- **Multicellular organisms need to have a coordinated system of structures so they can carry out vital processes, e.g. respiration, excretion, nutrition, etc.**
- **Differentiation is the process whereby cells become specialised.**
- **Stem cells have the potential to become almost any kind of cell.**
- **There are benefits, risks and objections to the use of stem cells.**

QUESTIONS

QUICK TEST

1. Which type of specialised animal cell helps protect the lungs against dust and microorganisms?
2. Which type of specialised plant cell is used to transport water through the stem, roots and leaves?
3. Name the regions of rapid cell division in plants.

EXAM PRACTICE

1. Name the process by which cells become specialised to carry out a particular function. **[1 mark]**
2. A fertilised egg cell is called a zygote. It is a type of stem cell.
   a) The zygote will divide and then produce specialised cells. Some of these cells will become neurones. Describe **one** adaptation of a neurone. **[1 mark]**
   b) Describe **two** applications of stem cell research. **[2 marks]**
   c) Explain why some people have objections to stem cell research. **[1 mark]**

DAY 1  40 Minutes

Microscopy and microorganisms

Microscopes

Microscopes:

- observe objects that are too small to see with the naked eye
- are useful for showing detail at cellular and sub-cellular level.

There are two main types of microscope: the **light microscope** and the **electron microscope**.

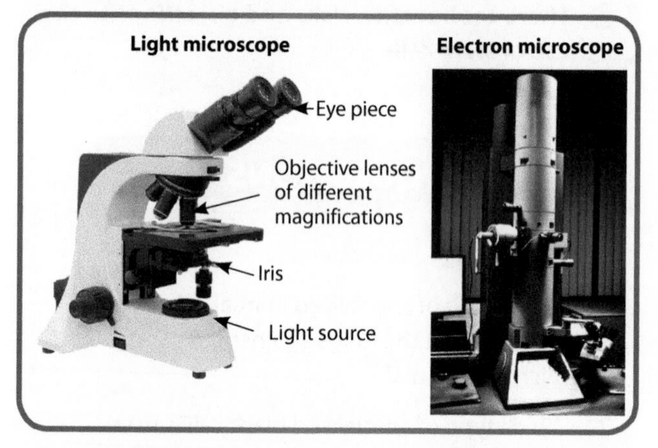

You will probably use a light microscope in your school.

The electron microscope was invented in 1931. It has increased our understanding of sub-cellular structures because it has much higher magnifications and resolution than a light microscope.

These are white blood cells, as seen through a transmission electron microscope. The black structures are nuclei.

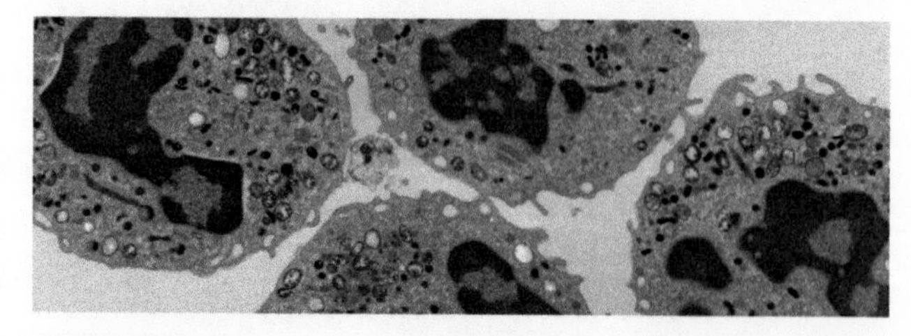

Comparing light and electron microscopes

Light microscope	Electron microscope
Uses light waves to produce images	Uses electrons to produce images
Low resolution	High resolution
Magnification up to ×1500	Magnification up to ×500 000 (2D) and ×100 000 (3D)
Able to observe cells and larger organelles	Able to observe small organelles
2D images only	2D and 3D images produced

HT Here are some common units used in microscopy.

Measure	Scale	Symbol
1 metre		m
1 millimetre	$\frac{1}{1\,000}$ (a thousandth) of a metre ($\times 10^{-3}$)	mm
1 micrometre	$\frac{1}{1\,000\,000}$ (a millionth) of a metre ($\times 10^{-6}$)	µm
1 nanometre	$\frac{1}{1\,000\,000\,000}$ (a thousand millionth or a billionth) of a metre ($\times 10^{-9}$)	nm
1 picometre	$\frac{1}{1\,000\,000\,000\,000}$ (a million millionth or a trillionth) of a metre ($\times 10^{-12}$)	pm

Magnification and resolution

Magnification measures how many times an object under a microscope has been made larger.

You may be asked to carry out calculations using magnification. To calculate the magnifying power of a microscope, use this formula:

$$\textbf{magnification} = \frac{\textbf{Size of image}}{\textbf{Size of real object}}$$

Example

A light microscope produces an image of a cell which has a diameter of 1500 µm. The cell's actual diameter is 50 µm. Calculate the magnifying power of the microscope.

$$\text{magnification} = \frac{1500}{50} = \times 30$$

Resolution is the smallest distance between two points on a specimen that can still be told apart.

The diagram below shows the limits of resolution for a light microscope. When looking through it you can see fine detail down to 200 nanometres. Electron microscopes can see detail down to 0.05 nanometres!

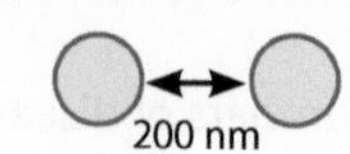

Staining techniques are used in light and electron microscopy to make organelles more visible.

- **Methylene blue** is used to stain the nuclei of animal cells for viewing under the light microscope.
- **Heavy metals** such as cadmium can be used to stain specimens for viewing under the electron microscope.

Culturing microorganisms

To view microorganisms under the microscope, pure, uncontaminated samples are grown or **cultured** in the laboratory. These cultures can be used for research, e.g. investigating the action of disinfectants or antibiotics.

In order to grow, microorganisms need:

- **food** in the form of a growth medium; in the laboratory, this is usually nutrient agar or broth
- **warmth** – a temperature of 25°C is used in school laboratories, as it lowers the possibility of **pathogens** growing
- **moisture**.

Aseptic technique

Aseptic technique is needed to culture microorganisms in the laboratory so that uncontaminated cultures are produced, which reduces the risk of scientists being infected by pathogens.

Bacterial growth

Under ideal conditions, a bacterial culture will grow quickly over a period of days by **binary fission**. In this process, a cell divides into two in as little time as twenty minutes!

The bacteria continue to divide in this way: 1, 2, 4, 8, 16, 32, 64, etc. So after a day, the numbers are in their millions. As they grow, the individual cells merge to produce circular **colonies** that are visible to the naked eye. You can count them or calculate their area to estimate the growth rate.

SUMMARY

- **There are two main types of microscope: light microscope and electron microscope.**
- **Magnification measures how many times an object under a microscope has been enlarged.**
- **To view microorganisms under a microscope, samples are cultivated using aseptic technique.**

HT Calculating the area of a bacterial colony

Bacterial colonies grow **radially**, i.e. in circles. To calculate a rate of growth, the area of the colony can be worked out at two points in time.

Example: A colony of *M. luteus* is observed growing on agar in a petri dish. The diameter is measured with a ruler and is 20 mm. Exactly 24 hours later, the colony's diameter is 30 mm. What is its growth rate?

- Area of colony for first measurement = πr^2 where $\pi = 3.14$ and $r = ½ \times 20$
  Area = $3.14 \times 10^2 = 314\,mm^2$
- Area of colony for second measurement = $3.14 \times 15^2 = 706.5\,mm^2$
- Rate of growth = $\frac{706.5 - 314}{24}$
  = $\mathbf{16.35\,mm^2 h^{-1}}$

QUESTIONS

QUICK TEST

1. Name the two types of microscope.
2. What three factors do microorganisms need in order to grow?

EXAM PRACTICE

1. Some bacterial colonies are grown on an agar plate. The radius of eight colonies are measured. The results are shown in the table below.

Colony	1	2	3	4	5	6	7	8
Radius/mm	10	8	2	3	2	7	5	4

Using the formula πr^2, calculate the **mean** cross sectional area of the colonies.
Show your working. **[2 marks]**

2. a) A student wishes to see a specimen at a higher power magnification. Which part of the light microscope must she adjust to do this? **[1 mark]**

   b) Explain why the student will not be able to observe the finer detail of organelles such as mitochondria. **[2 marks]**

   c) The same student wishes to obtain a 3D micrograph of a skin specimen surface. Which type of microscope should they use and why? **[2 marks]**

Cell division

Chromosomes

Chromosomes are found in the nucleus of eukaryotic cells. They are made of DNA and carry a large number of genes. Chromosomes exist as pairs called **homologues**.

For cells to duplicate exactly, it is important that all of the genetic material is duplicated. Chromosomes take part in a sequence of events that ensures the genetic code is transmitted precisely and appears in the new **daughter cells**.

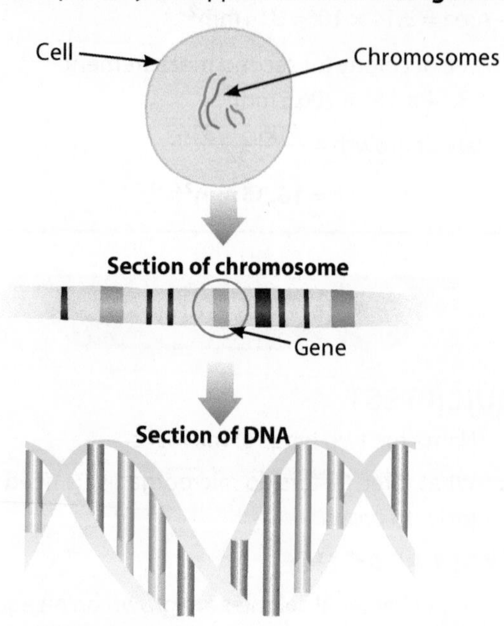

DNA replication

During the cell cycle, the genetic material (made of the **polymer** molecule, DNA) is doubled and then divided between the identical daughter cells. This process is called **DNA replication**.

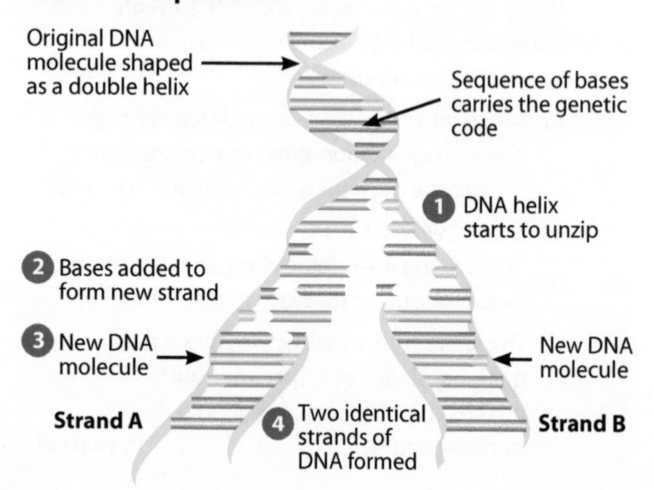

Mitosis

Mitosis is where a **diploid** cell (one that has a complete set of chromosomes) divides to produce two more diploid cells that are genetically identical. Most cells in the body are diploid.

Humans have a diploid number of 46.

Mitosis produces new cells:

- for growth
- to repair damaged tissue
- to replace old cells
- for asexual reproduction.

Before the cell divides, the DNA is duplicated and other organelles replicate, e.g. mitochondria and ribosomes. This ensures that there is an exact copy of all the cell's content.

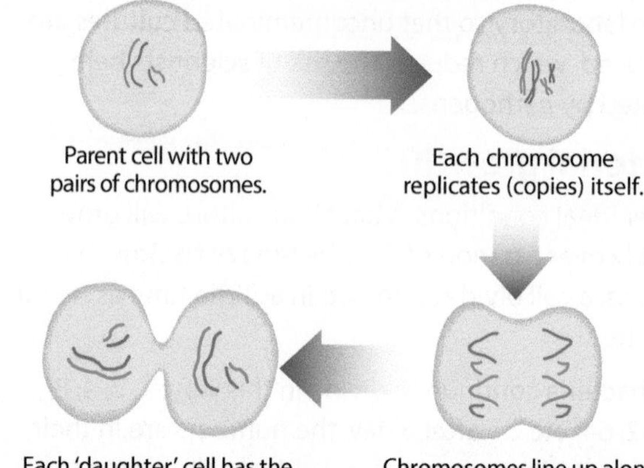

Meiosis

Meiosis takes place in the testes and ovaries of sexually reproducing organisms and produces gametes (eggs or sperm). The gametes are called **haploid** cells because they contain half the number of chromosomes as a diploid cell. This chromosome number is restored during **fertilisation**.

Humans have a haploid number of 23.

Meiosis – the cell divides twice to produce four cells with genetically different sets of chromosomes

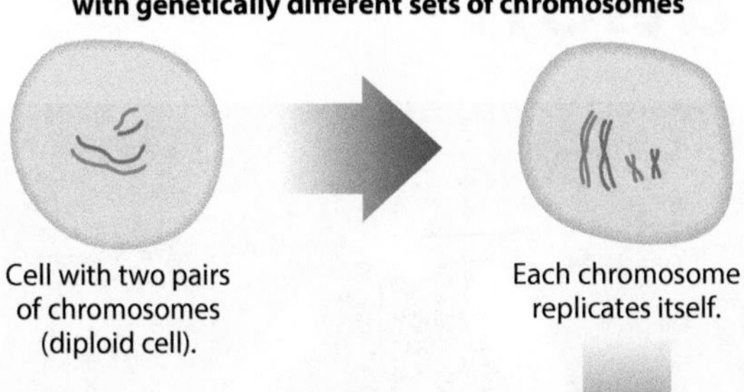

Cell with two pairs of chromosomes (diploid cell).

Each chromosome replicates itself.

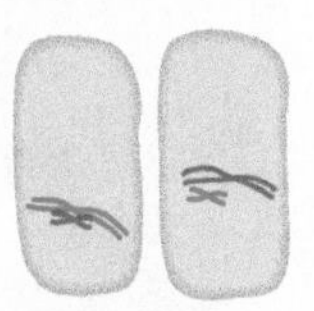

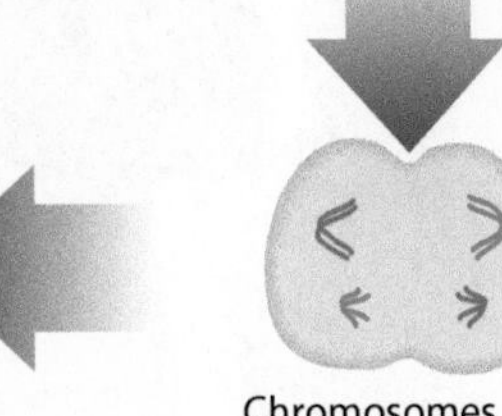

Cell divides for the first time.

Chromosomes part company and move to opposite poles.

Copies now separate and the second cell division takes place.

Four haploid cells (gametes), each with half the number of chromosomes of the parent cell.

Cancer

Cancer is a non-communicable disease caused by mutations in living cells.

Cancerous cells:

- divide in an uncontrolled way
- form tumours.

Benign tumours do not spread from the original site of cancer in the body. **Malignant tumour cells** invade neighbouring tissues. They spread to other parts of the body and form **secondary tumours**.

Making healthy lifestyle choices is one way to reduce the likelihood of cancer. These include:

- not smoking tobacco products (cigarettes, cigars, etc.)
- not drinking too much alcohol (causes cancer of the liver, gut and mouth)
- avoiding exposure to UV rays (e.g. sunbathing, tanning salons)
- eating a healthy diet (high fibre reduces the risk of bowel cancer) and doing moderate exercise to reduce the risk of obesity.

SUMMARY

- **Multicellular organisms grow and reproduce using cell division and cell enlargement.**
- **Plants can also grow via differentiation into leaves, branches, etc.**
- **Cell division begins from the moment of fertilisation when the zygote replicates itself exactly through mitosis. Later in life, an organism may use cell division to produce sex cells (gametes) in a different type of division called meiosis.**
- **The different stages of cell division make up the cell cycle of an organism.**

QUESTIONS

QUICK TEST

1. What name is given to sex cells?
2. Name the 46 structures in the human nucleus that carry genetic information.
3. How many chromosomes are in a human haploid cell?

EXAM PRACTICE

1. Amelia is looking at some cells through a microscope. The cells resulted from one original parent cell. The diagram below is a drawing of what she can see.

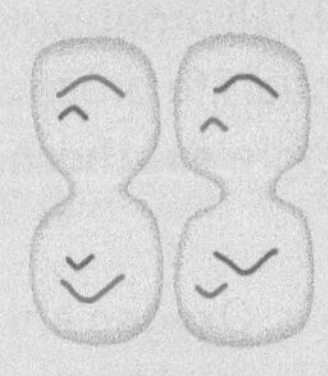

a) Which type of cell division is shown here? **[1 mark]**

b) Explain your answer to part a). **[2 marks]**

Metabolism – respiration

Metabolism

Metabolism is the sum of all the chemical reactions that take place in the body.

The two types of metabolic reaction are:

- building reactions (**anabolic**)
- breaking-down reactions (**catabolic**).

Anabolic Reactions

Anabolic reactions require the input of energy. Examples include:

- converting glucose to starch in plants, or glucose to glycogen in animals
- the synthesis of lipid molecules

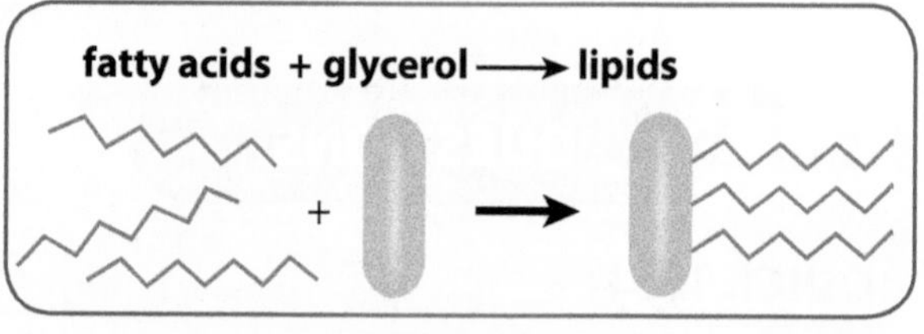

- the formation of **amino acids** in plants (from glucose and nitrate ions) which, in turn, are built up into proteins.

Catabolic reactions

Catabolic reactions release energy. Examples include:

- breaking down amino acids to form **urea**, which is then excreted
- respiration.

Catabolic reactions produce waste energy in the form of heat (an **exothermic** reaction), which is transferred to the environment.

Respiration

Respiration continuously takes place in all organisms – the need to release energy is an essential life process. The reaction gives out energy and is therefore **exothermic**.

Aerobic respiration

Aerobic respiration takes place in cells. Oxygen and glucose molecules react and release energy. This energy is stored in a molecule called **ATP**.

glucose + oxygen ⟶ carbon dioxide + water + energy (locked in ATP)

HT The symbol equation for aerobic respiration is:

$$C_6H_{12}O_6 + 6O_2 \longrightarrow 6CO_2 + 6H_2O + \text{energy released}$$

Energy is used in the body for many processes, including:

- muscle contraction (for movement)
- active transport
- transmitting nerve impulses
- synthesising new molecules
- maintaining a constant body temperature.

Anaerobic respiration

Anaerobic respiration takes place in the absence of oxygen and is common in muscle cells. It quickly releases a **smaller** amount of energy than aerobic respiration through the **incomplete breakdown** of glucose.

glucose ⟶ lactic acid + energy released

In plant and yeast cells, anaerobic respiration produces different products.

glucose ⟶ ethanol + carbon dioxide + energy released

HT The symbol equation for anaerobic respiration in plant and yeast cells is:

$$C_6H_{12}O_6 \longrightarrow 2C_2H_5OH + 2CO_2 + \text{energy released}$$

This reaction is used extensively in the brewing and wine-making industries. It is also the initial process in the manufacture of spirits in a distillery.

Response to exercise

In animals, anaerobic respiration takes place when muscles are working so hard that the lungs and circulatory system cannot deliver enough oxygen to break down all the available glucose through aerobic respiration. In these circumstances the **energy demand** of the muscles is high.

Anaerobic respiration and recovery

Anaerobic respiration releases energy much faster over short periods of time. It is useful when short, intense bursts of energy are required, e.g. a 100 m sprint.

However, the incomplete oxidation of glucose causes **lactic acid** to build up. Lactic acid is toxic and can cause pain, cramp and a sensation of fatigue.

The lactic acid must be broken down quickly and removed to avoid cell damage and prolonged muscle fatigue.

- During exercise the body's heart rate, breathing rate and breath volume increase so that sufficient oxygen and glucose is supplied to the muscles, and so that lactic acid can be removed.
- This continues after exercise when deep breathing or panting occurs until all the lactic acid is removed. This repayment of oxygen is called **oxygen debt**.

HT
- Lactic acid is transported to the liver where it is converted back to glucose.
- Oxygen debt is the amount of **extra** oxygen that the body needs after exercise to react with the lactic acid and remove it from the cells.

WS You may be asked to present data as graphs, tables, bar charts or histograms. For example, you could show data of breathing and heart rates on a line graph.

What does the line graph below tell you about breathing and pulse rates during recovery? Why is a line graph a good way to present the data?

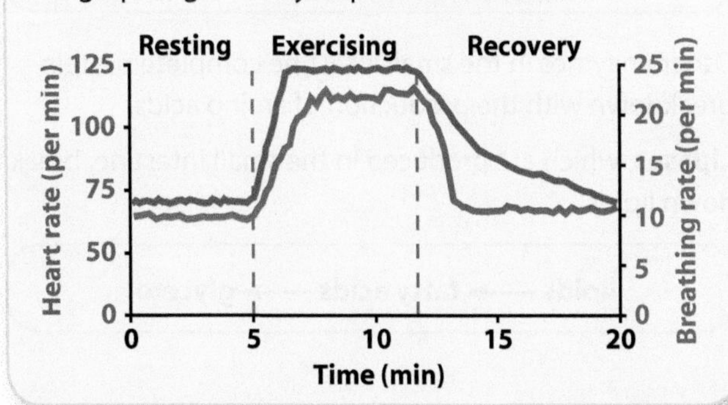

SUMMARY

- **Respiration takes place continuously in all organisms.**
- **Aerobic respiration is when oxygen and glucose molecules react and release energy.**
- **Anaerobic respiration takes place without oxygen, in muscle cells, and can cause lactic acid build up.**
- **Other organisms carrying out anaerobic respiration provide ethanol.**

QUESTIONS

QUICK TEST

1. Which type of respiration releases most energy – aerobic or anaerobic?
2. What is an exothermic reaction?
3. Give the product of anaerobic respiration in humans.
4. What type of reaction is respiration – anabolic or catabolic?

EXAM PRACTICE

1. The human muscle cell and a yeast cell can each carry out anaerobic respiration.

   Both produce a toxic waste product.

   a) Write down the word equations for both yeast anaerobic respiration and human anaerobic respiration. **[2 marks]**

   b) Compare and contrast the two types of respiration in terms of how they deal with their toxic waste products. **[3 marks]**

Metabolism – enzymes

Enzyme facts

Enzymes:

- are specific, i.e. one enzyme catalyses one reaction
- have an active site, which is formed by the precise folding of the enzyme molecule
- can be denatured by high temperatures and extreme changes in pH
- have an optimum temperature at which they work – for many enzymes this is approximately 37°C (body temperature)
- have an optimum pH at which they work – this varies with the site of enzyme activity, e.g. pepsin works in the stomach and has an optimum pH of 1.5 (acidic), salivary amylase works best at pH 7.3 (alkaline).

Enzyme activity

Enzyme molecules work by colliding with **substrate** molecules and forcing them to break up or to join with others in synthesis reactions. The theory of how this works is called the **lock and key theory**.

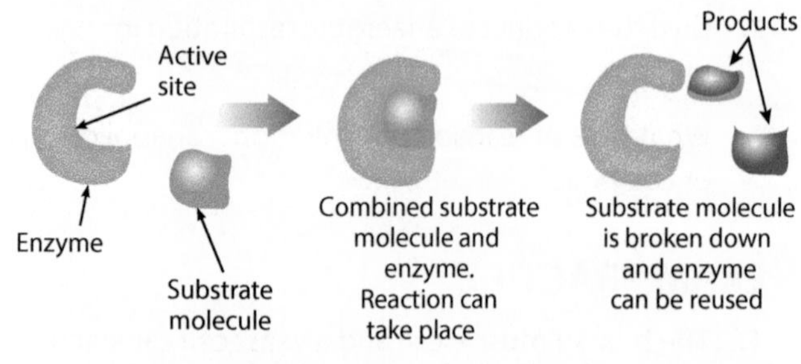

High temperatures denature enzymes because excessive heat vibrates the atoms in the protein molecule, putting a strain on bonds and breaking them. This changes the shape of the active site.

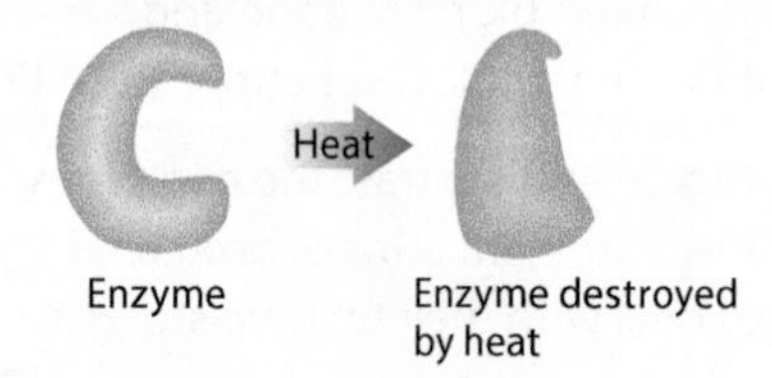

In a similar way, an extreme pH alters the active site's shape and prevents it from functioning.

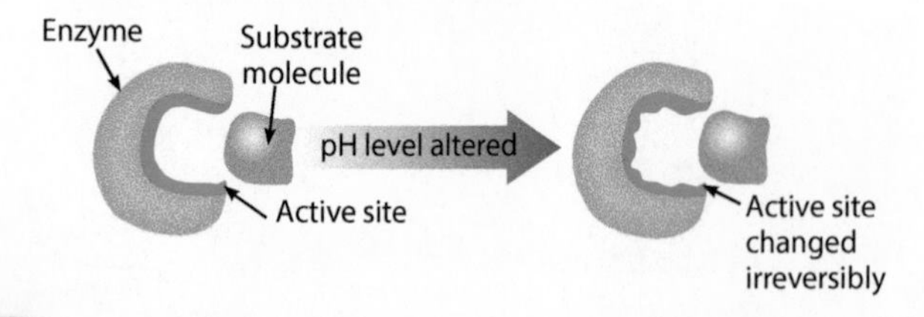

At lower than the optimal temperature, an enzyme still works but much more slowly. This is because the low **kinetic energy** of the substrate and enzyme molecules lowers the number of collisions that take place. When they do collide, the energy is not always sufficient to create a bond between them.

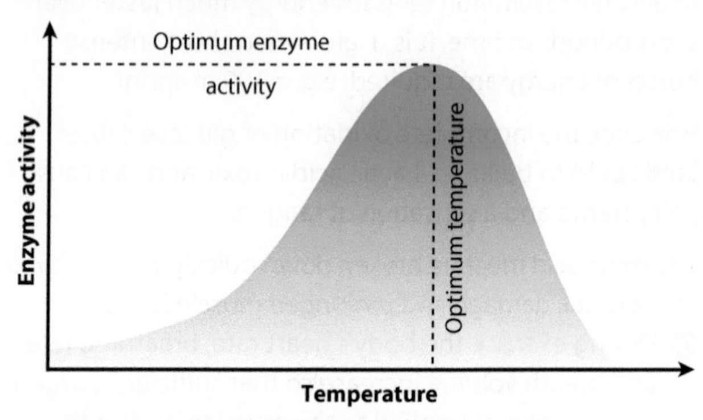

Enzymes in the digestive system

Enzymes in the digestive system help break down large nutrient molecules into smaller ones so they can be absorbed into the blood across the wall of the small intestine.

Enzyme types in the digestive system include carbohydrases, proteases and lipases.

Carbohydrases break down carbohydrates, e.g. **amylase**, which is produced in the mouth and small intestine.

starch ⟶ maltose

Other carbohydrases break down complex sugars into smaller sugars.

Proteases break down protein, e.g. the enzyme `pepsin, which is produced in the stomach.

protein ⟶ peptides ⟶ amino acids

Other enzymes in the small intestine complete protein breakdown with the production of amino acids.

Lipases, which are produced in the small intestine, break down lipids.

lipids ⟶ fatty acids ⟶ glycerol

Bile

Bile is a digestive chemical. It is produced in the liver and stored in the gall bladder.

- Its alkaline pH neutralises hydrochloric acid that has been produced in the stomach.
- It emulsifies fat, breaking it into small droplets with a large surface area.
- Its action enables lipase to break down fat more efficiently.

What happens to digested food?

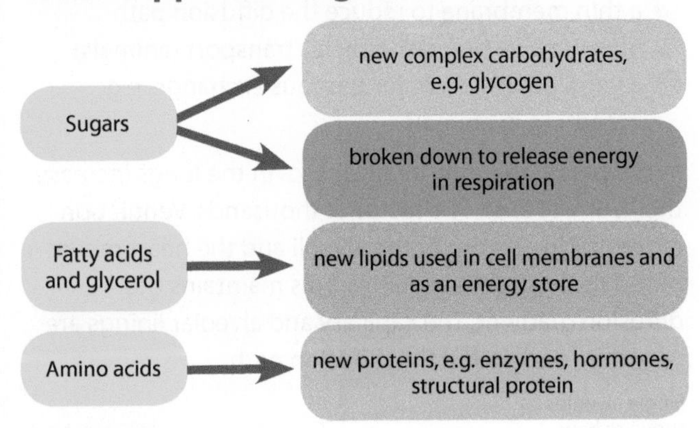

Calorimetry

Enzyme-controlled metabolic reactions such as respiration release energy. The amount of energy contained in a substrate can be measured using **calorimetry**.

The substrate (food) is placed in a calorimeter and burned in pure oxygen. The energy released heats up the surrounding water and the temperature rise is measured.

The energy released per gram can be calculated.

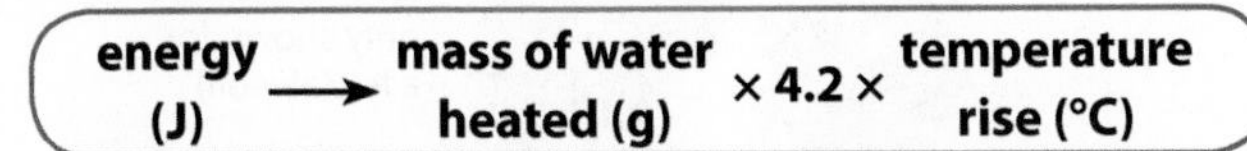

WS During your course you will investigate how certain factors affect the rate of enzyme activity. These include temperature, pH and substrate concentration.

Design an investigation to discover how temperature affects the activity of amylase. Here are some guidelines.

- Iodine solution turns from a red-brown colour to blue-black in the presence of starch.
- You can measure amylase activity by timing how long it takes for iodine solution to stop turning blue-black.
- A water bath can be set up with a thermometer. Add cold or hot water to regulate the temperature.
- Identify the independent, dependent and control variables in the investigation. Write out your method in clear steps.

SUMMARY

- **Enzymes are large proteins that act as biological catalysts.**
- **They speed up chemical reactions, including reactions that take place in living cells, e.g. respiration, photosynthesis and protein synthesis.**
- **Enzymes function according to the 'lock and key' theory.**

QUESTIONS

QUICK TEST

1. Which two factors affect the rate of enzyme activity?
2. What type of enzyme breaks down lipids?
3. Which molecules act as building blocks for protein polymers?

EXAM PRACTICE

1. Enzymes can only function in optimum pH and temperature conditions.

    a) Explain how a temperature of 45°C will affect the shape and function of the enzyme amylase when digesting starch. **[3 marks]**

    b) Explain how a pH of 2 allows the stomach protease enzyme to function efficiently.

    Use 'lock and key' theory to aid your explanation. **[3 marks]**

Cell transport

Diffusion

Living cells need to obtain oxygen, glucose, water, mineral ions and other dissolved substances from their surroundings. They also need to excrete waste products, such as carbon dioxide or urea. These substances pass through the cell membrane by **diffusion**.

Diffusion:

- is the (net) movement of particles in a liquid or gas from a region of high concentration to one of low concentration (down a **concentration gradient**)
- happens due to the random motion of particles past each other
- stops once the particles have completely spread out
- is passive, i.e. requires no input of energy
- can be increased in terms of rate by making the concentration gradient steeper, the diffusion path shorter, increasing the temperature or increasing the surface area over which the process occurs, e.g. having a folded cell membrane.

A protist called amoeba can absorb oxygen through diffusion.

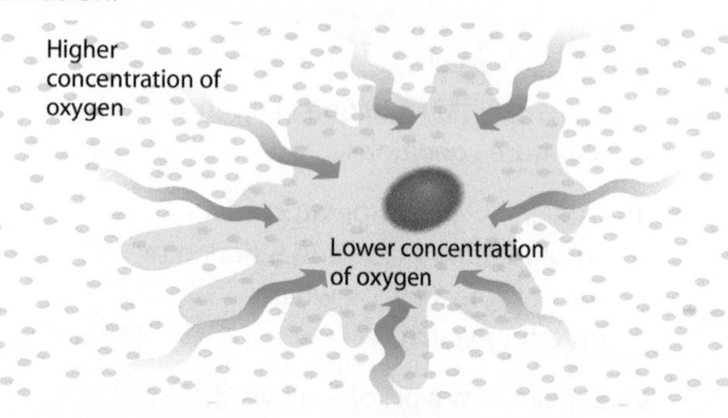

You may be asked to calculate rates of diffusion using **Fick's law**.

$$\textbf{rate of diffusion} \propto \frac{\textbf{surface area} \times \textbf{concentration difference}}{\textbf{thickness of membrane}}$$

$\propto$ means proportional to

Surface area to volume ratios

A unicellular organism (such as a protist) can absorb materials by diffusion directly from the environment. This is because it has a **large surface area to volume ratio**.

However, for a large, **multicellular organism**, the diffusion path between the environment and the inner cells of the body is long. Its large size also means that the **surface area to volume ratio** is **small**.

Adaptations

Multicellular organisms therefore need transport systems and specialised structures for exchanging materials, e.g. mammalian lungs and a small intestine, fish gills, and roots and leaves in plants. These increase diffusion efficiency in animals because they have:

- a large surface area
- a thin membrane to reduce the diffusion path
- an extensive blood supply for transport (animals)
- a ventilation system for gaseous exchange, e.g. breathing in animals.

In mammals, the individual air sacs in the lungs increase their surface area by a factor of thousands. Ventilation moves air in and out of the alveoli and the heart moves blood through the capillaries. This maintains the diffusion gradient. The capillary and alveolar linings are very thin, decreasing the diffusion path.

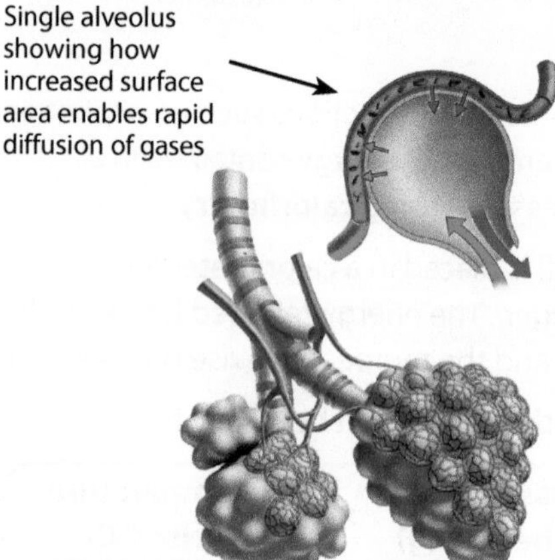

Osmosis

Osmosis is a special case of diffusion that involves the movement of water only.

There are two ways of describing osmosis.

1. The net movement of **water** from a region of **low** solute concentration to one of **high** concentration.
2. The movement of water down a **water potential gradient**.

Osmosis:

- occurs across a **partially permeable membrane**, so solute molecules cannot pass through (only water molecules can)
- occurs in all organisms
- is passive (in other words, requires no input of energy)
- allows water movement into root hair cells from the soil and between cells inside the plant

- can be demonstrated and measured in plant cells using a variety of tissues, e.g. potato chips, daffodil stems.

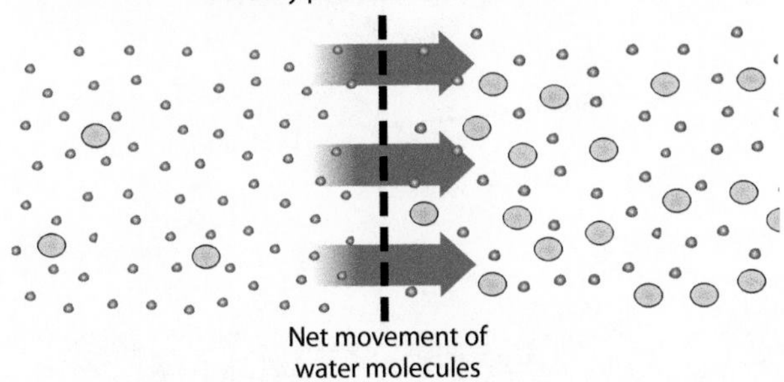

Active transport

Substances are sometimes absorbed against a concentration gradient, i.e. from a low to a high concentration.

Active transport:

- requires the release of energy from respiration
- takes place in the small intestine in humans, where sugar is absorbed into the bloodstream
- allows plants to absorb mineral ions from the soil through root hair cells.

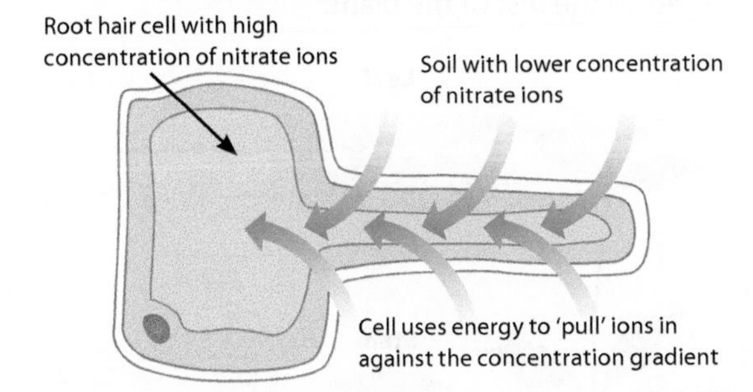

WS During your course, you may investigate the effect of salt or sugar solutions on plant tissue.

Here is one experiment you could do.

1. Immerse raw potato cut into chips of equal length in sugar solutions of various concentration.
2. You will see the potato chips change in length depending on whether individual cells have lost or gained water.

Can you **predict** what would happen to the potato chips immersed in:

- concentrated sugar (e.g. 1 molar)
- medium concentration sugar (e.g. 0.5 molar)
- water (0 molar)?

SUMMARY

- **Living cells obtain oxygen, glucose, water and other substances from their surroundings, and excrete waste products, through diffusion.**
- **Osmosis is a special type of diffusion involving only water.**
- **Active transport requires the release of energy from respiration.**

QUESTIONS

QUICK TEST

1. What is Fick's Law used for?
2. Some plant tissue is placed in a highly concentrated salt solution. Explain why water leaves the cells.

EXAM PRACTICE

1. Root hair cells absorb ions from the soil and water that surround them.

   Explain how the ions can enter, despite the fact that this is against a concentration gradient. **[2 marks]**

2. A student places a thin piece of rhubarb epidermis in a strong sugar solution and observes the cells under a microscope.

   `She notices that the cytoplasm has pulled away from the cell wall.

   Explain this change. **[4 marks]**

Plant tissues, organs and systems

Leaves

As this cross-section of a leaf shows, leaf tissues are adapted for efficient photosynthesis. The epidermis covers the upper and lower surfaces of the leaf and protects the plant against pathogens.

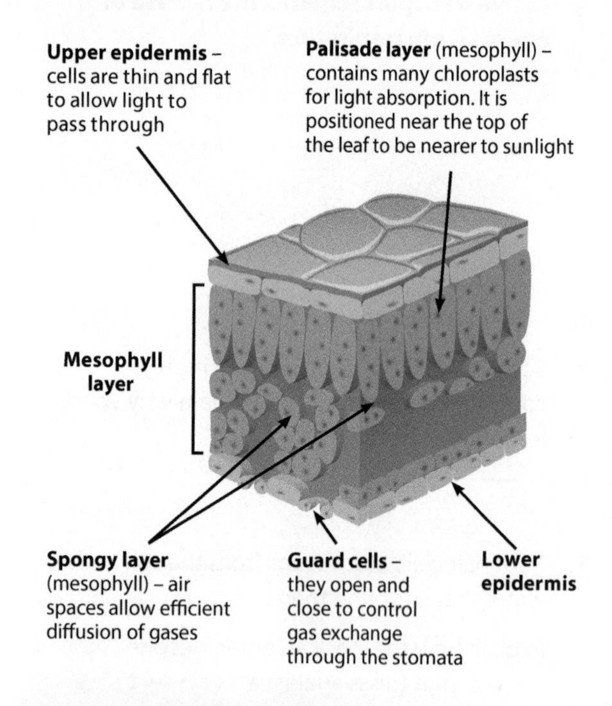

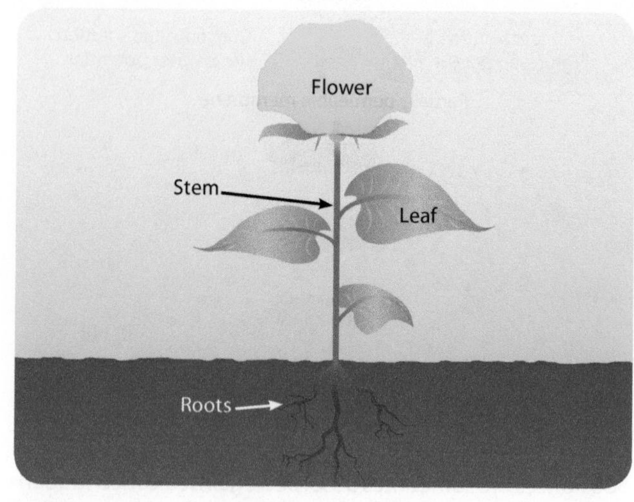

Stem and roots

Veins in the stem, roots and leaves contain tissues that transport water, carbohydrate and minerals around the plant.

- **Xylem tissue** transports water and mineral ions from the roots to the rest of the plant.
- **Phloem tissue** transports dissolved sugars from the leaves to the rest of the plant.

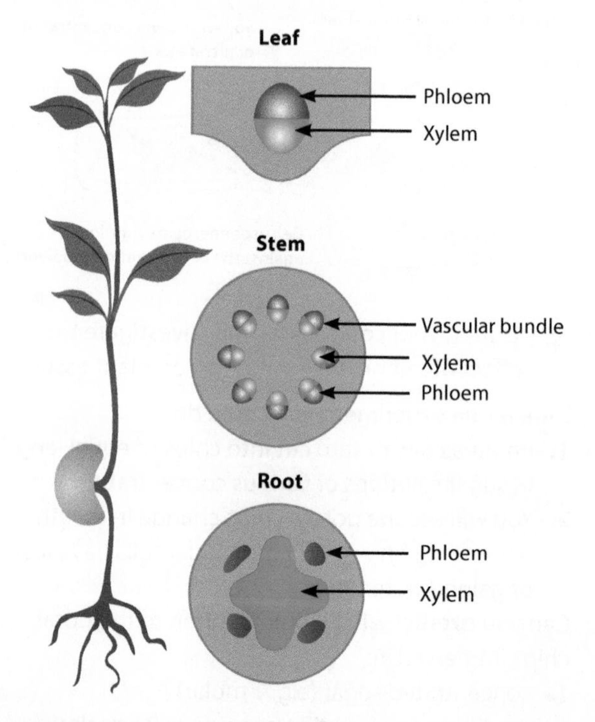

Meristem tissue is found at the growing tips of shoots and roots.

Xylem, phloem and root hair cells

Xylem, phloem and root hair cells are adapted to their function.

Part of plant	Appearance	Function	How they are adapted to their function
Xylem	Hollow tubes made from dead plant cells (the hollow centre is called a lumen)	Transport water and mineral ions from the roots to the rest of the plant in a process called transpiration	The cellulose cell walls are thickened and strengthened with a waterproof substance called lignin
Phloem	Columns of living cells	Translocate (move) cell sap containing sugars (particularly sucrose) from the leaves to the rest of the plant, where it is either used or stored	Phloem have pores in the end walls so that the cell sap can move from one phloem cell to the next
Root hair cells	Long and thin; have hair-like extensions	Absorb minerals and water from the soil	Large surface area

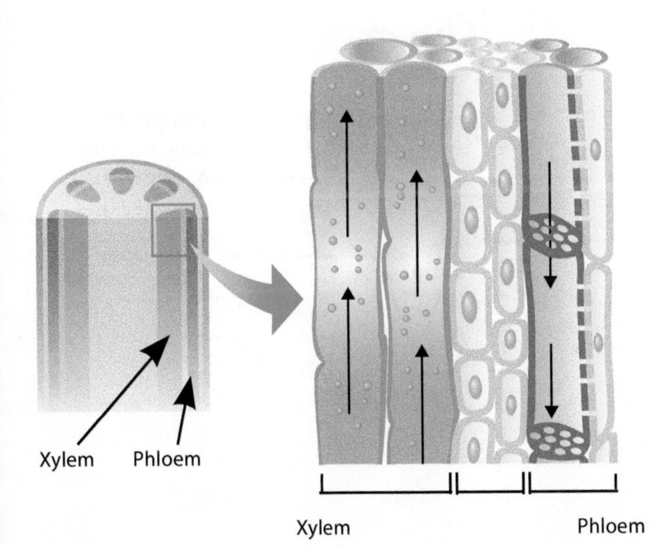

SUMMARY

- **A plant's system is made up of organs and tissues that enable it to be a photosynthetic organism.**
- **Roots absorb water and minerals. They anchor plants in the soil.**
- **The stem transports water and nutrients to leaves. It holds leaves up to the light for maximum absorption of energy.**
- **The leaf is the organ of photosynthesis.**
- **The flower makes sexual reproduction possible through pollination.**

QUESTIONS

QUICK TEST

1. What part of the plant is the main organ for photosynthesis?
2. What is the name of the waterproof substance that strengthens xylem?

EXAM PRACTICE

1. **a)** State one structural difference between xylem and phloem tissue. **[1 mark]**

   b) Small, herbivorous insects called aphids are found on plant stems. They have piercing mouthparts that can penetrate down to the phloem.

   Explain the reasons for this behaviour. **[2 marks]**

Transport in plants

Transpiration

The movement of water through a plant, from roots to leaves, takes place as a transpiration stream. Once water is in the leaves, it diffuses out of the stomata into the surrounding air. This is called (evapo)transpiration.

Water evaporates from the spongy mesophyll through the stomata. → Water passes by osmosis from the xylem vessels in the leaf into the spongy mesophyll cells to replace what has been lost. → This movement 'pulls' the column of water in that xylem vessel upwards. → Water enters root hair cells by osmosis to replace water that has entered the xylem.

Measuring rate of transpiration

A leafy shoot's rate of transpiration can be measured using a **potometer**.

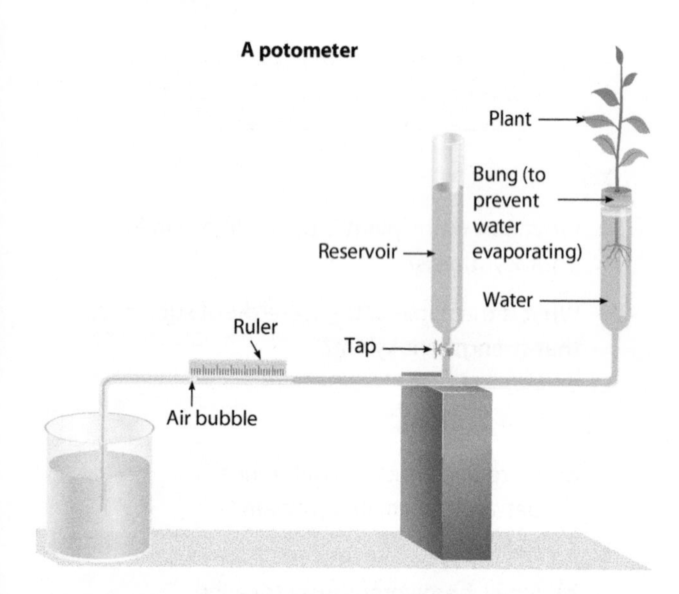

The shoot is held in a tube with a bung around the top to prevent any water from evaporating (this would give a false measurement of the water lost by transpiration) and to prevent air bubbles entering the apparatus.

As the plant transpires, it takes up water from the tube to replace what it has lost. All the water is then pulled up, moving the air bubble along.

The distance the air bubble moves can be used to calculate the plant's rate of transpiration for a given time period.

The experiment can be repeated, varying a different factor each time, to see how each factor affects the rate of transpiration.

Factors affecting rate of transpiration

Evaporation of water from the leaf is affected by **temperature**, **humidity**, **air movement** and **light intensity**.

- **Increased temperature** increases the kinetic energy of molecules and removes water vapour more quickly away from the leaf.
- **Increased air movement** removes water vapour molecules.
- **Increased light intensity** increases the rate of photosynthesis. This in turn draws up more water from the transpiration stream, which maintains high concentration in the spongy mesophyll.
- **Decreasing atmospheric humidity** lowers water vapour concentration outside of the stoma and so maintains the concentration gradient.

How water vapour exits the leaf

Sunshine

This gradient determines how quickly water vapour diffuses

Water vapour diffuses outwards

Guard cells

Opening and closing of stomata

Guard cells control the amount of water vapour that evaporates from the leaves and the amount of carbon dioxide that enters them.

- When light intensity is high and photosynthesis is taking place at a rapid rate, the sugar concentration rises in photosynthesising cells, e.g. palisade and guard cells.
- Guard cells respond to this by increasing the rate of water movement in the transpiration stream. This in turn provides more water for photosynthesis.

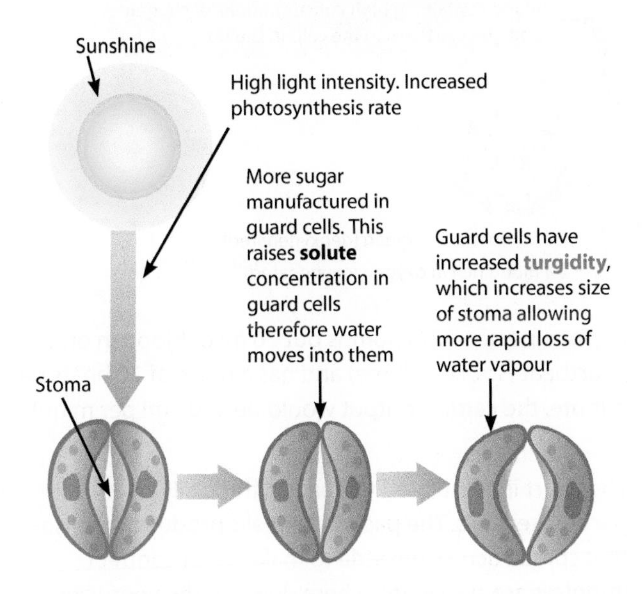

WS A **hypothesis** is an idea or explanation that you test through study and experiments. It should include a reason. For example: desert plants have fewer stomata than temperate plants **because** they need to minimise water loss.

- In an experiment investigating the factors that affect the rate of transpiration, a student plans to take measurements of weight loss or gain from a privet plant.
- Construct hypotheses for each of these factors: **temperature**, **humidity**, **air movement** and **light intensity**.
- The first has been done for you: As temperature increases, the plant will lose mass/water more quickly **because** diffusion occurs more rapidly.

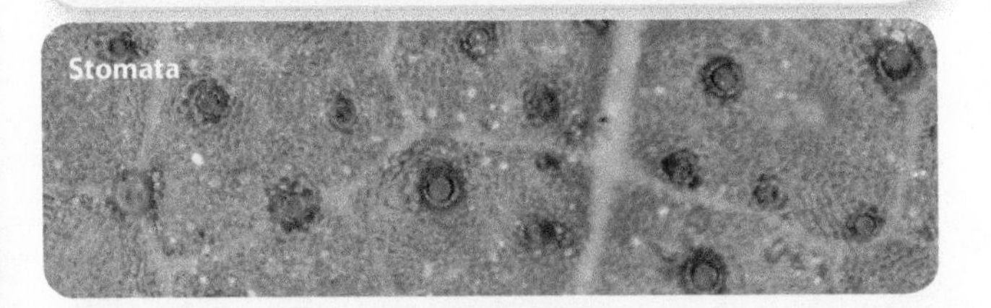

SUMMARY

- **A plant's system is made up of organs and tissues that enable it to be a photosynthetic organism.**
- **Roots absorb water and minerals. They anchor plants in the soil.**
- **The stem transports water and nutrients to leaves. It holds leaves up to the light for maximum absorption of energy.**
- **The leaf is the organ of photosynthesis.**

QUESTIONS

QUICK TEST

1. What effect would decreasing air humidity have on transpiration?
2. Which cells control the opening and closing of the stomata?
3. What equipment is used to measure a leafy shoot's rate of transpiration?

EXAM PRACTICE

1. A student set up a plant in a potometer and noted the position of the air bubble at 2.1 cm on the scale. He left the apparatus in still air at 20°C for exactly 60 minutes. He noted the new position of the air bubble on the scale as 2.3 cm. He then immediately placed a fan close to the plant and switched it on. He noted the temperature was still 20°C. After another 60 minutes he saw that the bubble was at 5.7 cm.

   a) Calculate the rate of water absorption in cm per minute for both periods of time.

   Show your working. **[4 marks]**

   b) Explain the difference between both sets of results. **[3 marks]**

Transport in humans 1

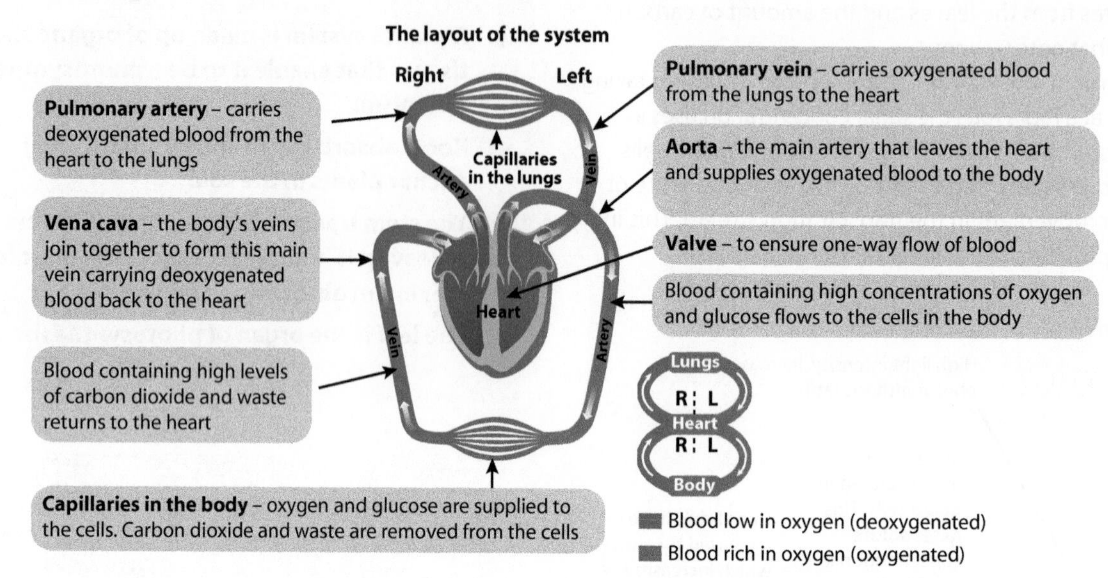

Blood circulation

Blood moves around the body in a **double circulatory system**. In other words, blood moves twice through the heart for every full circuit. This ensures maximum efficiency for absorbing oxygen and delivering materials to all living cells.

The heart

The heart is made of powerful muscles that contract and relax rhythmically in order to continuously pump blood around the body. The **heart muscle** is supplied with nutrients (particularly glucose) and oxygen through the coronary artery.

The sequence of events that takes place when the heart beats is called the **cardiac cycle**.

- The heart relaxes and blood enters both atria from the veins.
- The atria contract together to push blood into the ventricles, opening the atrioventricular valves.
- The ventricles contract from the bottom, pushing blood upwards into the arteries. The backflow of blood into the ventricles is prevented by the **semilunar valves.**

The left side of the heart is more muscular than the right because it has to pump blood further round the body. The right side only has to pump blood to the lungs and back.

A useful measurement for scientists and doctors to take is **cardiac output**. This is calculated using:

cardiac output = stroke volume × heart rate

So, for a person who pumps out 70 ml of blood in one heartbeat (stroke volume) and has a pulse of 70 beats per minute, the cardiac output would be 4900 ml per minute.

Controlling the heartbeat

The heart is stimulated to beat rhythmically by pacemaker cells. The pacemaker cells produce impulses that spread across the atria to make them contract. Impulses are spread from here down to the ventricles, making them contract, pushing blood up and out.

Nerves connecting the heart to the brain can increase or decrease the pace of the pacemaker cells in order to regulate the heartbeat.

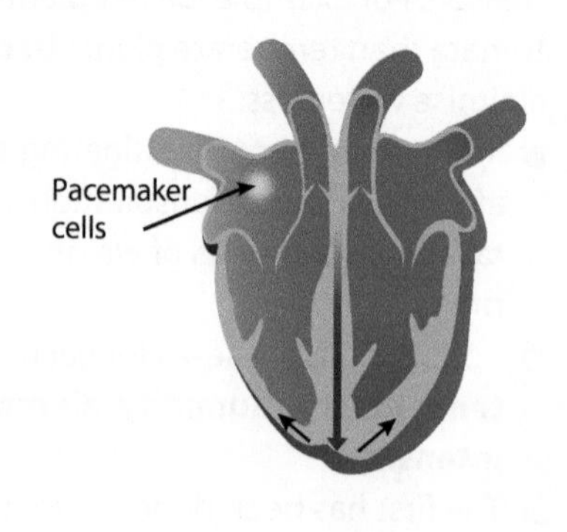

If a person has an irregular heartbeat, they can be fitted with an artificial, electrical pacemaker.

Blood vessels

Blood is carried through the body in three types of vessel.

- **Arteries** have thick walls made of elastic fibres and muscle fibres to cope with the high pressure. The **lumen** (space inside) is small compared to the thickness of the walls. There are no valves.

- **Veins** have thinner walls. The lumen is much bigger compared to the thickness of the walls and there are valves to prevent the backflow of blood.
- **Capillaries** are narrow vessels with walls only one cell thick. These microscopic vessels connect arteries to veins, forming dense networks or **beds**. They are the only blood vessels that have permeable walls to allow the exchange of materials.

Artery

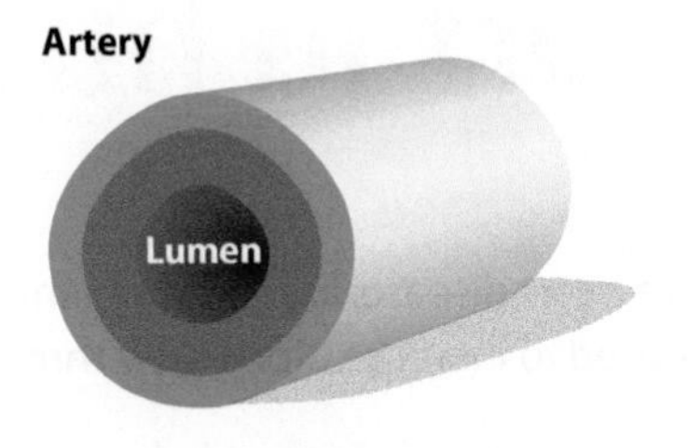

Vein

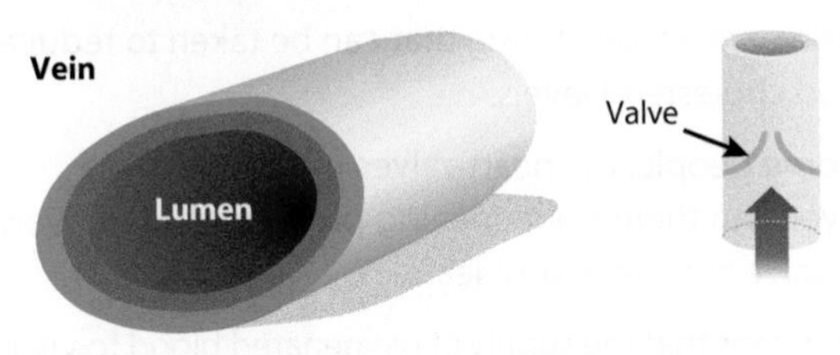

Capillary

Note: capillaries are much smaller than veins or arteries

Coronary heart disease

Coronary heart disease (CHD) is a non-communicable disease. It results from the build-up of **cholesterol**, leading to plaques laid down in the coronary arteries. This restricts blood flow and the artery may become blocked with a blood clot or **thrombosis**. The heart muscle is deprived of glucose and oxygen, which causes a **heart attack**.

The likelihood of plaque developing increases if you have a high fat diet. The risk of having a heart attack can be reduced by:

- eating a balanced diet and not being overweight
- not smoking tobacco
- lowering alcohol intake
- reducing salt levels in your diet
- reducing stress levels.

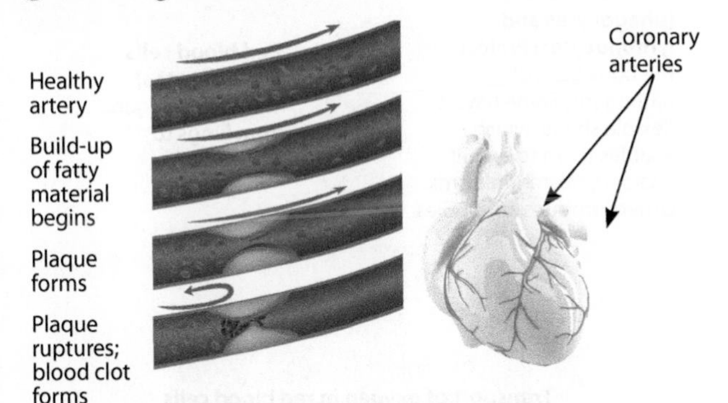

SUMMARY

- **Blood moves around the body in a double circulatory system.**
- **The heart is made of strong muscles that contract and relax to pump blood around the body.**
- **There are three types of blood vessels: arteries, veins and capillaries.**
- **Coronary heart disease can occur due to build-up of cholesterol in arteries.**

QUESTIONS

QUICK TEST

1. Which blood vessels contain valves?
2. Capillaries are smaller than veins. True or false?
3. Which vessel carries oxygenated blood from the lungs to the heart?

EXAM PRACTICE

1. a) Explain why arteries have thick, elastic muscle walls. **[1 mark]**

   b) Why do veins have valves? **[1 mark]**

2. Compared with reptiles and amphibians, humans have a double circulation.

   a) What is meant by this term? **[1 mark]**

   b) What is the advantage of having a double circulation? **[1 mark]**

Transport in humans 2

Remedying heart disease

For patients who have heart disease, artificial implants called **stents** can be used to increase blood flow through the coronary artery.

Statins are a type of drug that can be taken to reduce blood cholesterol levels.

In some people, the heart valves may deteriorate, preventing them from opening properly. Alternatively, the valve may develop a leak.

This means that the supply of oxygenated blood to vital organs is reduced. The problem can be corrected by surgical replacement using a **biological** or **mechanical** valve.

When complete heart failure occurs, a heart transplant can be carried out. If a donor heart is unavailable, the patient may be kept alive by an artificial heart until one can be found. Mechanical hearts are also used to give the biological heart a rest while it recovers.

Blood as a tissue

Blood transports digested food and oxygen to cells and removes the cells' waste products. It also forms part of the body's defence mechanism.

The four components of blood are:

- platelets
- plasma
- white blood cells
- red blood cells.

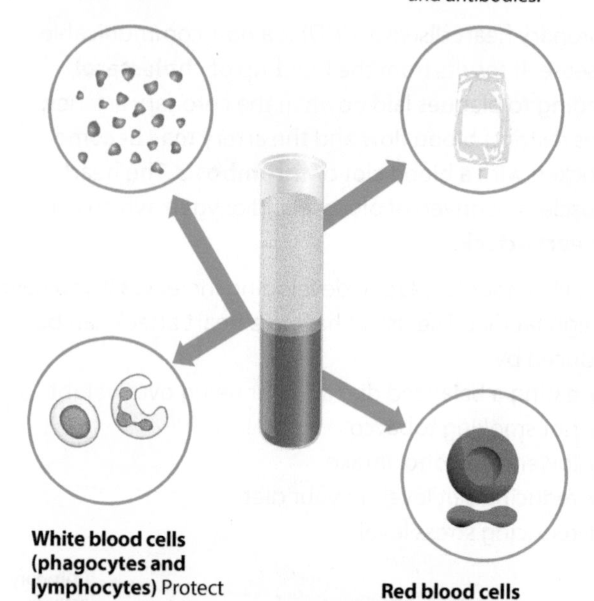

Oxygen transport

Red blood cells are small and have a biconcave shape. This gives them a large surface area to volume ratio for absorbing oxygen. When the cells reach the lungs, they absorb and bind to the oxygen in a molecule called haemoglobin.

haemoglobin + oxygen ⇌ oxyhaemoglobin

Blood is then pumped around the body to the tissues, where the reverse of the reaction takes place. Oxygen diffuses out of the red blood cells and into the tissues.

Transport of oxygen in red blood cells

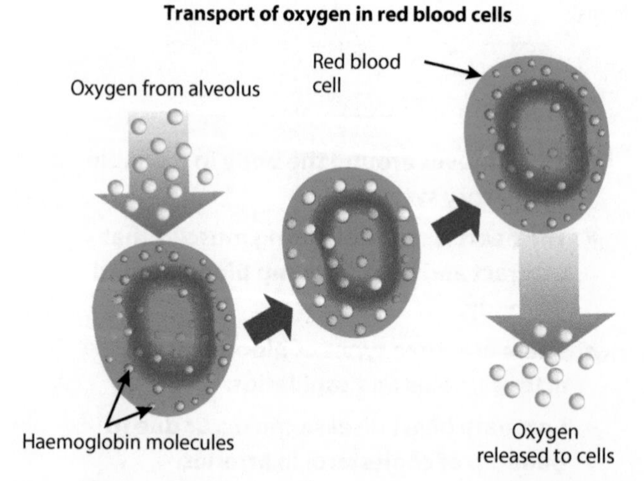

The lungs

Humans, like many vertebrates, have lungs to act as a **gaseous exchange surface**.

Other structures in the thorax enable air to enter and leave the lungs (ventilation).

- The trachea is a flexible tube, surrounded by rings of cartilage to stop it collapsing. Air is breathed in via the mouth and passes through here on its way to the lungs.
- Bronchi are branches of the trachea.
- The alveoli are small air sacs that provide a large surface area for the exchange of gases.
- Capillaries form a dense network to absorb maximum oxygen and release carbon dioxide.

In the alveoli, **oxygen** diffuses down a concentration gradient. It moves across the thin layers of cells in the alveolar and capillary walls, and into the red blood cells.

For **carbon dioxide**, the gradient operates in reverse. The carbon dioxide passes from the blood to the alveoli, and from there it travels back up the air passages to the mouth.

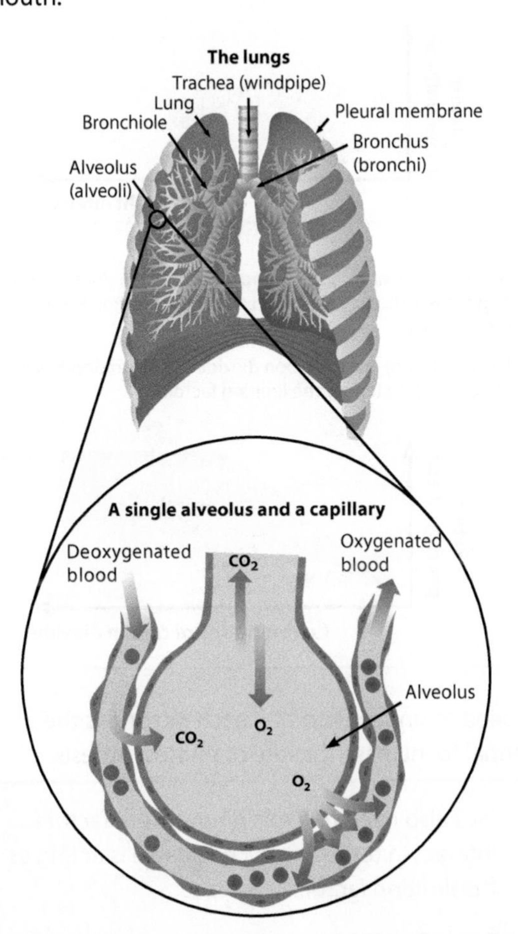

WS Scientists make observations, take measurements and gather data using a variety of instruments and techniques. Recording data is an important skill.

Create a table template that you could use to record data for the following experiment:

An investigation that involves measuring the resting and active pulse rates of 30 boys and 30 girls, together with their average breathing rates.

Make sure that:

- you have the correct number of columns and rows
- each variable is in a heading
- units are in the headings (so they don't need to be repeated in the body of the table).

SUMMARY

- **There are four components of blood: platelets, plasma, white blood cells and red blood cells.**
- **Oxygen is transported around the body by red blood cells.**
- **The lungs act as a gaseous exchange surface.**

QUESTIONS

QUICK TEST

1. Name the four components of blood.
2. What is produced when haemoglobin combines with oxygen?

EXAM PRACTICE

1. Describe the function of white blood cells. **[2 marks]**
2. A smoker's lungs develop a layer of tar lining the inside of the alveoli.

   Explain why this will make the smoker breathless.

   Use ideas about diffusion in your explanation. **[3 marks]**

Photosynthesis

Photosynthesis:

- is an endothermic reaction
- requires chlorophyll to absorb the sunlight; this is found in the **chloroplasts** of photosynthesising cells, e.g. palisade cells, guard cells and spongy mesophyll cells
- produces **glucose**, which is then respired for energy release or converted to other useful molecules for the plant
- produces **oxygen** that has built up in the atmosphere over millions of years; oxygen is vital for respiration in all organisms.

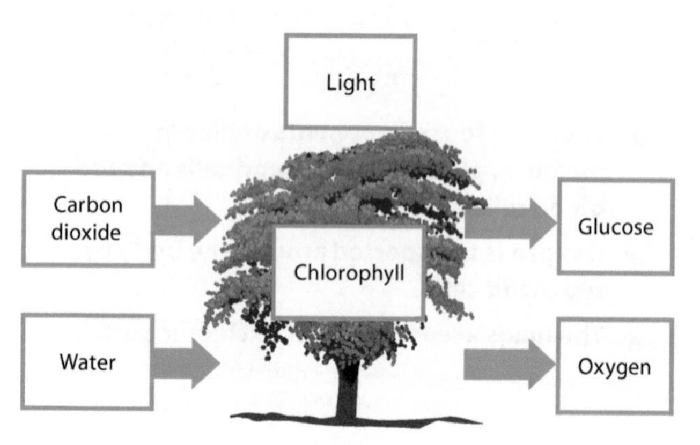

$$\text{carbon dioxide} + \text{water} \xrightarrow[\text{chloroplast}]{\text{light energy}} \text{glucose} + \text{oxygen}$$

HT $$6CO_2 + 6H_2O \xrightarrow[\text{chloroplast}]{\text{light energy}} C_6H_{12}O_6 + 6O_2$$

Rate of photosynthesis

The rate of photosynthesis can be affected by:

- temperature
- light intensity
- carbon dioxide concentration
- amount of chlorophyll.

In a given set of circumstances, **temperature**, **light intensity** and **carbon dioxide concentration** can act as limiting factors.

Temperature

1. As the temperature rises, so does the rate of photosynthesis. This means temperature is limiting the rate of photosynthesis.
2. As the temperature approaches 45°C, the enzymes controlling photosynthesis start to be denatured. The rate of photosynthesis decreases and eventually declines to zero.

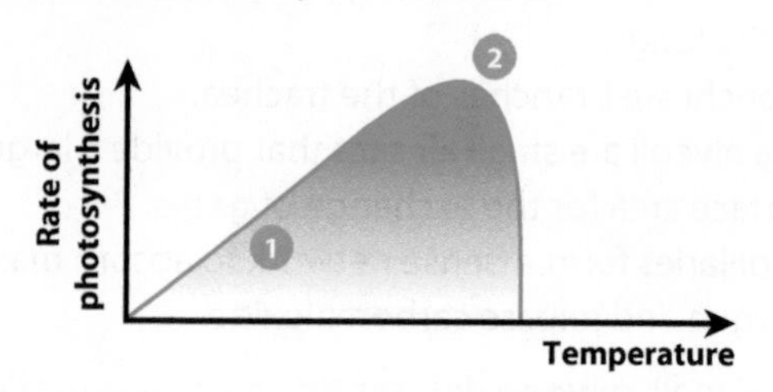

Light intensity

1. As the light intensity increases, so does the rate of photosynthesis. This means light intensity is limiting the rate of photosynthesis.
2. Eventually, the rise in light intensity has no effect on photosynthesis rate. Light intensity is no longer the limiting factor; carbon dioxide or temperature must be.

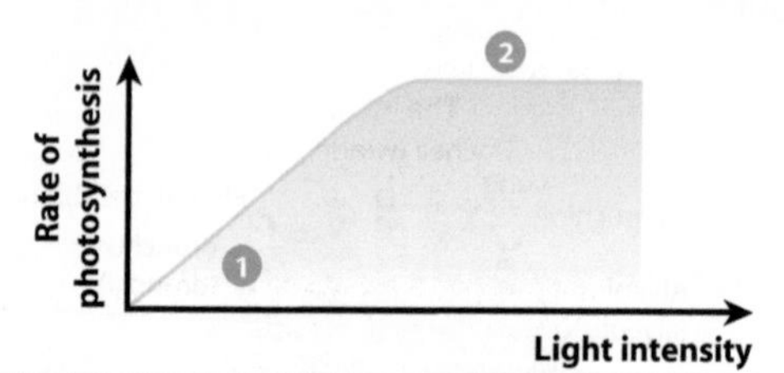

Carbon dioxide concentration

1. As carbon dioxide concentration increases, so does the rate of photosynthesis. Carbon dioxide concentration is the limiting factor.
2. Eventually, the rise in carbon dioxide concentration has no effect – it is no longer the limiting factor.

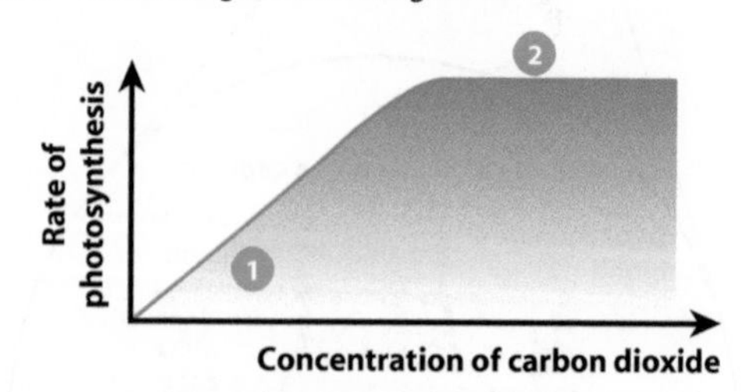

You need to understand that each factor has the potential to increase the rate of photosynthesis.

HT You also need to explain how these factors interact in terms of which variable is acting as the limiting factor.

HT The inverse law

The effect of light intensity on photosynthesis can be investigated by placing a lamp at varying distances from a plant. As the lamp is moved further away from the plant the light intensity decreases, as shown in the graph.

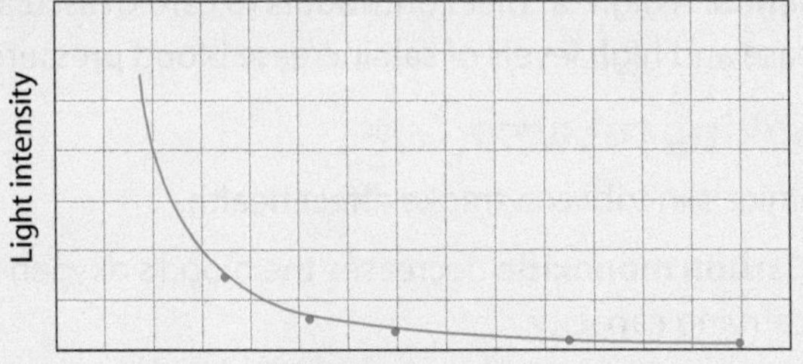

There is an **inverse relationship** between the two variables. The graph can be used to convert distances to light intensity, or light intensity can be calculated using the formula:

$$\textbf{light intensity} = \frac{1}{d^2}$$

- *d* is the distance from the lamp.

This formula produces a dimensionless quantity of light intensity, so no units need to be stated. Instruments that measure absolute light intensity may use units such as 'lux' or the 'candela'.

Commercial applications

Farmers and market gardeners can increase their crop yields in greenhouses. They do this by:

- making the temperature optimum for growth using heaters
- increasing light intensity using lamps
- installing fossil-fuel burning stoves to increase carbon dioxide concentration (and increase temperature).

If applied carefully, the cost of adding these features will be offset by increased profit from the resulting crop.

SUMMARY

- **Plants can photosynthesise, i.e. make food molecules in the form of carbohydrate from carbon dioxide and water. As such, they are the main producers of biomass.**
- **Sunlight energy is needed for photosynthesis.**
- **The rate of photosynthesis is affected by light intensity, temperature and carbon dioxide concentration.**
- **Other products such as starch and cellulose are made from glucose.**

Uses of glucose in plants

The glucose produced from photosynthesis can be used immediately in respiration, but some is used to synthesise larger molecules: starch, cellulose, protein and lipids.

Starch is insoluble. So it is suitable for storage in leaves, stem or roots.

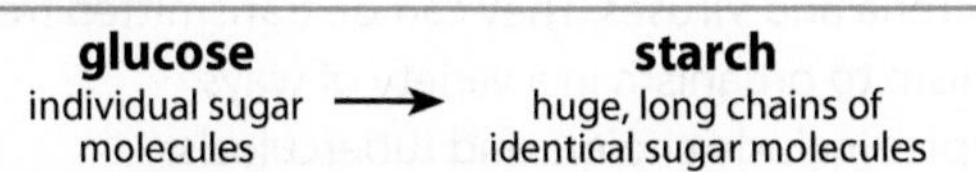

Cellulose is needed for cell walls.

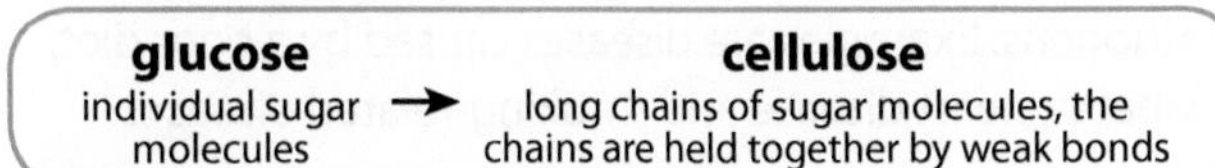

Protein is used for the growth and repair of plant tissue, and also to synthesise enzyme molecules.

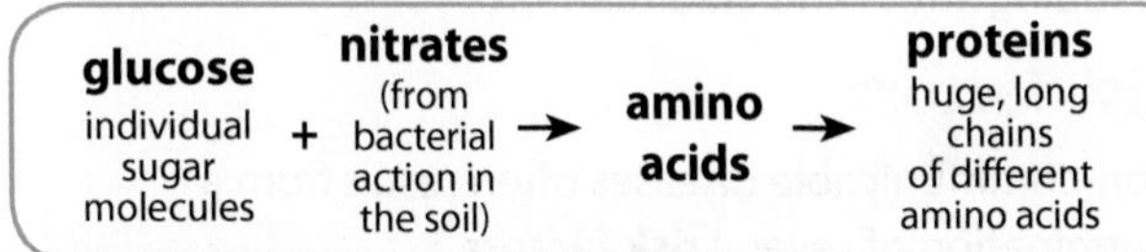

Lipids are needed in cell membranes, and for fat and oil storage in seeds.

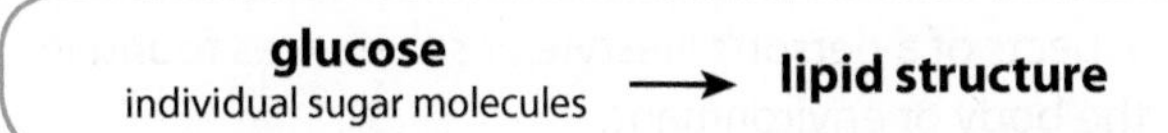

QUESTIONS

QUICK TEST

1. Give two reasons why photosynthesis is seen as the opposite of respiration.
2. What is cellulose needed for?

EXAM PRACTICE

HT **1.** The graph shows how the rate of photosynthesis changes in a plant with changing light intensity.

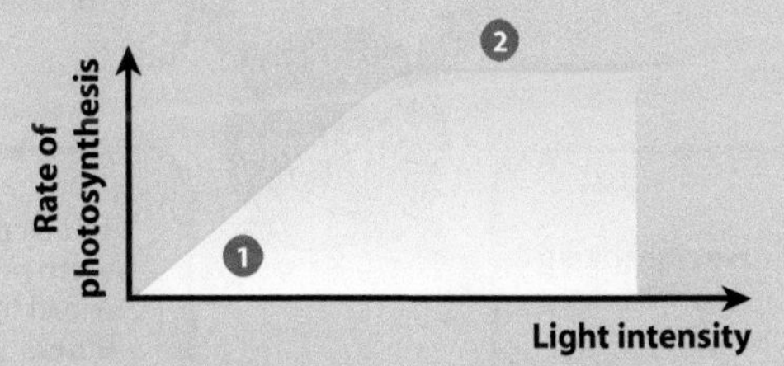

A market gardener understands this and wants to use the information to increase the yield of her tomatoes.

a) Use the graph to explain how she could accomplish this. **[4 marks]**

b) Explain why she should not invest too much money in her solution. **[2 marks]**

Non-communicable diseases

Diseases

Communicable diseases are caused by pathogens such as bacteria and viruses. They can be transmitted from organism to organism in a variety of ways.
Examples include cholera and tuberculosis.

Non-communicable diseases are not primarily caused by pathogens. Examples are diseases caused by a poor diet, diabetes, heart disease and smoking-related diseases.

Health is the state of physical, social and mental well-being. Many factors can have an effect on health, including stress and life situations.

Risk factors

Non-communicable diseases often result from a combination of several **risk factors**.

- Risk factors produce an increased likelihood of developing that particular disease. They can be aspects of a person's lifestyle or substances found in the body or environment.
- Some of these factors are difficult to quantify or to establish as a definite **causal connection**. So scientists have to describe their effects in terms of probability or likelihood.

The symptoms observed in the body may result from communicable and non-communicable components interacting.

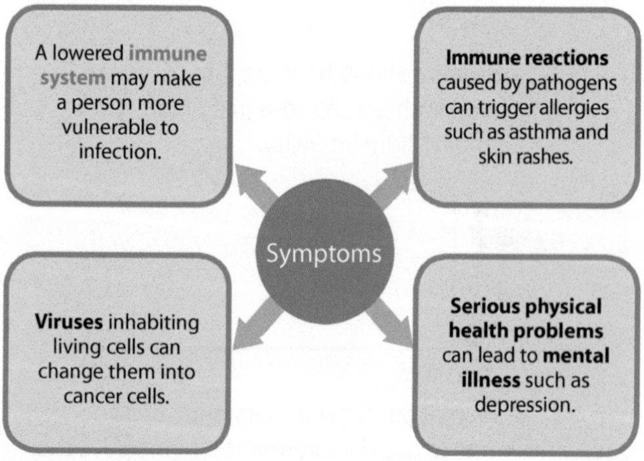

Symptoms

- A lowered immune system may make a person more vulnerable to infection.
- **Immune reactions** caused by pathogens can trigger allergies such as asthma and skin rashes.
- **Viruses** inhabiting living cells can change them into cancer cells.
- **Serious physical health problems** can lead to **mental illness** such as depression.

Poor diet

People need a **balanced diet**. If a diet does not include enough of the main food groups, malnutrition might result. Lack of correct vitamins leads to diseases such as **scurvy** and **rickets**. Lack of the mineral, iron, results in **anaemia**. A high fat diet contributes to cardiovascular disease and high levels of salt increase blood pressure.

Smoking tobacco

Chemicals in tobacco smoke affect health.

- **Carbon monoxide** decreases the blood's oxygen-carrying capacity.
- **Nicotine** raises the heart rate and therefore blood pressure.
- **Tar** triggers cancer.
- **Particulates** cause **emphysema** and increase the likelihood of **lung infections**.

Weight/lack of exercise

Obesity and lack of exercise both increase the risk of developing **type 2 diabetes** and cardiovascular disease.

One way to show whether someone is underweight or overweight for their height is to calculate their **body mass index** (**BMI**), using the following formula:

$$\text{BMI} = \frac{\text{mass (kg)}}{\text{height (m)}^2}$$

Recommended BMI chart

BMI	What it means
<18.5	Underweight – too light for your height
18.5–25	Ideal – correct weight range for your height
25–30	Overweight – too heavy
30–40	Obese – much too heavy. Health risks!

Example

Calculate a man's BMI if he is 1.65 m tall and weighs 68 kg.

$$\text{BMI} = \frac{\text{mass (kg)}}{\text{height (m)}^2} = \frac{68}{1.65^2} = \frac{68}{2.7} = \mathbf{25}$$

The recommended BMI for his height (1.65 m) is 18.5–25, so he is just a healthy weight.

There are drawbacks to using BMI as a way of assessing people's health. For example:

- teenagers go through a rapid growth phase

- a person could have a well-developed muscle system – this would increase their body mass but not make them obese.

Some scientists say a more accurate method is using the waist/hip ratio. A tape measure is used to measure the circumference of the hips and the waist (at its widest). The waist measurement is divided by the hips measurement. The following chart can then be used.

Waist to hip ratio (WHR)		
Male	**Female**	**Health risk based solely on WHR**
0.95 or below	0.80 or below	Low
0.96 to 1.0	0.81 to 0.85	Moderate
1.0+	0.85+	High

Alcohol

Drinking excess alcohol can impair brain function and lead to **cirrhosis** of the liver. It also contributes to some types of cancer and cardiovascular disease.

Smoking and drinking alcohol during pregnancy

Unborn babies receive nutrition from the mother via the placenta. Substances from tobacco, alcohol and other drugs can pass to the baby and cause **lower birth weight**, **foetal alcohol syndrome** and **addiction**.

Carcinogens

Exposure to **ionising radiation** (e.g. X-rays, gamma rays) can cause cancerous tumours. Overexposure to UV light can cause skin cancer. Certain chemicals such as mercury can also increase the likelihood of cancer.

SUMMARY

- **Diseases can be communicable or non-communicable.**
- **Communicable diseases are caused by pathogens. Non-communicable diseases include heart disease, diabetes and smoking-related illnesses.**
- **Symptoms of diseases are often due to an interaction between communicable and non-communicable factors.**
- **Risk factors include poor diet, smoking tobacco, lack of exercise and alcohol abuse.**

WS Interpreting complex data in graphs doesn't need to be difficult. This line graph shows data about smoking and lung cancer. Look for different patterns in it. For example:

- males have higher smoking rates in all years
- female cancer rates have increased overall since 1972.

Can you see any other patterns?

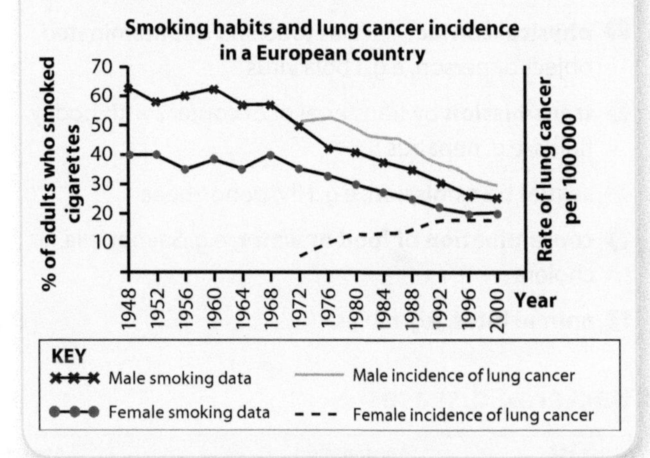

QUESTIONS

QUICK TEST

1. What does BMI stand for?
2. What is a pathogen?

EXAM PRACTICE

1. Write down two conditions that are non-communicable. **[2 marks]**
2. **a)** A woman's height is 1.55m and she weighs 64kg. What is her BMI? **[2 marks]**

   b) Using the BMI table on p.28, state whether you think she should take steps to lose weight.

   Give a reason for your answer. **[2 marks]**

Communicable diseases

How do pathogens spread?

Pathogens are disease-causing microorganisms from groups of bacteria, viruses, fungi and protists. All animals and plants can be affected by pathogens. They spread in many ways, including:

- **droplet infection** (sneezing and coughing), e.g. flu
- **physical contact**, such as touching a contaminated object or person, e.g. Ebola virus
- **transmission** by transferral of or contact with bodily fluids, e.g. hepatitis B
- **sexual transmission**, e.g. HIV, gonorrhoea
- **contamination of food or water**, e.g. Salmonella, cholera
- **animal bites**, e.g. rabies.

How do pathogens cause harm?

- Bacteria and viruses reproduce rapidly in the body.
- Viruses cause cell damage.
- Bacteria produce toxins that damage tissues.

These effects produce **symptoms** in the body.

How can the spread of disease be prevented?

The spread of disease can be prevented by:

- good hygiene, e.g. washing hands/whole body, using soaps and disinfectants
- destroying vectors, e.g. disrupting the life cycle of mosquitoes can combat malaria
- the isolation or quarantine of individuals
- vaccination.

Bacterial diseases

Disease	Transmission	Symptoms	Treatment/prevention
Tuberculosis	Droplet infection	Persistent coughing, which may bring up blood; chest pain; weight loss; fatigue; fever; night sweats; chills	Long course of antibiotics
Cholera	Contaminated water/food	Diarrhoea; vomiting; dehydration	Rehydration salts
Chlamydia	Sexually transmitted	May not be present, but can include discharge and bleeding from sex organs	Antibiotics Using condoms during sexual intercourse can reduce chances of infection
Helicobacter	Spread orally from person to person by saliva, or by fecal contamination	Abdominal pain; feeling bloated; nausea; vomiting; loss of appetite; weight loss	Proton pump inhibitors and antibiotics
Salmonella	Contaminated food containing toxins from pathogens – these could be introduced from unhygienic food preparation techniques	Vomiting; fever; diarrhoea; stomach cramps	Anti-diarrhoeals and antibiotics; vaccinations for chickens
Gonorrhoea	Sexually transmitted	Thick yellow or green discharge from vagina or penis; pain on urination	Antibiotic injection followed by antibiotic tablets; penicillin is no longer effective against gonorrhoea; prevention through use of condoms

Fungal diseases

Disease	Transmission	Symptoms	Treatment/prevention
Athlete's foot	Direct and indirect contact, e.g. skin-to-skin, bed sheets and towels (often spreads at swimming pools and in changing rooms)	Itchy, red, scaly, flaky and dry skin	Self-care and anti-fungal medication externally applied

Viral and protist diseases

Disease	Transmission	Symptoms and notes	Treatment/prevention
Measles (viral)	Droplets from sneezes and coughs	Fever; red skin rash; fatal if complications arise	No specific treatment; vaccine is a highly effective preventative measure
HIV (viral)	Sexually transmitted; exchange of body fluids; sharing of needles during drug use	Flu-like symptoms initially; late-stage AIDS produces complications due to compromised immune system	Anti-retroviral drugs
Ebola (viral)	Via body fluids; contaminated needles; bite from infected animal	Early stages: muscle pain; sore throat; diarrhoea Later stages: kidney/liver failure; internal bleeding	No direct treatments or tested vaccines are available at the time of writing; symptoms and infections are treated as they appear
Malaria (protist)	Via mosquito vector	Headache; sweats; chills and vomiting; symptoms disappear and reappear on a cyclical basis; further life-threatening complications may arise	Various anti-malarial drugs are available for both prevention and cure; prevention of mosquito breeding and use of mosquito nets

Viruses carry out their life cycle largely within a host cell. Viruses tend to follow one of two types of cycle.

- **Lysogenic**: the virus does not lyse (split) the host cell. Instead, the viral DNA becomes part of the host DNA. When the host cell divides, the viral DNA is also copied and is present in each of the daughter cells.
- **Lytic**: the virus replicates using the host DNA (as before). The new viruses that are formed destroy the infected cell and its membrane as they are released.

Influenza viral life cycle

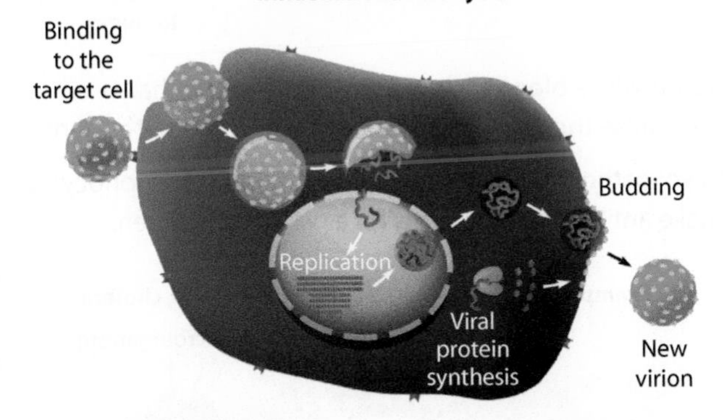

SUMMARY

- **Pathogens can be spread through physical contact, droplet infection, sexual transmission, contamination of food/water, transmission of bodily fluids, or animal bites.**
- **Bacterial diseases include salmonella and gonorrhoea.**
- **Viral and protist diseases include measles, HIV and malaria.**

QUESTIONS

QUICK TEST

1. Why does cholera spread quickly in areas that have poor sanitation?
2. Name a fungal disease.

EXAM PRACTICE

1. Microorganisms consist of bacteria, viruses, fungi and protista.

   Many cause harm to the human body.

   a) Write down the term that describes such organisms. **[1 mark]**

   b) Harmful microorganisms produce symptoms when they reproduce in large numbers.

   Write down **two** ways in which microorganisms do this. **[2 marks]**

2. Malaria kills many thousands of people every year. The disease is common in areas that have warm temperatures and stagnant water.

   Explain why this is. **[2 marks]**

Human defences

Non-specific defences

The body has a number of general or non-specific defences to stop pathogens multiplying inside it.

The skin covers most of the body – it is a **physical barrier** to pathogens. It also secretes antimicrobial peptides to kill microorganisms. If the skin is damaged, a clotting mechanism takes place in the blood preventing pathogens from entering the site of the wound.

Tears contain enzymes called **lysozymes**. Lysozymes break down pathogen cells that might otherwise gain entry to the body through tear ducts.

Hairs in the nose trap particles that may contain pathogens.

Tubes in the respiratory system (**trachea** and **bronchi**) are lined with special epithelial cells. These cells either produce a sticky, liquid mucus that traps microorganisms or have tiny hairs called cilia that move the mucus up to the mouth where it is swallowed.

The stomach produces **hydrochloric acid**, which kills microorganisms.

Phagocytes are a type of white blood cell. They move around in the bloodstream and body tissues searching for pathogens. When they find pathogens, they engulf and digest them in a process called **phagocytosis**.

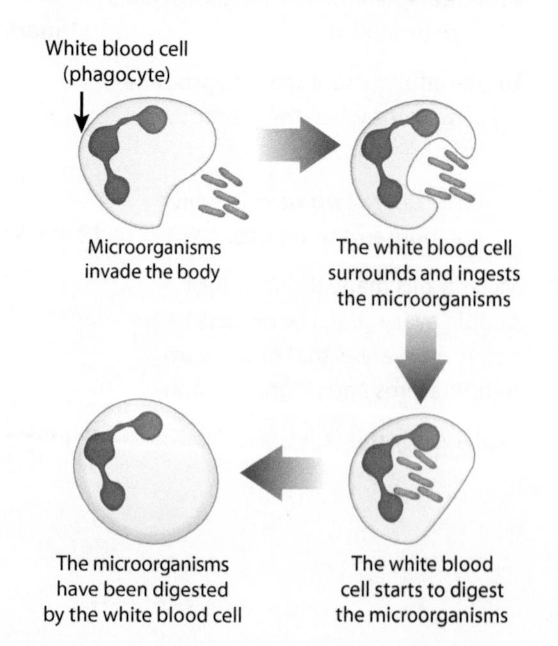

Specific defences

White blood cells called **lymphocytes** recognise molecular markers on pathogens called antigens. They produce antibodies that lock on to the antigens on the cell surface of the pathogen cell. The immobilised cells are clumped together and engulfed by phagocytes.

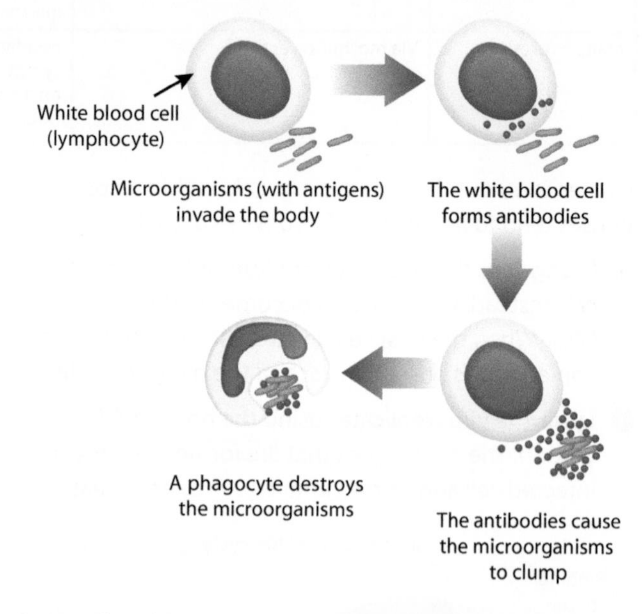

Some white blood cells produce **antitoxins** that neutralise the poisons produced from some pathogens.

Every pathogen has its own unique antigens. Lymphocytes make antibodies specifically for a particular antigen.

Example: Antibodies to fight TB will not fight cholera

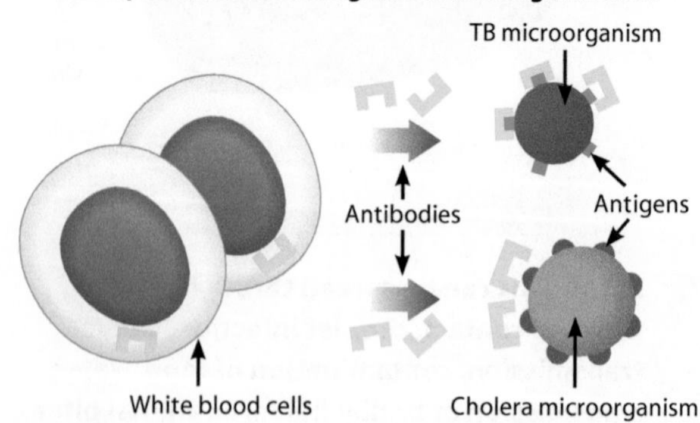

Active immunity

Once lymphocytes recognise a particular pathogen, the interaction is stored as part of the body's immunological memory through memory lymphocytes. These memory cells can produce the right antibodies much quicker if the same pathogen is detected again, therefore providing future protection against the disease. The process is called the secondary response and is part of the body's active immunity.

Active immunity can also be achieved through vaccination.

Memory lymphocytes and antibody production

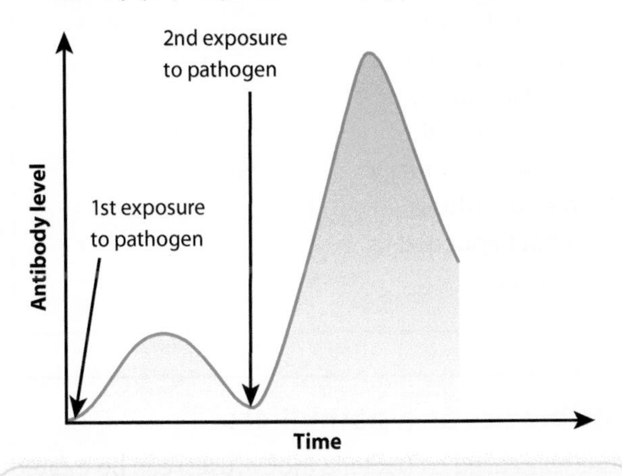

WS Investigating the growth of microorganisms involves culturing microorganisms.

This presents hazards that require a risk assessment. A risk assessment involves taking into account the severity of each hazard and the likelihood that it will occur.

Any experiment of this type involves thinking about risks in advance. Here is an example of a risk assessment table for this investigation.

Hazard	Infection from pathogen	Scald from autoclave (a specialised pressure cooker for superheating its contents)
Risk	High	High
How to lower the risk	• Observe aseptic technique. • Wash hands thoroughly before and after experiment. • Store plates at a maximum temperature of 25°C.	• Ensure lid is tightly secured. • Adjust heat to prevent too high a pressure. • Wait for autoclave to cool down before removing lid.

SUMMARY

- **The body has a number of non-specific defences, including the skin, tears, and stomach acid.**
- **The body has specific defences in the form of lymphocytes.**
- **Lymphocytes produce antibodies, which lock on to antigens.**
- **Memory cells remember the shape of antigens and provide a defence against further infection. This is active immunity.**

QUESTIONS

QUICK TEST

1. Describe the process of phagocytosis.
2. What is an antigen?
3. What enzyme is found in tears?

EXAM PRACTICE

1. The diagram shows a white blood cell producing small proteins as part of the body's immune system.

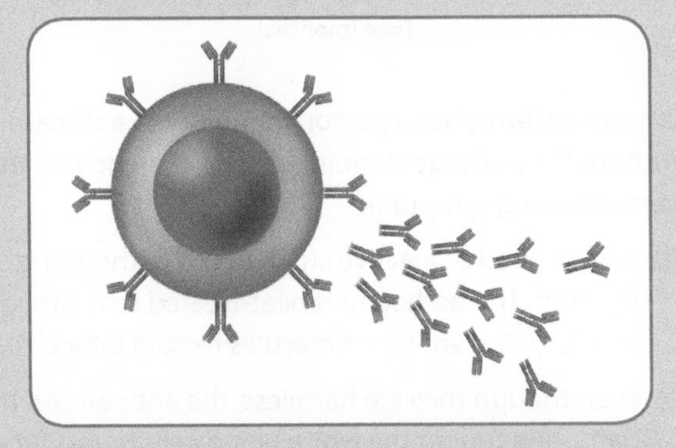

a) What is the name of these proteins? **[1 mark]**

b) i) What is the name of this type of white blood cell? **[1 mark]**

ii) Describe what happens to these components if the body is invaded again by the same pathogen. **[3 marks]**

Fighting disease

Vaccination

There are two types of vaccination: active and passive.

Passive immunisation

Antibodies are introduced into an individual's body, rather than the person producing them on their own. Some pathogens or toxins (e.g. snake venom) act very quickly and a person's immune system cannot produce antibodies quickly enough. So the person must be injected with the antibodies. However, this does not give long-term protection.

Active immunisation

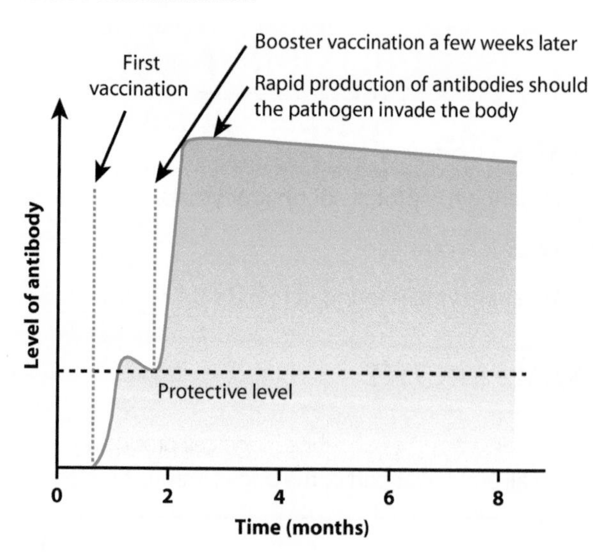

Immunisation gives a person immunity to a disease without the pathogens multiplying in the body, or the person having symptoms.

- A weakened or inactive strain of the pathogen is injected. The pathogen is heat-treated so it cannot multiply. The antigen molecules remain intact.
- Even though they are harmless, the antigens on the pathogen trigger the white blood cells to produce specific antibodies.
- As with natural immunity, **memory lymphocytes** remain sensitised. This means they can produce more antibodies very quickly if the same pathogen is detected again.

HT

Benefits of immunisation	Risks of immunisation
• It protects against diseases that could kill or cause disability (e.g. polio, measles). • If everybody is vaccinated and herd immunity is established, the disease eventually dies out (this is what happened to smallpox).	• A person could have an allergic reaction to the vaccine (small risk).

Antibiotics and painkillers

Diseases caused by bacteria (not viruses) can be treated using **antibiotics**, e.g. penicillin. Antibiotics are drugs that destroy the pathogen. Some bacteria need to be treated with antibiotics specific to them.

Antibiotics work because they **inhibit** cell processes in the bacteria but not the body of the host.

Viral diseases can be treated with **antiviral drugs**, e.g. swine flu can be treated with 'Tamiflu' tablets. It is a challenge to develop drugs that destroy viruses without harming body tissues.

Antibiotic resistance

Antibiotics are very effective at killing bacteria. However, some bacteria are **naturally resistant** to particular antibiotics. It is important for patients to follow instructions carefully and take the full course of antibiotics so that all the harmful bacteria are killed.

If doctors over-prescribe antibiotics, there is more chance of resistant bacteria surviving. These multiply and spread, making the antibiotic useless. **MRSA** is a bacterium that has become resistant to most antibiotics. These bacteria have been called 'superbugs'.

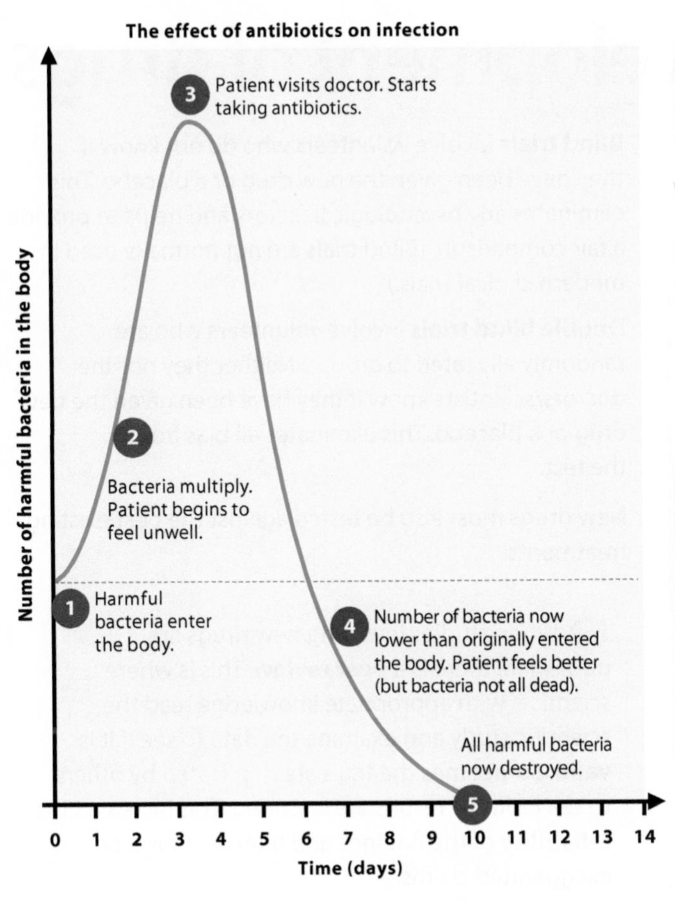

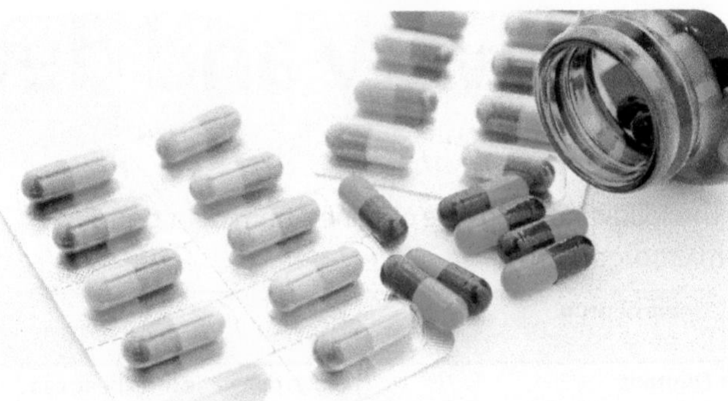

Painkillers

Painkillers or **analgesics** are given to patients to relieve symptoms of a disease, but they do not kill pathogens. Types of painkiller include paracetamol and ibuprofen. Morphine is another painkiller – it is a medicinal form of heroin used to treat extreme pain.

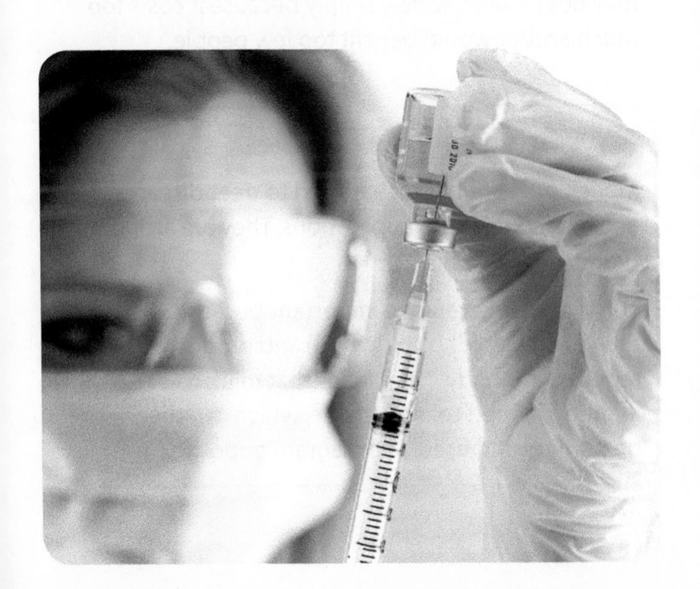

SUMMARY

- **There are two types of vaccination: passive immunisation and active immunisation.**
- **Diseases caused by bacteria can be treated using antibiotics, but some bacteria are resistant to antibiotics.**
- **Painkillers do not kill pathogens; they only relieve symptoms.**

QUESTIONS

QUICK TEST

1. What is an antibiotic?
2. Name the two types of vaccination.
3. What can doctors and patients do to reduce the risk of antibiotic-resistant bacteria developing?

EXAM PRACTICE

1. Explain how a vaccine works. **[4 marks]**
2. Bella is suffering from the flu and has asked her doctor for some antibiotics.
   a) How do antibiotics work? **[1 mark]**
   b) Explain why the doctor will not prescribe her antibiotics. **[2 marks]**
3. Explain why antibiotics are becoming increasingly less effective against 'superbugs' such as MRSA. **[3 marks]**

Discovery and development of drugs

Discovery of drugs

The following drugs are obtained from plants and microorganisms.

Name of drug	Use
Digitalis Foxgloves (common garden plants that are found in the wild)	Slows down the heartbeat; can be used to treat heart conditions
Aspirin From Willow trees (aspirin contains the active ingredient salicylic acid)	Mild painkiller
Penicillin Penicillium mould (discovered by Alexander Fleming)	Antibiotic

Modern pharmaceutical drugs are synthesised by chemists in laboratories, usually at great cost. The starting point might still be a chemical extracted from a plant.

New drugs have to be developed all the time to combat new and different diseases. This is a lengthy process, taking up to ten years. During this time the drugs are tested to determine:

- that they work
- that they are safe
- that they are given at the correct dose (early tests usually involve low doses).

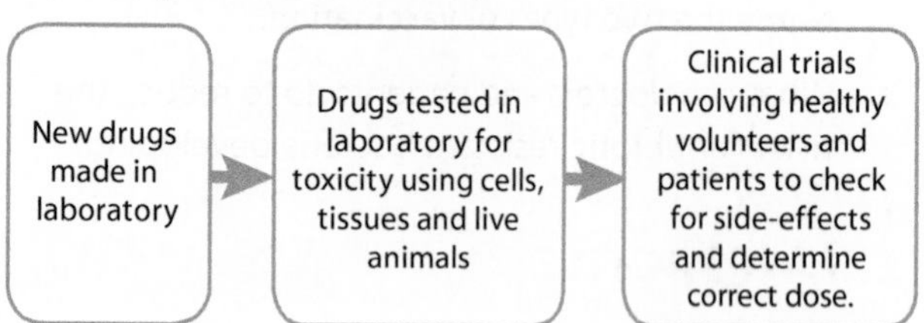

In addition to testing, computer models are used to predict how the drug will affect cells, based on knowledge about how the body works and the effects of similar drugs. There are many who believe this type of testing should be extended and that animal testing should be phased out.

Clinical trials

Clinical trials are carried out on healthy volunteers and patients who have the disease. Some are given the new drug and others are given a placebo. The effects of the drug can then be compared to the effects of taking the placebo.

Blind trials involve volunteers who do not know if they have been given the new drug or a placebo. This eliminates any psychological factors and helps to provide a fair comparison. (Blind trials are not normally used in modern clinical trials.)

Double blind trials involve volunteers who are randomly allocated to groups. Neither they nor the doctors/scientists know if they have been given the new drug or a placebo. This eliminates all bias from the test.

New drugs must also be tested against the best existing treatments.

WS When studies involving new drugs are published, there is a **peer review**. This is where scientists with appropriate knowledge read the scientific study and examine the data to see if it is **valid**. Sometimes the trials are duplicated by others to see if similar results are obtained. This increases the **reliability** of the findings and filters out false or exaggerated claims.

Once a **consensus** is agreed, the paper is published. This allows others to hear about the work and to develop it further.

In the case of pharmaceutical drugs, clinical bodies have to decide if the drug can be **licensed** (allowed to be used) and whether it is **cost-effective**. This can be controversial because a potentially life-saving drug may not be used widely simply because it costs too much and/or would benefit too few people.

HT Monoclonal antibodies

Monoclonal antibodies are used to treat diseases and in technological applications. They are artificially produced in the laboratory.

The hybridoma cells are important because they divide rapidly (by cloning) and produce the required antibody. This means that millions of identical cloned cells are made, which can then be purified and used. See diagram opposite.

HT

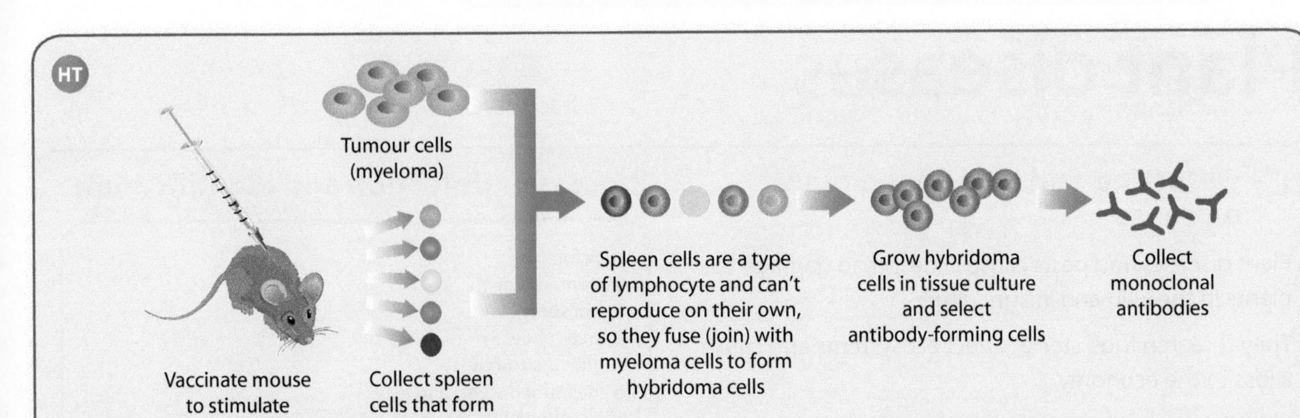

Uses of monoclonal antibodies

Use	How?	Advantages
Cancer treatment	Antibodies are made that react to an antigen found on a cancer cell, e.g. in a pancreatic cancer tumour. A toxic agent, such as a radioactive substance, is bound to the antibody and can then be injected into the patient.	The antibody enables the agent to be carried directly to cancer cells and does not harm other cells in the body (unlike chemotherapy and radiotherapy). The technique can also be used to locate blood clots.
Pregnancy testing kits	HCG is a hormone produced in pregnancy. Antibodies to the protein hormone are bound to a test strip and used to detect HCG in urine.	Rapid results (within minutes).
Measuring hormone levels in laboratories	Antibodies can be used to detect hormones or to detect pathogens and other chemicals in the blood.	Provides new information that wouldn't otherwise have been available.
Research	Used to locate a specific molecule in cells or tissues by binding to them with a fluorescent marker.	Marking of molecules made more specific.

When using monoclonal antibodies to treat cancer, more side-effects than expected can arise.

SUMMARY

- **Drugs used for treating illnesses and health conditions include antibiotics, analgesics and other chemicals that modify body processes and chemical reactions. In the past, these drugs came from plants and microorganisms.**
- **Clinical trials are carried out to ensure that drugs are safe to use.**
- HT **Monoclonal antibodies are a new technological advancement and have a variety of beneficial applications.**

QUESTIONS

QUICK TEST

1. What is a double blind trial?
2. The drug, Digitalis, is obtained from foxglove plants. True or false?
3. HT List two uses of monoclonal antibodies.

EXAM PRACTICE

1. **a)** New pharmaceutical drugs have to be extensively tested before their release on to the market. Explain why this is. **[2 marks]**

   b) Name one alternative to animal testing of drugs. **[1 mark]**

   HT **c)** Explain how monoclonal antibodies can be used in pregnancy testing kits. **[2 marks]**

Plant diseases

HT Detecting and identifying plant diseases

Plant diseases and pests cause widespread damage to plants in the wild and in agriculture.

They threaten food stocks, affect ecosystems and cause a loss to the economy.

Identifying and detecting plant diseases involves several stages.

Process of detection and identification

Look for:
- stunted growth
- leaf spots
- areas of decay
- growths/tumours
- malformed stems and leaves
- discolouration
- presence of pests.

↓

Eliminate environmental causes, e.g. abrasion, poor water supply, climatic factors.

↓

Study the distribution of affected plants and refer to garden manuals or websites.

↓

Carry out diagnostic testing in a laboratory to identify pathogens, e.g. observation; using monoclonal antibody testing kits; DNA testing.

Pathogens and pests that affect plants

Disease	Pathogen	Appearance/effect on plants	Treatment
Rose black spot	Fungal disease – the fungal spores are spread by water and wind	Purple/black spots on leaves; these then turn yellow and drop early, leading to a lack of photosynthesis and poor growth	Apply a fungicide and/or remove affected leaves
Tobacco mosaic virus (TMV)	Widespread disease that affects many plants (including tomatoes)	'Mosaic' pattern of discolouration; can lead to lack of photosynthesis and poor growth	
Ash dieback	Caused by the fungus Chalara	Leaf loss and bark lesions	
Barley powdery mildew	Erysiphe graminis	Causes powdery mildew to appear on grasses, including cereals	Fungicides and careful application of nitrogen fertilisers
Crown gall disease	Agrobacterium tumefaciens	Tumours or 'galls' at the crown of plants such as apple, raspberry and rose	Use of copper and methods of biological control

Pests	What they do	Appearance/effect on plants	Control
Invertebrates and particularly insects, e.g. many species of aphids	Feed on sap, leaves and storage organs; transmit pathogenic viruses		Chemical pesticides or biological control methods

Mineral ion deficiencies

Plants need **mineral ions** to build complex molecules. The ions are obtained from the soil via the roots in an active manner (requiring energy). In particular, plants need:

- **nitrates** to form **amino acids**, the building blocks of proteins. They are also needed to make nucleic acids such as DNA. Lack of nitrates in a plant leads to yellow leaves and stunted growth
- **magnesium** to form chlorophyll, which absorbs light energy for photosynthesis. Lack of magnesium results in chlorosis, which is a discolouration of the leaves.

Defence responses of plants

Plants defend themselves from herbivores and disease in a range of ways. These adaptations have evolved over millions of years to maximise a plant's survival.

Mechanical defences include:

- thorns and hairs to deter plant-eaters
- leaves that droop or curl on contact
- mimicry, to fool animals into avoiding them as food or laying their eggs on them. For example, passion flowers have structures that look like yellow eggs. A butterfly is less likely to lay eggs on a plant that has apparently already been used.

Physical defences include:

- tough, waxy leaf cuticles
- cellulose cell walls
- layers of dead cells around stems, e.g. bark. These fall off, taking pathogens with them.

Chemical defences include:

- producing antibacterial chemicals, e.g. mint and witch hazel
- producing toxins to deter herbivores, e.g. deadly nightshade, foxgloves and tobacco plants.

SUMMARY

- **Pathogens can affect plants. Diseases include Rose black spot and Tobacco mosaic virus (TMV).**
- **Plants need minerals for healthy functioning, especially nitrates and magnesium.**
- **Plants defend themselves from pests and pathogens in three ways: mechanical defences, physical defences and chemical defences.**

QUESTIONS

QUICK TEST

1. List three things that plant scientists look for when detecting whether a plant is diseased.
2. What type of disease is rose black spot?
3. What do plants need magnesium for?

EXAM PRACTICE

1. Black spot is a fungal disease that affects rose leaves.

   Fatima has noticed that roses growing in areas with high air pollution are less affected by the disease. She thinks that regular exposure to acid rain is stopping the disease from developing.

   Describe a simple experiment that Fatima could do to test her theory. **[3 marks]**

2. State one mechanical and one physical defence that plants use to defend themselves against herbivores or disease. **[2 marks]**

Homeostasis and negative feedback

Homeostasis

The body has automatic control systems to maintain a constant internal environment (**homeostasis**). These systems make sure that cells function efficiently.

Homeostasis balances inputs and outputs to ensure that optimal levels of temperature, pH, water, oxygen and carbon dioxide are maintained.

For example, even in the cold, homeostasis ensures that body temperature is regulated at about 37°C.

Control systems in the body may involve the nervous system, the endocrine system, or both. There are three components of control.

- **Effectors** cause responses that restore optimum levels, e.g. muscles and glands.
- **Coordination centres** receive and process information from the receptors, e.g. brain, spinal cord and pancreas.
- **Receptors** detect stimuli from the environment, e.g. taste buds, nasal receptors, the inner ear, touch receptors and receptors on retina cells.

WS When taking measurements, the quality of the measuring instrument and a scientist's skill is very important to achieve **accuracy**, **precision** and **minimal error**.

Adrenaline levels in blood plasma are measured by a chromatography method called HPLC. This is often coupled to a detector that gives a digital readout.

This digital readout displays the concentration of adrenaline as 6.32. This means that the instrument is precise up to $\frac{1}{100}$ of a unit.

A less precise instrument might only measure down to $\frac{1}{10}$ of a unit, e.g. 6.3 (one decimal place).

A bar graph of some data generated from HPLC is shown alongside. The graph shows the **average concentration** of three different samples of blood. The average is taken from many individual measurements.

The vertical error bars indicate the range of measurements (the difference between highest and lowest) obtained for each sample.

Sample C shows **the greatest precision** as the individual readings do not vary as much as the others. There is less error so we can be more confident that the average is closer to the **true value** and therefore more **accurate**.

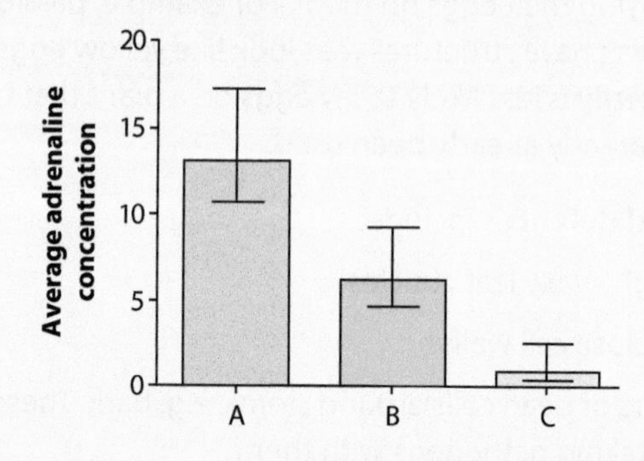

HT Negative feedback

Negative feedback occurs frequently in homeostasis. It involves the automatic reversal of a change in the body's condition.

In the body, examples of negative feedback include osmoregulation/water balance, balancing blood sugar levels, maintaining a constant body temperature and controlling metabolic rate.

Metabolism needs to be controlled so that chemical reactions in the body take place at an optimal rate. Negative feedback controls metabolic rate by using the hormones **thyroxine** and **adrenaline**.

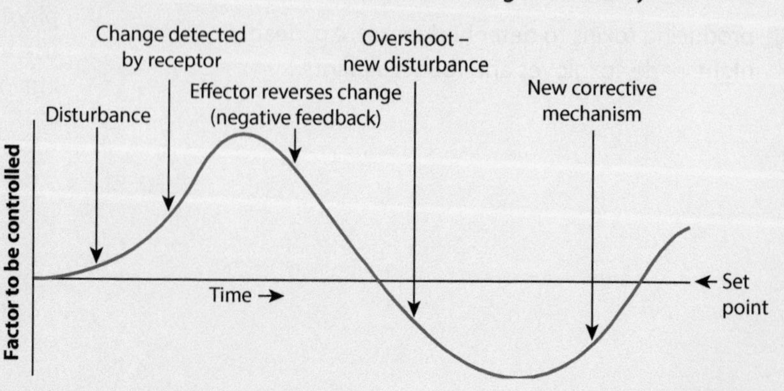

HT Thyroxine

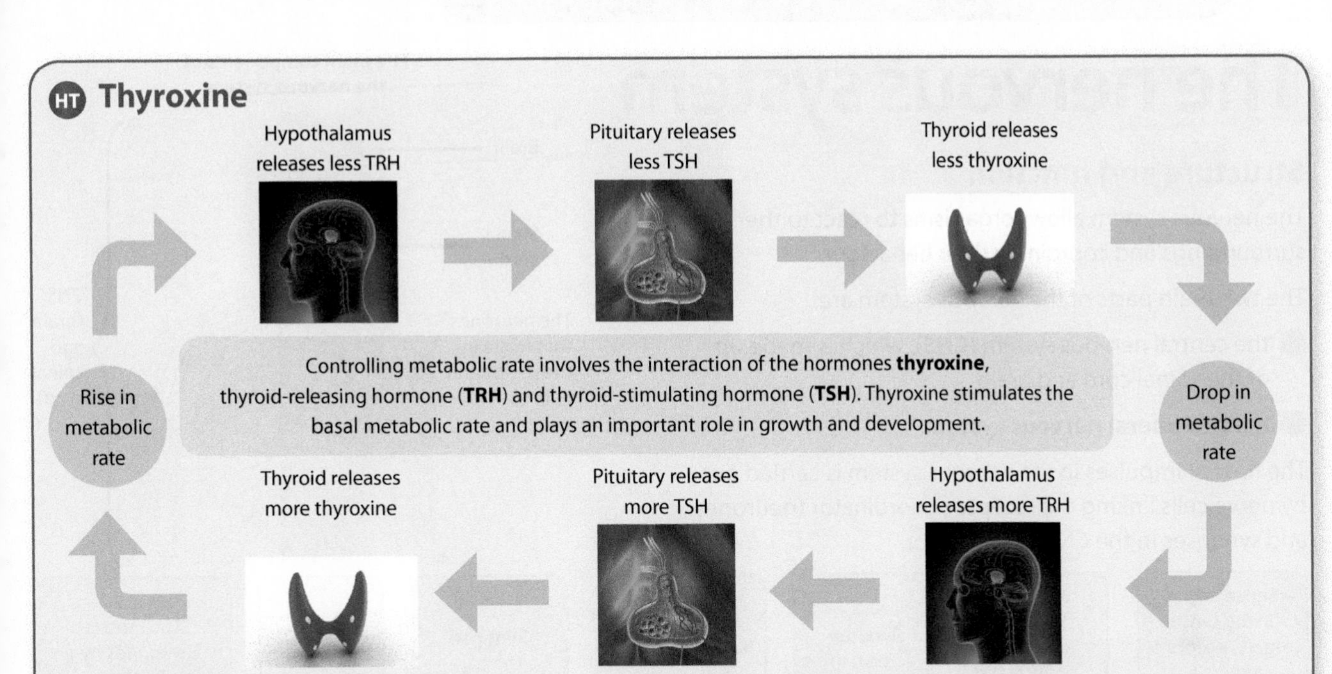

Controlling metabolic rate involves the interaction of the hormones **thyroxine**, thyroid-releasing hormone (**TRH**) and thyroid-stimulating hormone (**TSH**). Thyroxine stimulates the basal metabolic rate and plays an important role in growth and development.

Adrenaline

Adrenaline is sometimes called the 'flight or fight' hormone. During times of stress the adrenal glands produce adrenaline. It has a direct effect on muscles, the liver, intestines and many other organs to prepare the body for sudden bursts of energy. Specifically, adrenaline increases the heart rate so that the brain and muscles receive oxygen and glucose more rapidly.

SUMMARY

- **The body has control systems that work automatically to maintain a constant internal environment. This is homeostasis.**
- **Negative feedback is the automatic reversal of a change in the body's condition.**

QUESTIONS

QUICK TEST

1. At what temperature is the human body regulated?
2. (HT) Give two examples where negative feedback operates in the human body.

QUESTIONS

EXAM PRACTICE

1. (HT) Match the organs below, with the function they perform:

Pancreas	Releases TSH
Skin receptor	Detects pressure
Pituitary gland	Releases TRH
Hypothalamus	Produces insulin

[3 marks]

The nervous system

Structure and function

The nervous system allows organisms to react to their surroundings and coordinate their behaviour.

The two main parts of the nervous system are:

- the central nervous system (**CNS**), which is made up of the spinal cord and brain
- the **peripheral nervous system**.

The main components of the nervous system

Brain

Spinal cord

The neurones that make up the peripheral nervous system

CNS (brain and spinal cord)

The flow of impulses in the nervous system is carried out by nerve cells linking the receptor, coordinator (neurones and synapses in the CNS) and effector.

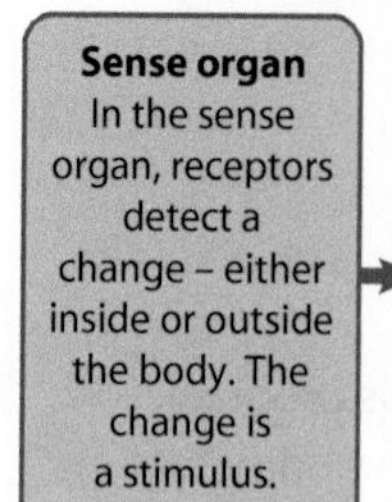

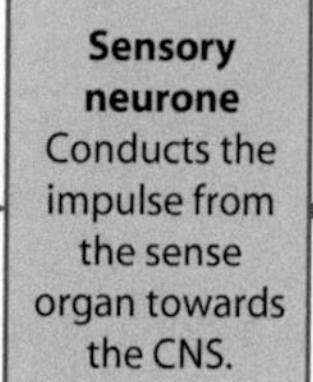

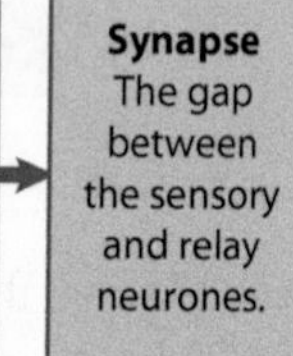

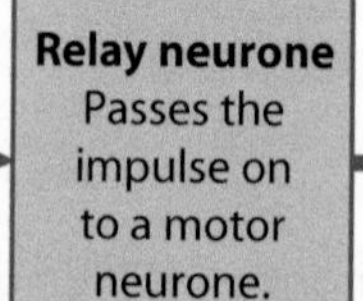

Synapse
The gap between the relay neurone and the motor neurone.

Motor neurone
Passes the impulse on to the muscle (or gland).

Muscle
The muscle responds by contracting, which results in a movement. Muscles and glands are examples of effectors.

Nerve cells or neurones are specially adapted to carry nerve impulses, which are electrical in nature. The impulse is carried in the long, thin part of the cell called the axon.

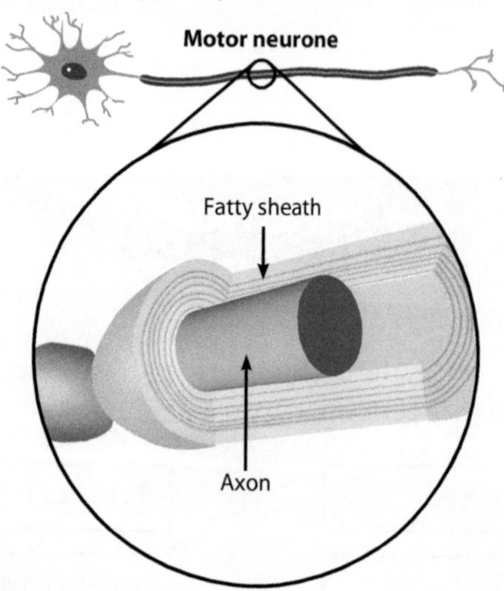

There are three types of neurone:

Sensory neurones carry impulses from receptors to the CNS.

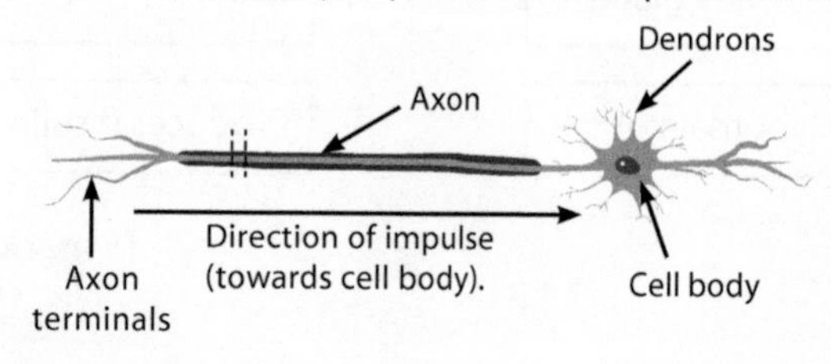

Relay neurones make connections between neurones inside the CNS.

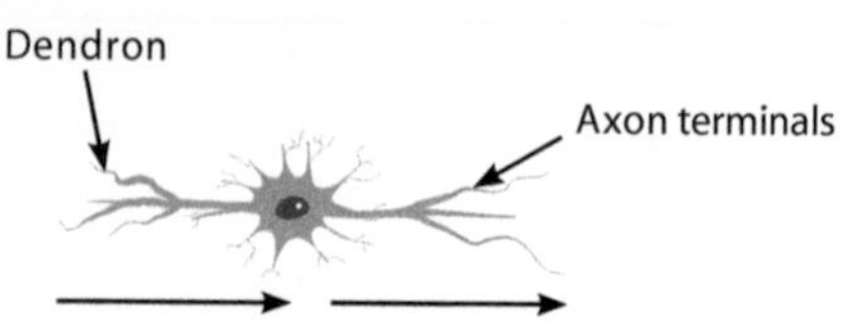

Impulse travels first towards, and then away from, cell body.

Motor neurones carry impulses from the CNS to muscles and glands.

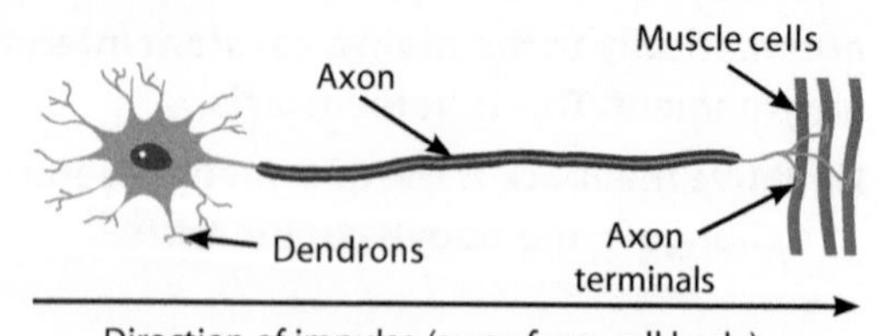

Direction of impulse (away from cell body).

Synapses

Synapses are junctions between neurones.

They play an important part in regulating the way impulses are transmitted. Synapses can be found between different neurones, neurones and muscles, and between dendrites (the root-like outgrowths from the cell body).

When an impulse reaches a synapse, a neurotransmitter is released by the neurone ('A' in the following diagram) into the gap that lies between the neurones. It travels by diffusion and binds to receptor molecules on the next neurone. This triggers a new electrical impulse to be released.

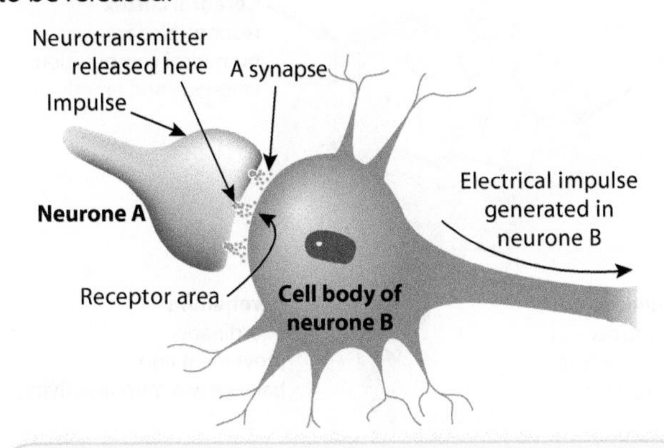

WS You may have to investigate the effect of factors on human reaction time.

For example, you could be asked to investigate a learned reflex by measuring how far up a ruler someone can catch it. The nearer to the zero the ruler is caught, the faster the reflex.

You could investigate factors such as:

- experience/practice at catching
- sound
- touch
- sight.

Can you design experiments to test these variables? Which factors will need to be kept the same?

SUMMARY

- **The nervous system is made up of the brain and spinal cord (CNS) and neurones.**
- **There are three types of neurone: sensory, relay and motor neurones.**
- **Synapses are junctions between neurones where a neurotransmitter is released.**
- **A reflex arc is the pathway taken by impulses around the body in response to a reflex action.**

QUESTIONS

QUICK TEST

1. Name the junctions between neurones.
2. How do reflex arcs aid survival of an organism?

Reflex arcs

Reflex actions:

- are involuntary/automatic
- are very rapid
- protect the body from harm
- bypass conscious thought.

The pathway taken by impulses around the body is called a reflex arc. Examples include:

- opening and closing the pupil in the eye
- the knee-jerk response
- withdrawing your hand from a hot plate.

Here is the arc pathway for a pain response.

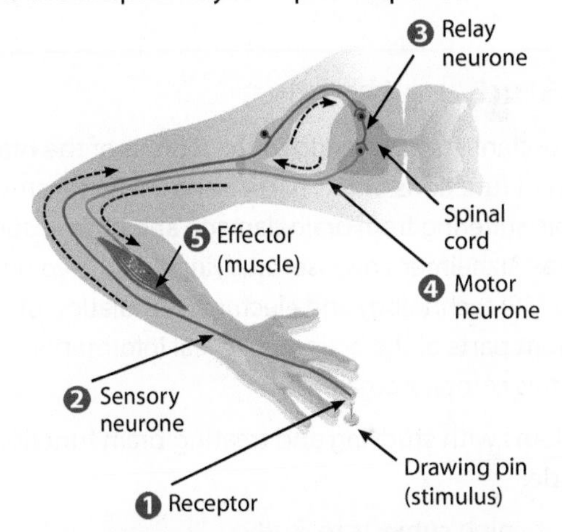

QUESTIONS

EXAM PRACTICE

1. The flow chart illustrates the events that occur during a reflex action.

   X → Receptor → Sensory neurone → Relay neurone → Brain and/or spinal cord → Motor neurone → Effector

   Paul accidentally puts his hand on a pin. Without thinking, he immediately pulls his hand away.

   a) Which component of a reflex arc is represented by the letter X? **[1 mark]**

   b) Give **two** reasons why this can be described as a reflex action. **[2 marks]**

   c) Use the features in the flow chart to describe what happens in this reflex action. **[4 marks]**

Nervous control

The brain

The brain is part of the central nervous system (**CNS**) and is studied by neuroscientists. The brain controls many activities in the body but is also responsible for memory, learning and behaviour.

The brain is made up of billions of neurones, each interlinked via synapses. There are four main parts of the brain.

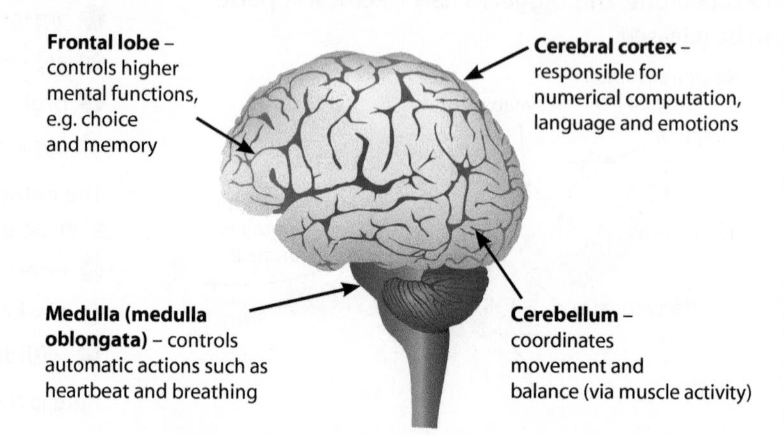

HT Studying the brain

Neuroscientists have mapped the regions of the brain to different functions. This is particularly useful for studying people suffering from brain damage and brain disorders, such as Alzheimer's disease. Studying healthy volunteers, using MRI technology and electrical stimulation of different parts of the brain, can reveal information about how this complex organ works.

Problems with studying and treating brain function include:

- obtaining subjects to study
- ethical issues relating to using human subjects
- difficulty when interpreting case studies
- difficulty in accessing areas of the brain where damage has occurred.

When treating nerve damage, the prospects can be limited because nervous and surrounding tissues are difficult to repair.

Scanning the brain

CT or **CAT** scans (computerised axial tomography) are X-ray tests producing cross-sectional images of the brain using computer programs. Many images can be taken at different angles to create 3D pictures.

PET stands for positron emission tomography. A PET scan can show what body tissues look like and how they are working. PET scanners work by detecting radiation given off by radiotracers, which the patient takes orally. The tracers accumulate in particular areas of the body. PET is often combined with CT and MRI technology to give 3D views of the brain.

Controlling body temperature

The thermoregulatory centre in the brain monitors and controls body temperature.

Enzymes work at around **37°C** (core temperature for humans). It is essential that this temperature does not fluctuate too much. Body heat is released by chemical reactions in the body (metabolism) and distributed via the bloodstream.

The skin plays a major role in temperature regulation. The structures responsible are found in the dermis and epidermis of the skin.

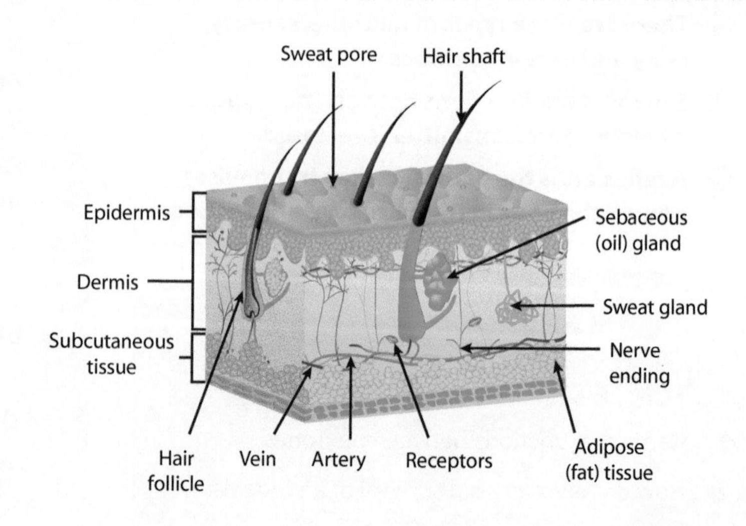

When core temperature falls

- There is reduced blood flow near the skin to limit heat loss to the environment. This is called vasoconstriction.
- Tiny involuntary contractions of muscles (shivering) generate heat.
- Sweating reduces.
- Body hairs rise due to the contraction of special muscles. Humans have less body hair than most mammals so there is minimal benefit.

When core temperature rises

- Blood flows closer to the skin surface because blood vessels widen. This is called vasodilation.
- Sweating increases. The sweat evaporates on the skin's surface, which absorbs heat from the skin in an endothermic change.
- Body hairs lower.

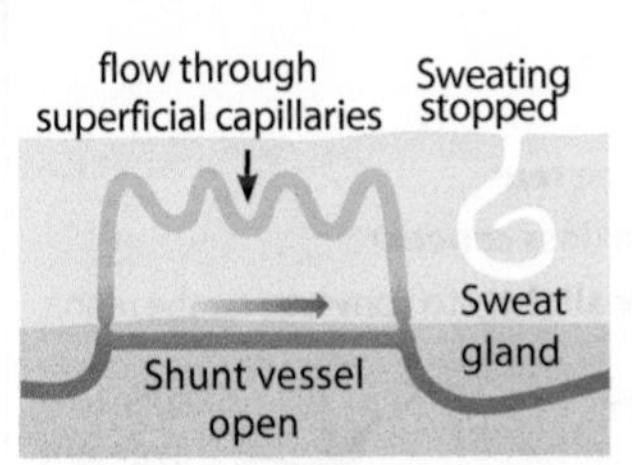

Core temperature falling

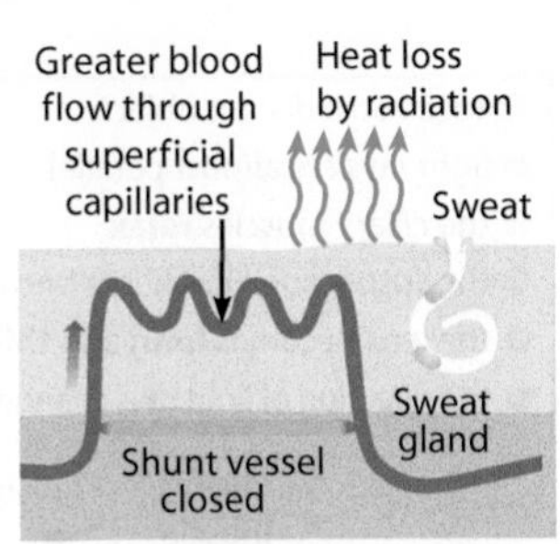

Core temperature rising

Negative feedback system

There is a negative feedback system for regulating body temperature.

In this control system we have:

- receptors – thermoreceptors in the skin
- a controller – hypothalamus in the brain
- effectors – a variety of organs and tissues.

If a control system breaks down it can be life-threatening.

- **Hypothermia** is when the body temperature falls well below 37°C. Symptoms include shivering, blue skin colour, disorientation and unconsciousness.
- **Heat stroke/dehydration** is when the body temperature rises well above 37°C. Symptoms include an altered mental state, nausea, flushed skin, and rapid breathing and pulse.

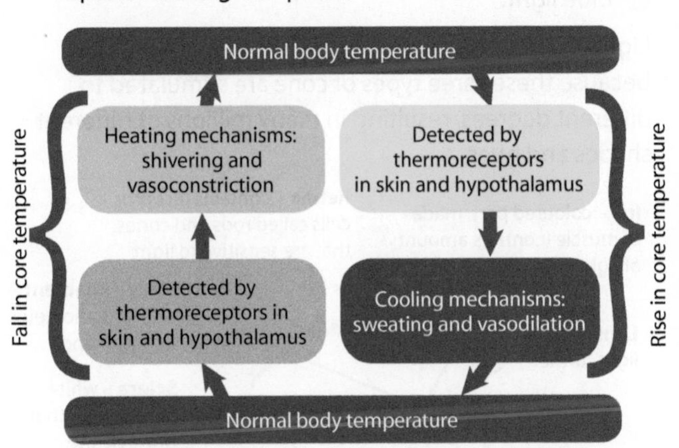

SUMMARY

- **The brain is part of the CNS and has four main parts: frontal lobe, cerebral cortex, medulla and cerebellum.**
- **The thermoregulatory centre in the brain controls and monitors body temperature.**
- **The skin plays a major role in keeping core temperature within acceptable limits.**

QUESTIONS

QUICK TEST

1. What does CNS stand for?
2. What is vasodilation?

QUESTIONS

EXAM PRACTICE

HT 1. Jenny has been sunbathing on a beach where the temperature hasn't fallen below 35°C.

Her body's thermoregulatory mechanisms have responded.

Describe how negative feedback has operated to try to ensure that her core temperature does not rise much above 37°C.

In your answer, explain how the thermoregulatory centre and the skin are involved. **[6 marks]**

The eye

Structure

The eye is a major sense organ in the human body. The receptors are on the **retina** – they consist of light-sensitive cells that detect the intensity and colour of light that enters the eye.

Light-sensitive cells are of two types – rods and cones. Rods are sensitive to low light intensities and enable vision in black and white. Cones are less sensitive and operate at higher light intensities. They enable us to see in colour. The three types of cone are those sensitive to:

- red light
- green light
- blue light.

Light entering the eye is perceived as a certain colour because these three types of cone are stimulated to different degrees, resulting in many millions of different shades and hues.

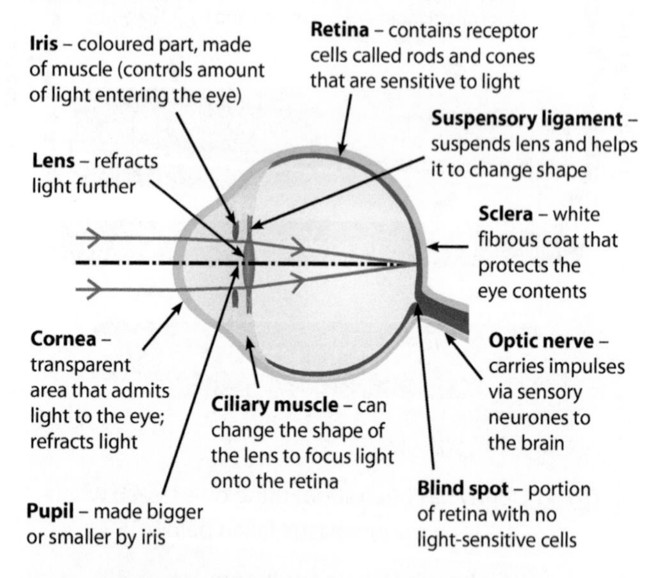

Accommodation

The function of the eye is to receive light rays and focus them to form a sharp image on the retina. The optical information is sent to the brain, which interprets it to give a sense of a 3D image, movement, colour and brightness. This interpretation by the occipital lobe in the brain is called **perception**.

The focusing process is called **accommodation**. It involves changing the shape of the lens in order to refract light.

To focus on a **near object**:

- light rays diverge significantly
- the ciliary muscles **contract**
- the suspensory ligaments loosen/go **slack**
- the lens becomes **short** and **fat** (**more convex**)
- the direction of light rays changes **greatly** to **converge** on the retina.

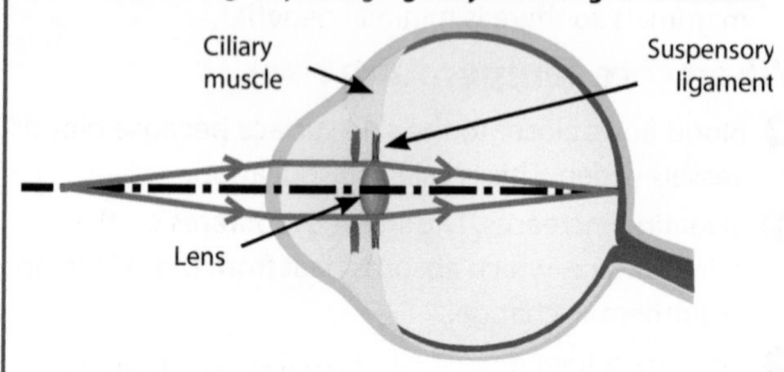

To focus on a distant object:

- light rays are almost parallel
- the ciliary muscles **relax**
- the suspensory ligaments become **taut**
- the lens becomes **long** and **thin** (**less convex**)
- the direction of light rays changes **slightly** to converge on the retina.

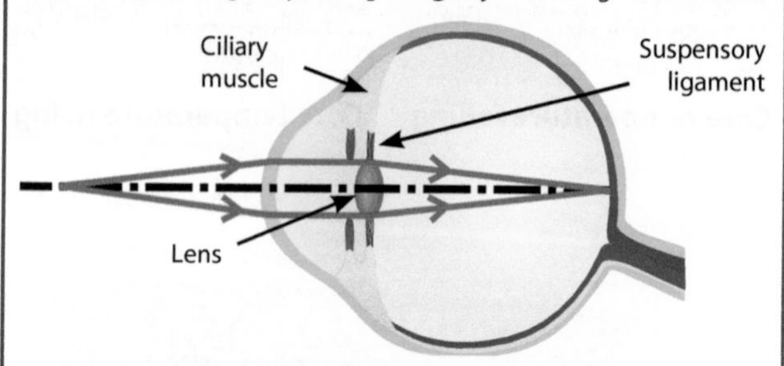

This is the front view of what happens during accommodation:

Near object

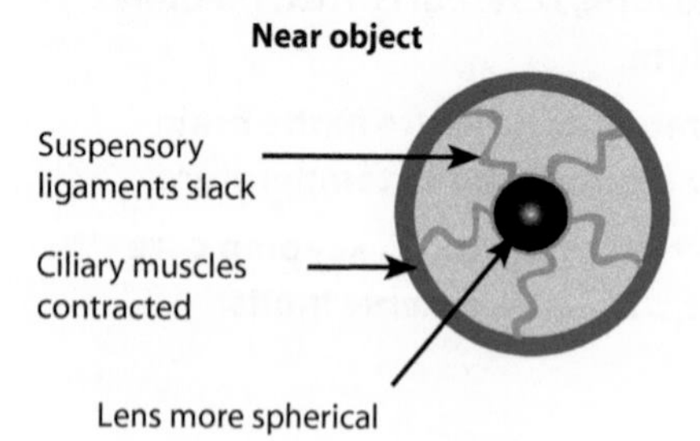

Distant object

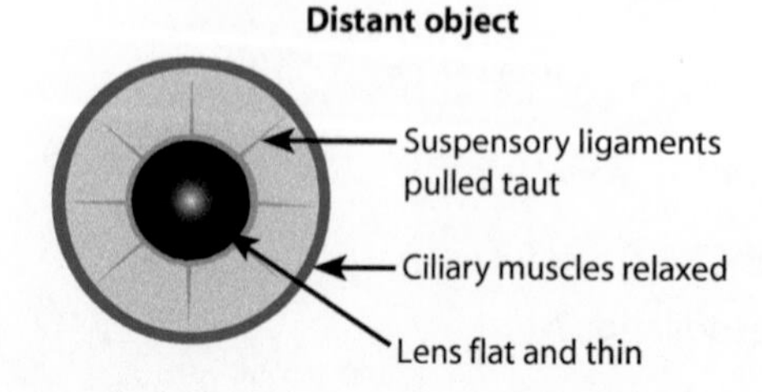

Defects

Common eye defects are:

- hyperopia (long sightedness)
- myopia (short sightedness)
- red–green colour blindness
- cataracts.

Hyperopia and myopia can be corrected using spectacles, contact lenses or laser surgery (to change the shape of the cornea).

Hyperopia is:

- caused by an eyeball that is too short, or a lens that stays too long and thin
- corrected by a **convex** lens.

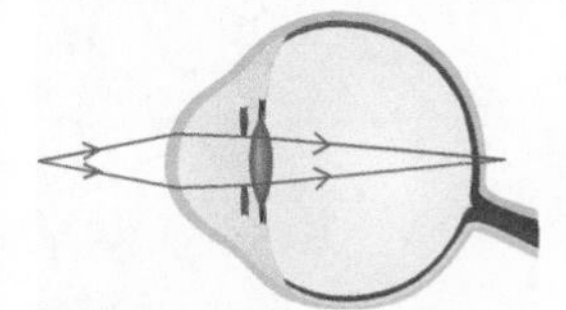
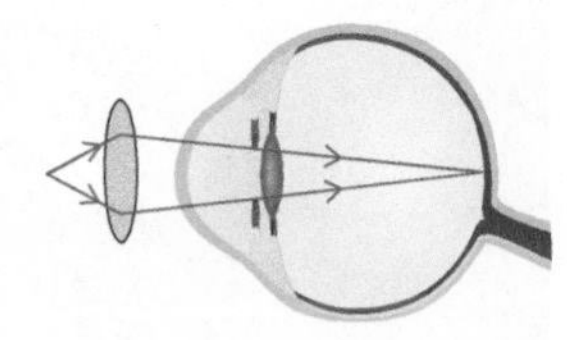

Myopia is:

- caused by an eyeball that is too long, or weak suspensory ligaments that cannot pull the lens into a thin shape
- corrected by a **concave** lens.

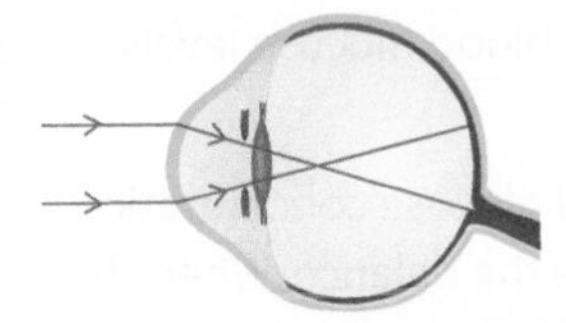
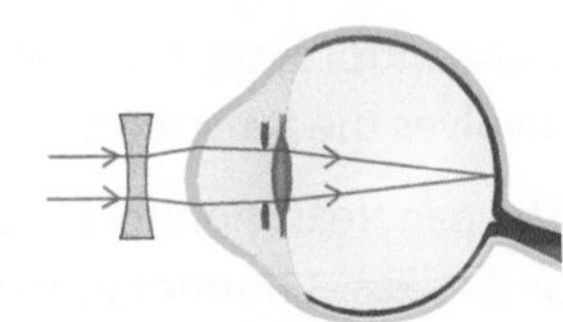

Red–green colour blindness is more common in boys as it is determined by a defective gene on the X chromosome. The result is one or more cone types being faulty, causing an inability to distinguish between different shades of red or green.

A **cataract** develops when the lens clouds over. This is caused by some of the protein in the lens clumping together. Using spectacles with strong bifocals is a temporary treatment. The only permanent solution is removal during surgery.

An alternative to spectacles is **contact lenses**. These are miniature lenses (either hard or soft) that can be placed over the front of the eye. They need to be kept clean if they are a permanent type, or disposed of if temporary. Some operations can replace the natural lens with a new one.

WS You will be expected to give examples of how scientific ideas can lead to technical applications. The development of optical instruments for helping vision is a good example.

Sir Isaac Newton developed laws of optics in the seventeenth century. This built on theories put forward by scientists previously. An understanding of how light is refracted and reflected, and of ray diagrams, has led to inventions such as telescopes, microscopes and spectacles.

SUMMARY

- **The eye receives light rays and focuses them to form an image on the retina.**
- **The focusing process is called accommodation.**
- **Common eye defects are hyperopia and myopia.**

QUESTIONS

QUICK TEST

1. Which part of the eye contains rods and cones?
2. What causes myopia?

EXAM PRACTICE

1. Martin is reading a book when he hears a plane fly over above him. He shields his eyes against the sun and soon observes it as it flies into the distance.

   a) Explain the change that occurs in the pupil of Martin's eye as he observes the plane.

   State why this change needs to take place. **[3 marks]**

   b) In order to see the plane clearly, Martin's eyes need to focus on it.

   Describe how muscles in the eye affect the lenses in order to change the pathway of light rays. **[3 marks]**

The endocrine system

Structure and function

The endocrine system is made up of glands that are ductless and secrete **hormones** directly into the bloodstream. The blood carries these chemical messengers to **target organs** around the body, where they cause an effect.

Hormones:

- are large protein molecules
- interact with the nervous system to exert control over essential biological processes
- act over a longer time period than nervous responses but their effects are slower to establish.

Endocrine gland	Hormone(s) produced
Pituitary gland	TSH, ADH, FSH, LH, etc.
Pancreas	Insulin, glucagon
Thyroid	Thyroxine
Adrenal gland	Adrenaline
Ovaries (female)	Oestrogen, progesterone
Testes (male)	Testosterone

The endocrine system

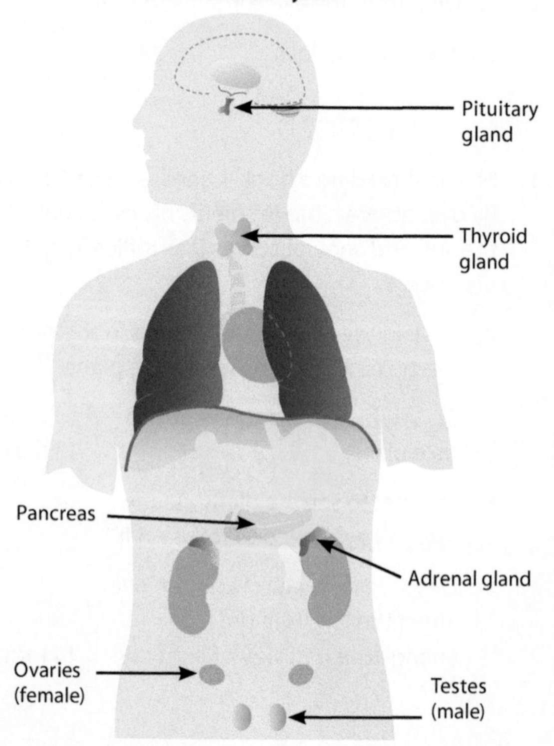

The pituitary gland

The pituitary is often referred to as the **master gland** because it secretes many hormones that control other processes in the body. Pituitary hormones often trigger other hormones to be released.

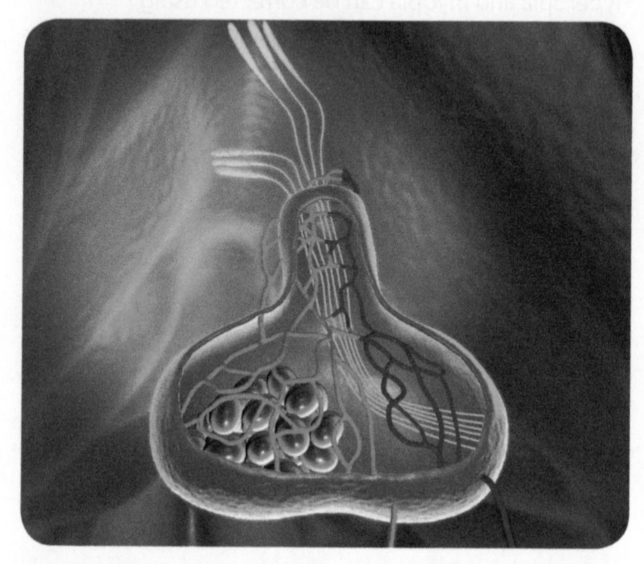

Controlling blood glucose concentration

The control system for balancing blood glucose levels involves the **pancreas**.

The pancreas monitors the blood glucose concentration and releases hormones to restore the balance. When the concentration is too high, the pancreas produces insulin that causes glucose to be absorbed from the blood by all body cells, but particularly those in the liver and muscles. These organs convert glucose to glycogen for storage until required.

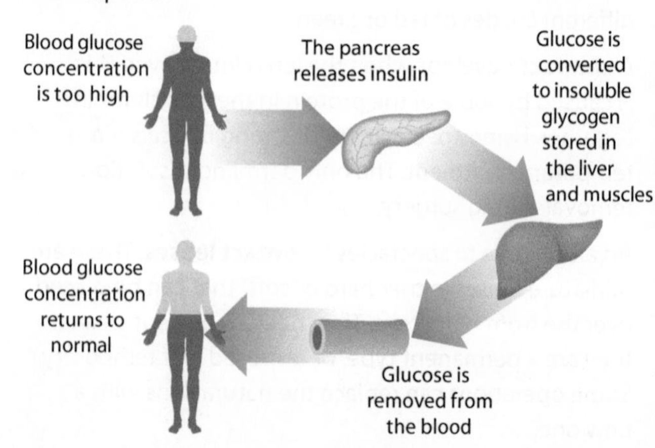

HT If blood glucose concentration is too low, the pancreas secretes glucagon. This stimulates the conversion of glycogen to glucose via enzymic systems. It is then released into the blood.

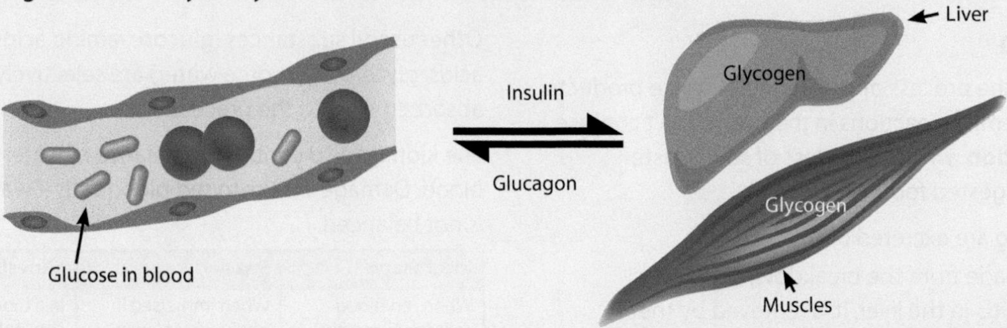

Diabetes

There are two types of diabetes.

Type I diabetes:

- is caused by the pancreas' inability to produce insulin
- results in dangerously high levels of blood glucose
- is controlled by delivery of insulin into the bloodstream via injection or a 'patch' worn on the skin
- is more likely to occur in people under 40
- is the most common type of diabetes in childhood
- is thought to be triggered by an auto-immune response where cells in the pancreas are destroyed.

Type II diabetes:

- is caused by fatty deposits preventing body cells from absorbing insulin; the pancreas tries to compensate by producing more and more insulin until it is unable to produce any more, which results in dangerously high levels of blood glucose
- is controlled by a low carbohydrate diet and exercise initially; it may require insulin in the later stages
- is more common in people over 40
- is a risk if you are obese.

SUMMARY

- **The endocrine system consists of glands that secrete hormones into the bloodstream.**
- **The pituitary gland secretes many hormones that control other processes.**
- **The pancreas monitors blood glucose concentration.**
- **Diabetes results from the body's inadequate production of insulin.**

QUESTIONS

QUICK TEST

1. Which hormone is produced in the thyroid?
2. Where is insulin produced in the body?
3. What effects does type II diabetes have on the body?

EXAM PRACTICE

1. a) When someone eats a meal, the pancreas responds by producing insulin.

   Describe the effects this has on the body? **[2 marks**

   b) People who have type I diabetes have trouble balancing their blood glucose concentrations.

   How is this condition treated? **[1 mark**

   c) How is type II diabetes caused and managed? **[2 marks**

Water and nitrogen balance

Excretion

Excretion is the process of getting rid of waste products made by chemical reactions in the body. Don't confuse it with **egestion**, which is the loss of solid waste (mainly undigested food).

The following are excreted products.

- Urea is made from the breakdown of excess amino acids in the liver. It is removed by the kidneys along with excess water and ions and transferred to the bladder as **urine** before being released.
- **Sweat** containing water, urea and salt is excreted by sweat glands onto the surface of the skin. Sweating aids the body's cooling process.
- **Carbon dioxide** and **water** are produced by respiration and leave the body from the lungs during exhalation.

The lungs and skin don't control the loss of substances. They are simply the organs by which these substances are removed.

The kidneys

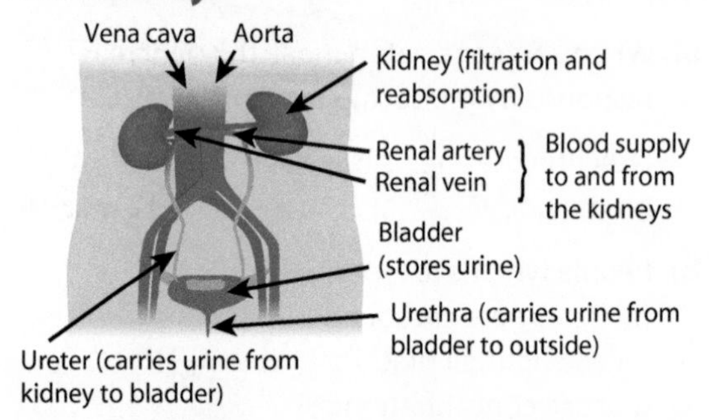

The kidneys filter the blood, allowing urea to pass to the bladder. The filtering is carried out within the kidney by thousands of tiny **kidney tubules**.

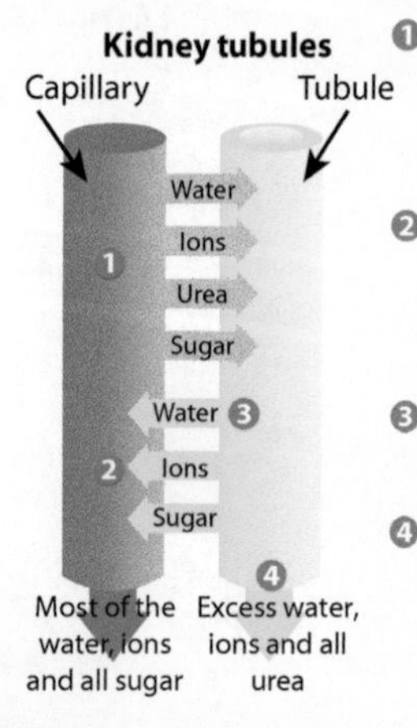

1. **Filtration**
Lots of water plus all the small molecules are squeezed out of the blood, under pressure, into the tubules.
2. **Selective reabsorption**
Useful substances, including glucose, ions and water, are reabsorbed into the blood from the tubules.
3. **Osmoregulation**
Amount of water in the blood and urine is adjusted here.
4. **Excretion of waste**
Excess water, ions and all the urea now pass to the bladder in the form of urine and are eventually released from the body.

Other useful substances (glucose, amino acids, fatty acids, glycerol and some water) are selectively re-absorbed early in the process.

The kidneys also control the balance of water in the blood. Damage occurs to red blood cells if water content is not balanced.

Ideal shape	Swollen	Shrivelled
When red blood cells (erythrocytes) are in solutions with equal concentration to their cytoplasm, they have an ideal, biconcave shape. This is because there is no net movement of water in or out.	When immersed in a solution of lower concentration (higher water concentration), the cells absorb water by osmosis. The weak cell membrane cannot resist the added water pressure and may burst.	In a more concentrated solution (lower water concentration), cells lose water by osmosis. They shrivel up and become **crenated** (have scalloped edges).

Kidney tubules (nephrons)

In terms of nephron structure, filtration, selective reabsorption and excretion of waste occur in the following regions.

Structure of the nephron

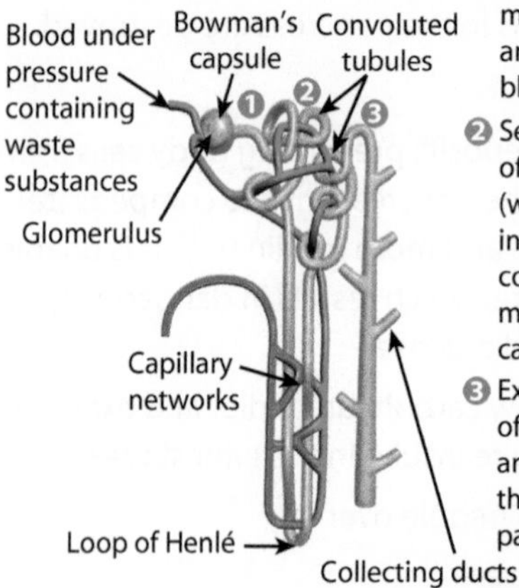

1. Filtration, where all small molecules and lots of water are squeezed out of the blood and into the tubules.
2. Selective reabsorption of useful substances (water, ions, glucose) back into the blood from the convoluted tubules. This may take energy in the case of glucose and ions.
3. Excretion of waste in the form of excess water, excess ions and all urea. These drain into the collecting tubules and pass to the bladder as urine.

Kidney failure

Kidneys may fail due to accidents or disease. A patient can survive with one kidney. If both kidneys are affected, two treatments are available.

Kidney transplant – involves a healthy person donating one kidney to replace two failed kidneys in another person.

Dialysis – offered to patients while they wait for the possibility of a kidney transplant. A dialysis machine removes urea and maintains levels of sodium and glucose in the blood.

This is what happens during dialysis.

1. Blood is taken from a person's vein and run into the dialysis machine, where it comes into close contact with a partially permeable membrane.
2. This separates the blood from the dialysis fluid.
3. The urea and other waste diffuse from the blood into the dialysis fluid. The useful substances remain and are transferred back to the body.

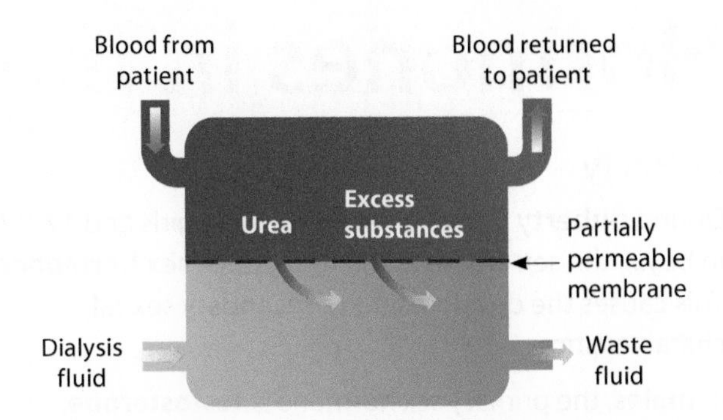

HT How urea is formed

Proteins obtained from the diet may produce a surplus of **amino acids** that need to be excreted safely.

1. First, the amino acids are deaminated in the **liver** to form ammonia.
2. Ammonia is toxic so is immediately converted to urea, which is then filtered out in the kidney.

Controlling water content

The osmotic balance of the body's fluids needs to be tightly controlled because if cells gain or lose too much water they do not function efficiently.

The amount of water re-absorbed by the kidneys is controlled by **anti-diuretic hormone (ADH)**. This is produced in the pituitary.

1. ADH directly increases the permeability of the kidney tubules to water.
2. When the water content of the blood is low (higher blood concentration), **negative feedback** operates to restore normal levels.

The effect of ADH on blood water content

Blood water level too low (salt concentration too high)

Detected by the pituitary gland

More ADH released into the blood by pituitary gland

More water reabsorbed into the blood from the renal tubules

Small amount of concentrated urine

Normal blood water level

Blood water level too high (salt concentration too low)

Detected by the pituitary gland

Less ADH released into the blood by pituitary gland

Less water reabsorbed into the blood from the renal tubules

Large amount of dilute urine

Normal blood water level

Pituitary gland

SUMMARY

- **Urea, sweat, carbon dioxide and water are the products of excretion.**
- **The kidneys remove urea from the body. When kidneys fail, a patient may receive dialysis or a kidney transplant.**
- **The kidney carries out water balance in tandem with the hormone ADH.**

QUESTIONS

QUICK TEST

1. List three substances that are selectively re-absorbed back into the bloodstream from the kidney.
2. Which organ releases ADH into the blood?

QUESTIONS

EXAM PRACTICE

1. A student is exercising.

   She produces carbon dioxide in her cells.

   a) What name is given to the process that converts substances into waste products? **[1 mark]**

   b) The student's liver produces urea.

   Which substances are changed to form urea? **[1 mark]**

Hormones in human reproduction

Puberty

During **puberty** (approximately 10–16 in girls and 12–17 in boys), the sex organs begin to produce **sex hormones**. This causes the development of secondary sexual characteristics.

In **males**, the primary sex hormone is **testosterone**.

During puberty, testosterone is produced from the testes and causes:

- production of sperm in testes
- development of muscles and penis
- deepening of the voice
- growth of pubic, facial and body hair.

In **females**, the primary sex hormone is **oestrogen**. Other sex hormones are **progesterone**, **FSH** and **LH**.

During puberty, oestrogen is produced in the ovaries and progesterone production starts when the menstrual cycle begins.

The secondary sexual characteristics are:

- ovulation and the menstrual cycle
- breast growth
- widening of hips
- growth of pubic and armpit hair.

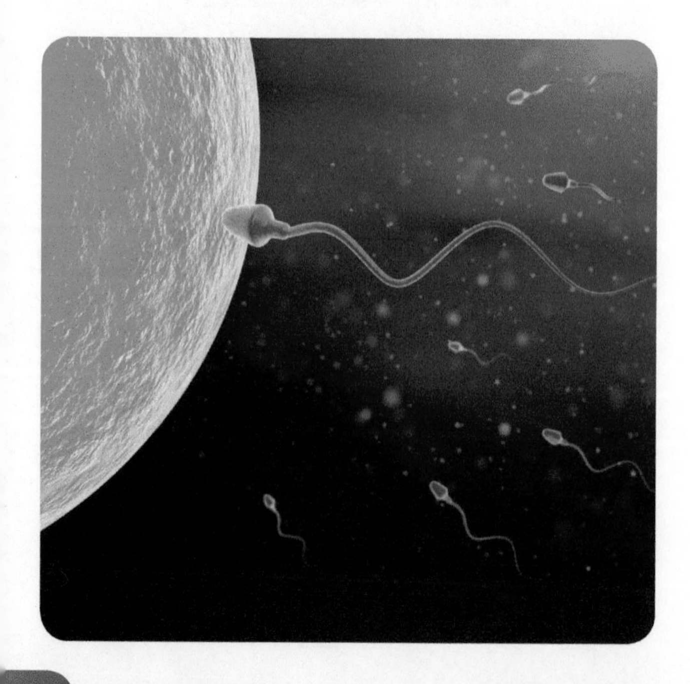

The menstrual cycle

A woman is fertile between the ages of approximately 13 and 50.

During this time, an egg is released from one of her ovaries each month and the lining of her uterus is replaced each month (approximately 28 days) to prepare for pregnancy.

The menstrual cycle

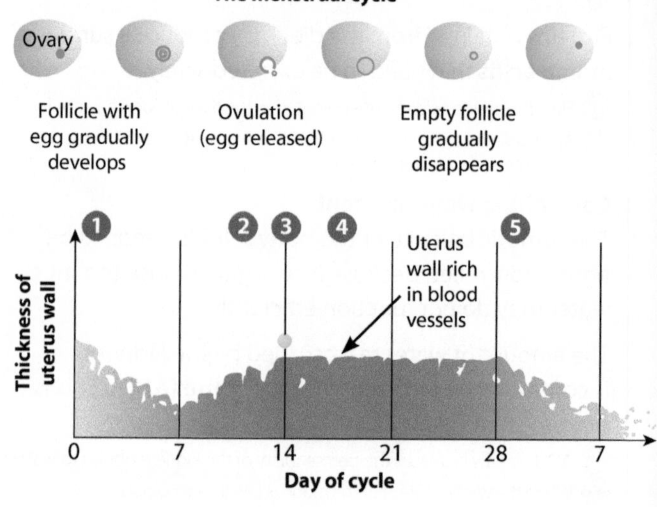

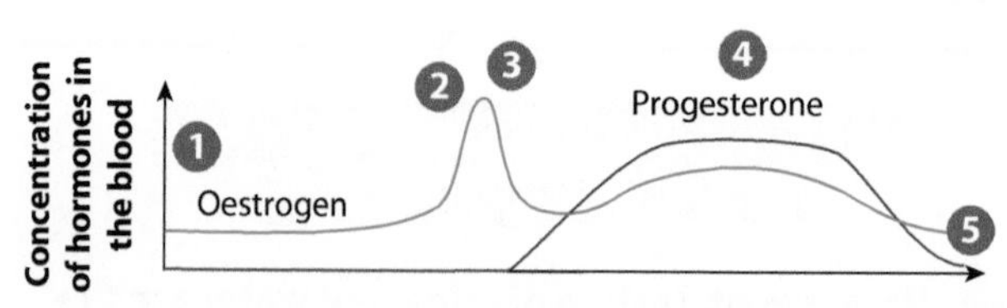

1. Uterus lining breaks down (i.e. a period).
2. Repair of the uterus wall.
   Oestrogen causes the uterus lining to gradually thicken.
3. Egg released by the ovary.
4. Progesterone and oestrogen make the lining stay thick, waiting for a fertilised egg.
5. No fertilised egg so cycle restarts.

As well as oestrogen and progesterone, the two other hormones involved in the cycle are:

- **FSH** or follicle stimulating hormone, which causes maturation of an egg in the ovary
- **LH** or luteinising hormone, which stimulates release of an egg.

HT Negative feedback in the menstrual cycle

The four female hormones interact in a complex manner to regulate the cycle.

- **FSH** is produced in the pituitary and acts on the ovaries, causing an egg to mature. It stimulates the ovaries to produce oestrogen.
- **Oestrogen** is secreted in the ovaries and inhibits further production of FSH. It also stimulates the release of LH and promotes repair of the uterus wall after menstruation.
- **LH** is produced in the pituitary. It also stimulates release of an egg.
- **Progesterone** is secreted by the empty follicle in the ovary (left by the egg). It maintains the lining of the uterus after ovulation has occurred. It also inhibits FSH and LH.

Female reproductive system

Low progesterone levels allow FSH from the pituitary gland to stimulate the maturation of an egg (in a follicle). This in turn stimulates oestrogen production.

High levels of oestrogen stimulate a surge in LH from the pituitary gland. This triggers ovulation in the middle of the cycle.

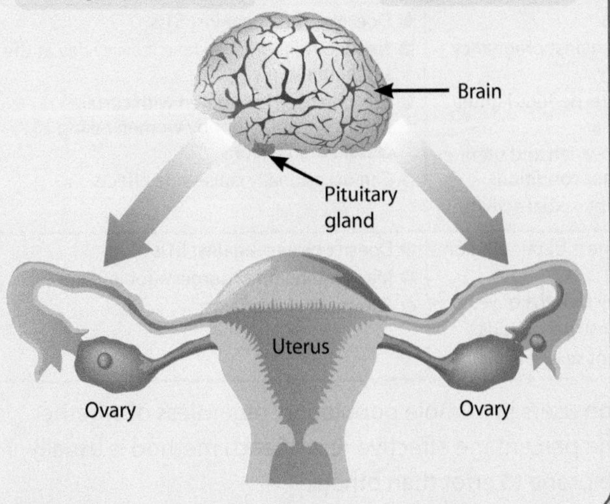

SUMMARY

- **During puberty the sex organs begin to produce sex hormones.**
- **Hormones play a vital role in regulating human reproduction, especially in the female menstrual cycle.**

QUESTIONS

QUICK TEST

1. Name the four hormones involved in the menstrual cycle.
2. Name one effect of testosterone in puberty.

EXAM PRACTICE

1. a) State the functions of oestrogen and progesterone in the menstrual cycle. **[2 marks]**

   HT b) Oestrogen also stimulates release of LH.

   What is the result of this surge? **[1 mark]**

2. From the list below, circle **three** characteristics that develop as a result of puberty.

   Breast development **Fertilisation**

   Sperm production **Secretion of ADH**

   Menstrual cycle

   [3 marks]

Contraception and infertility

Non-hormonal contraception

Contraceptive method	Method of action	Advantages	Disadvantages
• Barrier method – condom (male + female)	Prevents the sperm from reaching the egg	82% effective • Most effective against STIs	• Can only be used once • May interrupt sexual activity • Can break • Some people are allergic to latex
• Barrier method – diaphragm	Prevents the sperm from reaching the egg	88% effective • Can be put in place right before intercourse or 2–3 hours before • Don't need to take out between acts of sexual intercourse	• Increases urinary tract infections • Doesn't protect against STIs
• Intrauterine device	Prevents implantation – some release hormones	99% effective • Very effective against pregnancy • Doesn't need daily attention • Comfortable • Can be removed at any time	• Doesn't protect against STIs • Needs to be inserted by a medical practitioner • Higher risk of infection when first inserted • Can have side effects such as menstrual cramping • Can fall out and puncture the uterus (rare)
• Spermicidal agent	Kills or disables sperm	72% effective • Cheap	• Doesn't protect against STIs • Needs to be reapplied after one hour • Increases urinary tract infections • Some people are allergic to spermicidal agents
• Abstinence • Calendar method	Refraining from sexual intercourse when an egg is likely to be in the oviduct	76% effective • Natural • Approved by many religions • Woman gets to know her body and menstrual cycles	• Doesn't protect against STIs • Calculating the ovulation period each month requires careful monitoring and instruction • Can't have sexual intercourse for at least a week each month
• Surgical method	Vasectomy and female sterilisation	99% effective • Very effective against pregnancy • One-time decision providing permanent protection	• No protection against STIs • Need to have minor surgery • Permanent

Hormonal contraception

Contraceptive method	Method of action	Advantages	Disadvantages
• Oral contraceptive	Contains hormones that inhibit FSH production, so eggs fail to mature	91% effectiveness • Very effective against pregnancy if used correctly • Makes menstrual periods lighter and more regular • Lowers risk of ovarian and uterine cancer, and other conditions • Doesn't interrupt sexual activity	• Doesn't protect against STIs • Need to remember to take it every day at the same time • Can't be used by women with certain medical problems or by women taking certain medications • Can occasionally cause side effects
• Hormone injection • Skin patch • Implant	Provides slow release of progesterone; this inhibits maturation and release of eggs	91–99% effectiveness depending on method used • Lasts over many months or years • Light or no menstrual periods • Doesn't interrupt sexual activity	• Doesn't protect against STIs • May require minor surgery (for implant) • Can cause side effects

The percentage figures in the contraception tables are based on users in a whole population, regardless of whether they use the method correctly. If consistently used correctly, the percentage effectiveness of each method is usually higher. Some methods, such as the calendar method, are more prone to error than others.

HT Infertility treatment

Infertility treatment is used by couples who have problems conceiving.

Reasons for infertility

No eggs being released from the ovaries.
Endometriosis, which occurs when the tissue that lines the inside of the uterus enters other organs of the body, such as the abdomen and fallopian tubes; this reduces the maturation rate and release of eggs.
Male infertility/low sperm count.
Uterine fibroids.
Complete or partial blocking and/or scarring of the fallopian tubes.
Reduced number and quality of eggs.

Methods of treatment

Treating infertility is known as ART (assisted reproductive technology). There are a number of methods of treatment.

- Fertility drugs containing FSH and LH are given to women who do not produce enough FSH themselves. They may then become pregnant naturally.
- Clomifene therapy prevents the production of oestrogen and so inhibits negative feedback.
- In vitro fertilisation (IVF) is a method in which the potential mother is given FSH and LH to stimulate the production of several eggs. Sperm is collected from the father. The sperm and eggs are then introduced together outside the body in a petri dish. One or two growing embryos can then be transplanted into the woman's uterus.

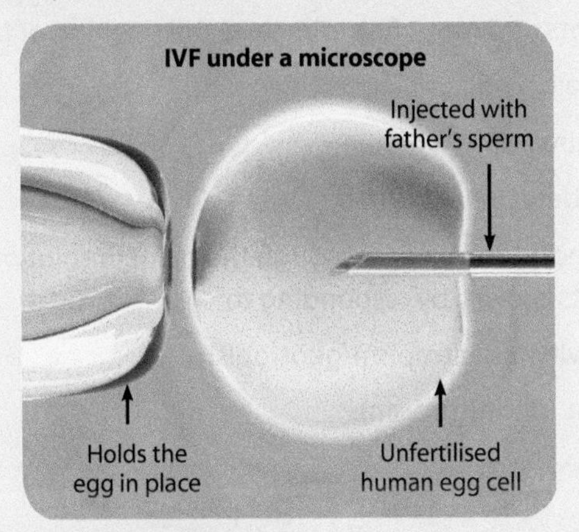

Disadvantages of IVF

It is very expensive.
It can be mentally and physically stressful.
Success rates are only approximately 40% (at the time of writing).
There is an increased risk of multiple births.

SUMMARY

- **Fertility and the possibility of pregnancy can be controlled using non-hormonal and hormonal methods of contraception.**
- **Infertility has a number of causes and may be treated using fertility drugs and surgical procedures.**

QUESTIONS

QUICK TEST

1. Which non-hormonal contraceptive is the least effective? What is the benefit of this method?
2. Why are condoms effective against the spread of HIV?

EXAM PRACTICE

1. State one advantage and one disadvantage of using a condom barrier method of contraception. **[2 marks]**
2. (HT) IVF is one method of treating infertility.

   Explain how this is carried out. **[3 marks]**

Plant hormones

Tropisms and general control

Plants, as well as animals, respond to changes in their environment.

Plant hormones are chemicals that control the growth of shoots and roots, flowering and the ripening of fruits.

Two major groups of hormones are:

- auxins
- gibberellins.

These hormones move through the plant in solution and affect its growth by responding to:

- gravity (gravitropism/geotropism)
- light (phototropism).

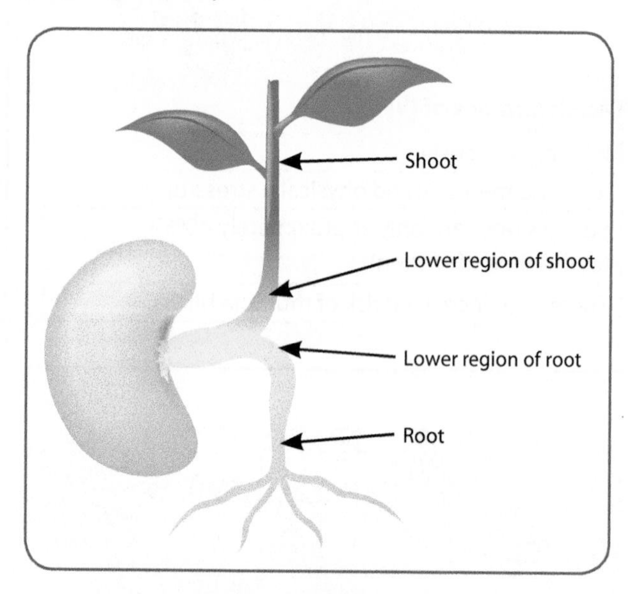

Response to gravity

- **Shoots** grow against gravity (negative gravitropism).
- **Roots** grow in the direction of gravity (positive gravitropism).

In the roots, auxins inhibit growth in the lower region, which makes the roots grow downwards. The roots anchor the plant in the soil and seek out water and minerals for absorption.

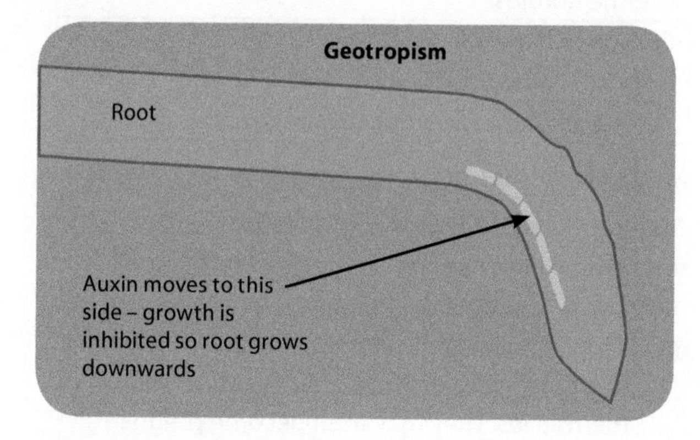

Response to light

- **Roots** grow away from the light (negative tropism).
- **Shoots** grow towards the light (positive tropism).

In the shoot, auxin is made in the tip. When light reaches the tip from all directions, the hormone is equally distributed. When light is **uni-directional** (in this case from the right), the auxin moves to the shaded side, causing the cells here to **elongate**. This makes the shoot bend towards the light.

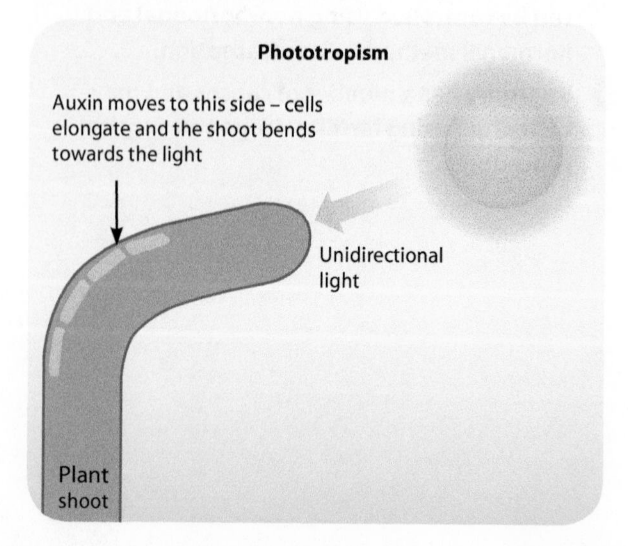

HT Commercial uses of hormones

Auxins

- **Rooting powder** consists of an auxin that encourages the growth of roots in stem cuttings, so that many new plants can be grown from a parent.
- **Tissue culture** technique also requires auxins to promote growth.
- **Selective weed killers**, used in agriculture, contain auxins that disrupt the growth patterns of their target plants (which are often broad-leaved rather than narrow-leaved). Therefore the crop plant is not harmed.

Gibberellins

Gibberellins are important in initiating seed germination. They can be used to:

- end seed dormancy
- promote flowering
- increase fruit size
- produce seedless fruit (parthenocarpy).

Ethene

Ethene is a hydrocarbon that controls cell division. It can be used in the food industry to control ripening of fruit while it is being transported or stored.

WS One of your required practicals is likely to be investigating the effect of light on the growth of newly germinated shoots.

This is one possible investigation. What other investigations could you carry out?

Experiment to show that shoots grow towards light

1. Cut a hole in the side of a box. Put three cuttings into the box. The cuttings detect light coming from the hole and will grow towards it.
2. Cut a hole in the side of another box. Put three cuttings with foil-covered tips in the box. These shoots can't detect the light so they grow straight up.

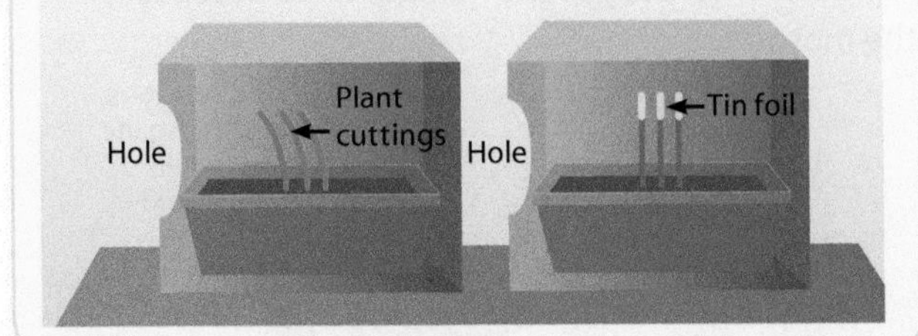

SUMMARY

- **Plant hormones control the growth of shoots and roots, flowering, and ripening of fruits.**
- **There are two major groups of plant hormones: auxins and gibberellins.**
- **The hormones respond to gravity and light.**
- HT **Commercial uses of plant hormones include plant propagation, weed killers and improved fruit production.**

QUESTIONS

QUICK TEST

1. What name is given to the process where shoots grow towards the light?
2. HT Which hormone could a farmer apply to her crop to kill weeds?
3. HT Which hormone could be used to cause plants to flower early?

EXAM PRACTICE

1. Which substance does rooting powder contain that stimulates root growth?

   Tick (✓) one box. **[1 mark]**

   Enzymes ☐
   Nitrates ☐
   Plant hormones ☐
   Vitamins ☐

2. The diagram shows a bean seedling that has been grown horizontally in laboratory conditions.

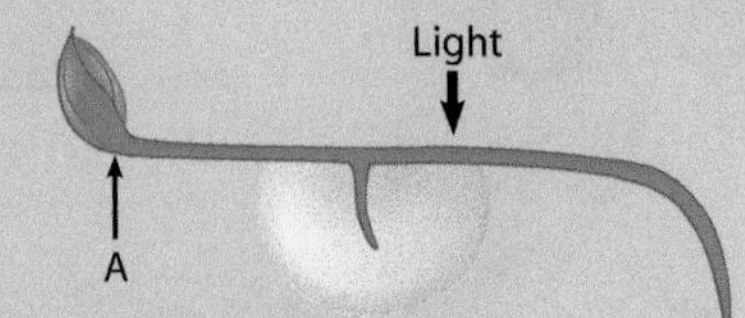

   Explain what has happened to cause the growth in the shoot at point 'A' in the diagram. **[2 marks]**

Sexual and asexual reproduction

Sexual reproduction

Sexual reproduction is where a male **gamete** (e.g. sperm) meets a female gamete (e.g. egg). This fusion of the two gametes is called fertilisation and may be internal or external.

Gametes are produced by meiosis in the sex organs.

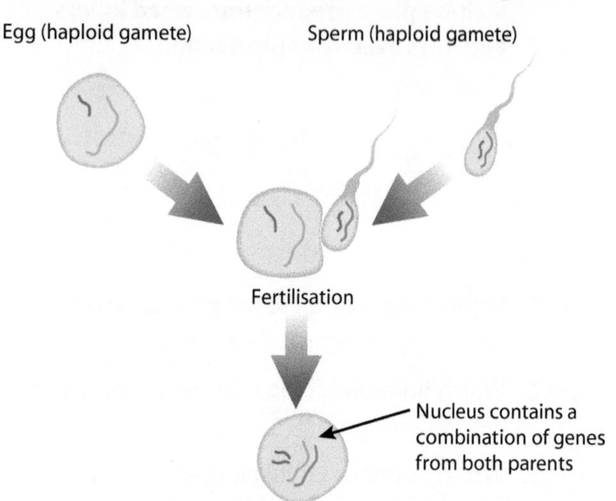

Asexual reproduction

Asexual reproduction does not require different male and female cells. Instead, genetically identical clones are produced from mitosis. These may just be individual cells, as in the case of yeast, or whole multicellular organisms, e.g. aphids.

Many organisms can reproduce using both methods, depending on the environmental conditions.

A yeast budding

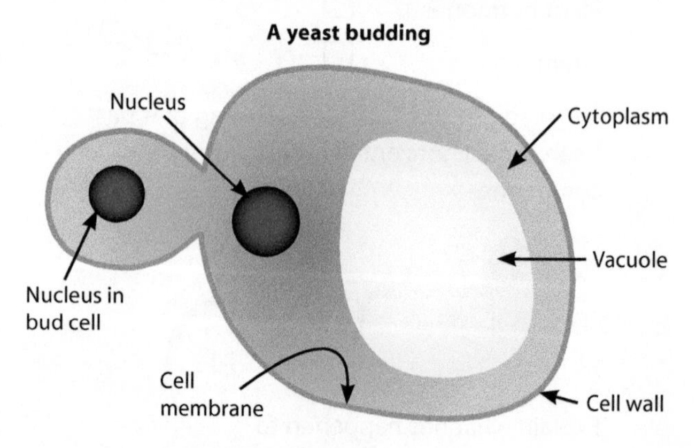

Comparing sexual and asexual reproduction

Advantages of sexual reproduction
• Produces **variation** in offspring through the process of meiosis, where genes are 'shuffled.' Variation is increased by **random fusion of gametes**. • Survival advantage gained when the environment changes because different genetic types have more chance of producing well-adapted offspring. • Humans can make use of sexual reproduction through **selective breeding**. This enhances food production.
Disadvantages of sexual reproduction
• Relatively slow process. • Variation can be a disadvantage in stable environments. • More resources required than for asexual reproduction, e.g. energy, time. • Results of selective breeding are unpredictable and might lead to genetic abnormalities from 'in-breeding'.

Advantages of asexual reproduction
• Only one parent required. • Fewer resources (energy and time) need to be devoted to finding a mate. • Faster than sexual reproduction – survival advantage of producing many offspring in a short period of time. • Many identical offspring of a well-adapted individual can be produced to take advantage of favourable conditions.
Disadvantages of asexual reproduction
• Offspring may not be well adapted in a changing environment.

Different reproductive strategies

Malarial parasite

The **plasmodium** is a **protist** that causes malaria. It can reproduce asexually in the human host but sexually in the mosquito.

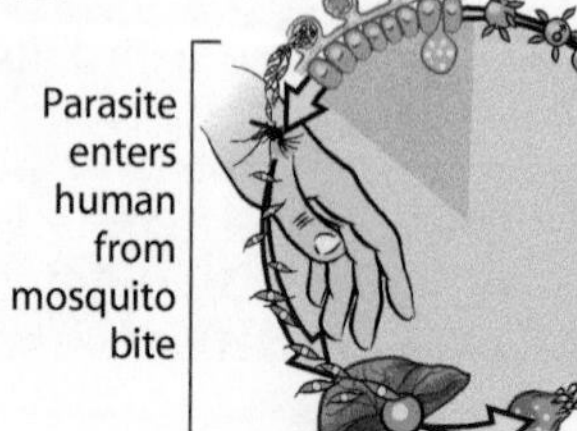

Strawberry plants and daffodils

These plants can reproduce sexually using flowers, or asexually using **runners** (strawberry) or **bulb division** (daffodils).

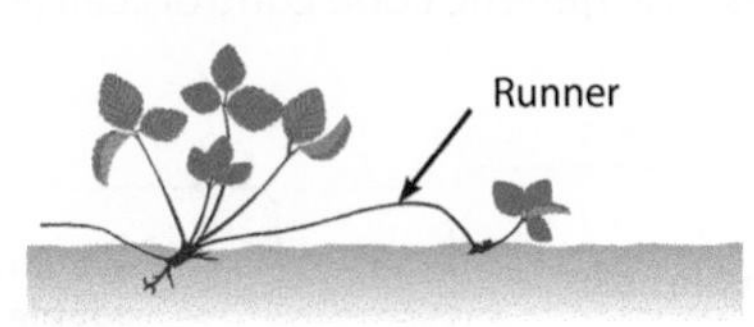

Fungi

Many fungi, such as toadstools, mushrooms and moulds, can reproduce sexually (giving variation) or asexually by **spores**.

WS During your course, you will learn about historical developments in science and technology. It is important to appreciate how understanding develops from previous studies and their publication in scientific journals. Scientists build on each other's work and applications develop from this.

Malaria is a disease that kills many people every year. Treatment via anti-malarial drugs has been available for years and techniques are available to disrupt the parasite's life cycle. At the time of writing, scientists are close to developing a vaccine.

All these developments have come from an understanding of the complex interaction between the plasmodium and mosquito's life cycles. Studies were carried out by many different people and took about ten years to reach their conclusions. Finally, a Scottish physician called Sir Ronald Ross produced evidence for the complete life cycle of the malaria parasite in mosquitoes. For this work, he received the 1902 Nobel Prize in Medicine.

SUMMARY

- **One of the basic characteristics of life is reproduction. This is the means by which a species continues. If sufficient offspring are not produced, the species becomes extinct.**
- **Reproduction may be sexual or asexual.**
- **Some organisms reproduce using both means.**

QUESTIONS

QUICK TEST

1. What name is given to the fusion of two gametes?
2. How do strawberry plants reproduce asexually?
3. How do fungi reproduce asexually?

EXAM PRACTICE

1. State one advantage and one disadvantage of reproducing by sexual means. **[2 marks]**
2. **a)** The malarial parasite can live in two different organisms.

   Suggest a survival advantage in this. **[1 mark]**

   b) Name one other type of organism that combines asexual and sexual reproduction. **[1 mark]**

DAY 5  40 Minutes

DNA

DNA and the genome

The nucleus of each cell contains a complete set of genetic instructions called the **genetic code**. The information is carried as genes, which are small sections of DNA found on **chromosomes**. The genetic code controls cell activity and, consequently, characteristics of the whole organism.

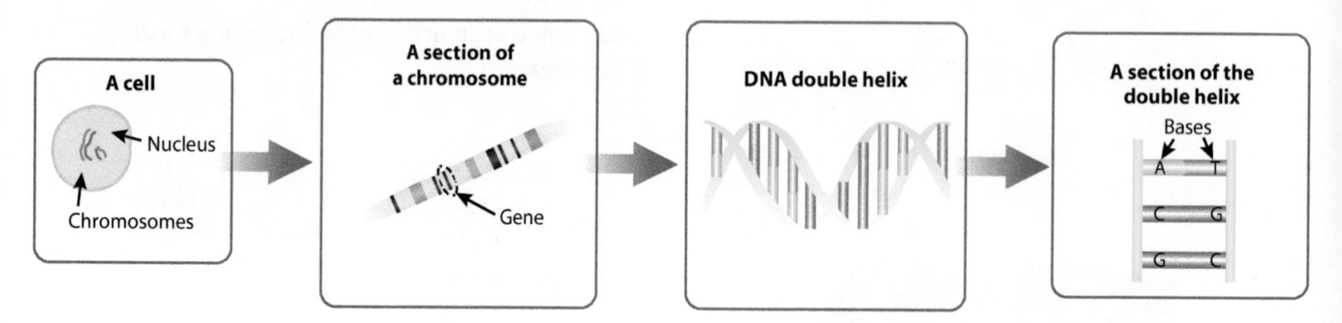

DNA facts

- DNA is a polymer.
- It is made of two strands coiled around each other called a **double helix**.
- The genetic code is in the form of nitrogenous **bases**.
- Bases bond together in pairs forming hydrogen bond cross-links.
- The structure of DNA was discovered in 1953 by **James Watson** and **Francis Crick**, using experimental data from **Rosalind Franklin** and **Maurice Wilkins**.
- A single gene codes for a particular sequence of **amino acids**, which, in turn, make up a single **protein**.

The human genome

The **genome** of an organism is the entire genetic material present in an adult cell of an organism.

The Human Genome Project (HGP)

The HGP was an international study. Its purpose was to map the complete set of genes in the human body.

HGP scientists worked out the code of the human genome in three ways. They:

- determined the sequence of all the bases in the genome
- drew up maps showing the locations of the genes on chromosomes
- produced linkage maps that could be used for tracking inherited traits from generation to generation, e.g. for genetic diseases. This could then lead to targeted treatments for these conditions.

The results of the project, which involved collaboration between UK and US scientists, were published in 2003. Three billion base pairs were determined.

The mapping of the human genome has enabled anthropologists to work out historical human migration patterns. This has been achieved by collecting and analysing DNA samples from many people across the globe. The study is called the **Genographic Project**.

World map showing suggested migration patterns of early hominids

DNA structure and base sequences

The four bases in DNA are A, T, C and G. The code is 'read' on one strand of DNA. Three consecutive bases (a **triplet**) code for one particular amino acid. The sequence of these triplets determines the structure of a whole protein.

The bases are attached to a sugar phosphate **backbone**. These form a basic unit called a nucleotide.

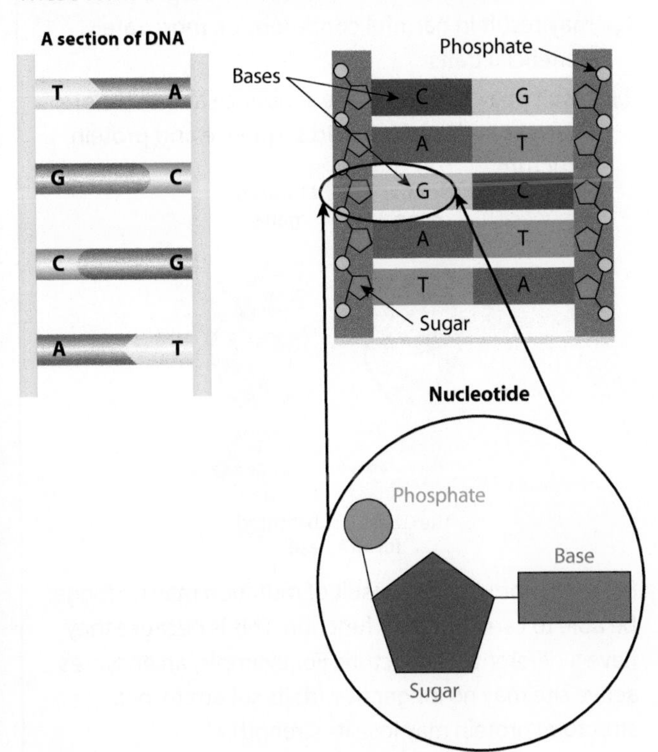

SUMMARY

- **Each cell in the body contains a nucleus with chromosomes.**
- **Genes are small sections of DNA found on chromosomes. The four bases in DNA are A, T, C and G.**
- **The Human Genome Project has produced a complete map of all the genes in the body.**

QUESTIONS

QUICK TEST

1. Explain how the Human Genome Project has advanced medical science.
2. What are the four bases in DNA?

EXAM PRACTICE

1. Explain how scientists have used knowledge about the human genome to contribute to the **Genographic project**. **[2 marks]**
2. Describe how a nucleotide from the DNA molecule is made up. **[3 marks]**

The genetic code

HT DNA structure and base sequences

In complementary strands of DNA, T always bonds with A, and C with G.

Proteins are synthesised in structures called **ribosomes**, which are located in the cytoplasm of the cell. In order for the DNA code in the nucleus to be translated as new protein by the ribosomes, the following process occurs.

1. DNA unzips and exposes the bases on each strand.
2. A molecule of **messenger RNA** (**mRNA**) is constructed from one of these template strands.
3. The mRNA, which carries a complementary version of the gene, travels out of the nucleus to the ribosome in cytoplasm.
4. In the ribosome, the mRNA is 'read'. **Transfer RNA** (**tRNA**) molecules carry individual amino acids to add to a growing protein (polypeptide).
5. The new polypeptide folds into a unique shape and is released into the cytoplasm.

In the nucleus, transcription takes place.

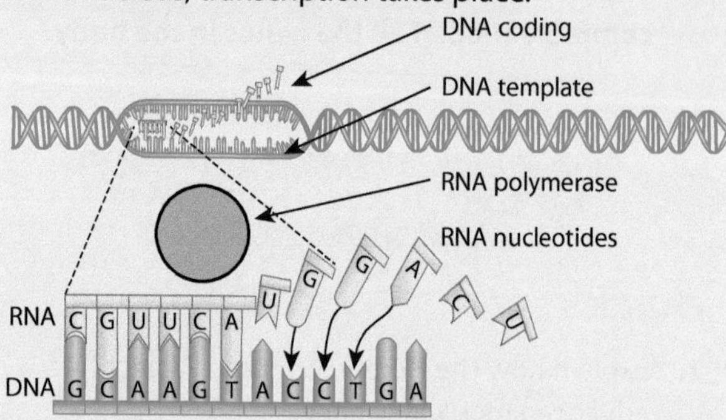

In the ribosome, translation takes place.

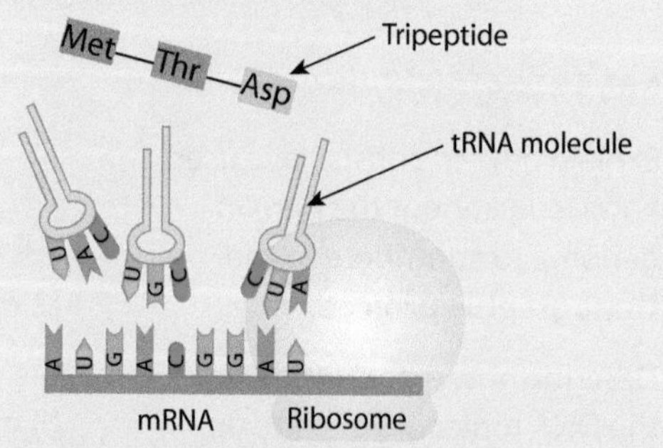

The proteins that are produced carry out a specific function. Examples include enzymes, hormones or structural protein such as **collagen**.

Gene switching

Every cell contains a complete set of genes for the whole body, but only some of these genes are used in any one cell.

- Some genes are not **expressed**. They are said to be 'switched off'. Switching off genes is accomplished through **non-coding** DNA. Most DNA is non-coding.
- The genes that are 'switched on' eventually determine the function of a cell.

Mutations

Mutations (genetic variants):

- are changes to the structure of a DNA molecule
- occur continuously during the cell division process or as a result of external influences, e.g. exposure to radioactive materials or emissions such as X-rays or UV light
- usually have a neutral effect as amino acids may still be produced or the proteins produced work in the same way
- may result in harmful conditions or, more rarely, beneficial traits
- result in a change in base sequence and therefore changes in the amino acid sequence and protein structure.

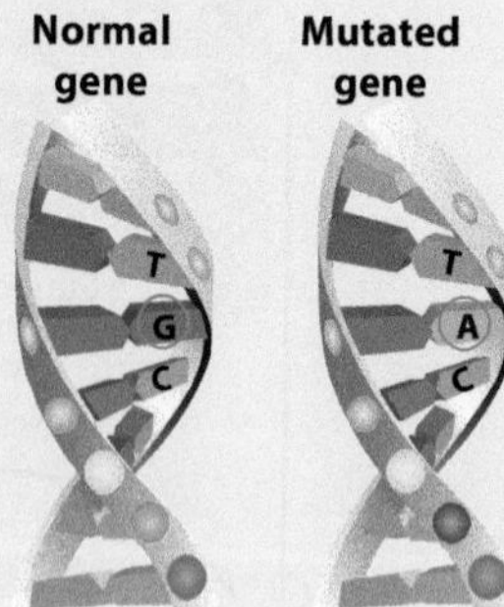

The G base is substituted for an A base

Proteins produced as a result of mutation may no longer be able to carry out their function. This is because they have a different 3D structure. For example, an enzyme's active site may no longer fit with its substrate, or a structural protein may lose its strength.

Changes in the base sequence may be passed on to daughter cells when cell division occurs. This in turn may lead to offspring having genetic conditions.

Monohybrid crosses

Most characteristics or **traits** are the result of multiple alleles interacting but some are controlled by a single gene. Examples include fur colour in mice and the shape of ear lobes in humans.

These genes exist as pairs called alleles on homologous chromosomes.

Alleles are described as **dominant** or **recessive**.

- A dominant allele controls the development of a characteristic even if it is present on only one chromosome in a pair.
- A recessive allele controls the development of a characteristic only if a dominant allele is not present, i.e. if the recessive allele is present on both chromosomes in a pair.

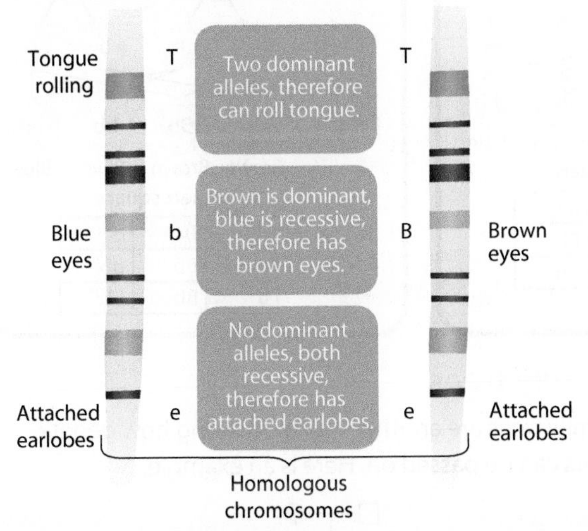

If both chromosomes in a pair contain the same allele of a gene, the individual is described as being **homozygous** for that gene or condition.

If the chromosomes in a pair contain different alleles of a gene, the individual is **heterozygous** for that gene or condition.

The combination of alleles for a particular characteristic is called the **genotype**. For example, the genotype for a homozygous dominant tongue-roller would be TT. The fact that this individual is able to roll their tongue is termed their **phenotype**.

Other examples are:

- bb (genotype), blue eyes (phenotype)
- EE or Ee (genotype), unattached/pendulous ear lobes (phenotype).

When a characteristic is determined by just one pair of alleles, as with eye colour and tongue rolling, it is called **monohybrid inheritance**.

WS During your course, you will be expected to recognise, draw and interpret scientific diagrams.

In this module, the way complementary strands of DNA are arranged can be worked out once you know that base T bonds with A and base C bonds with G.

Can you write out the complementary (DNA) strand to this sequence?

____

A T T A C G T G A G C C

SUMMARY

- HT **Bases in DNA always bond T to A and G to C.**
- HT **Mutations are genetic variants and may result in harmful conditions or occasionally beneficial traits.**
- **Alleles can be dominant or recessive.**

QUESTIONS

QUICK TEST

1. Name the two different types of alleles.
2. HT What are mRNA and tRNA?

EXAM PRACTICE

1. Explain how a recessive allele could control the development of a characteristic. **[2 marks]**
2. HT Compare and contrast the locations and roles of transfer RNA and messenger RNA. **[4 marks]**

Inheritance and genetic disorders

Genetic diagrams

Genetic diagrams are used to show all the possible combinations of alleles and outcomes for a particular gene. They use:

- capital letters for dominant alleles
- lower-case letters for recessive alleles.

For eye colour, brown is dominant and blue is recessive. So B represents a brown allele and b represents a blue allele.

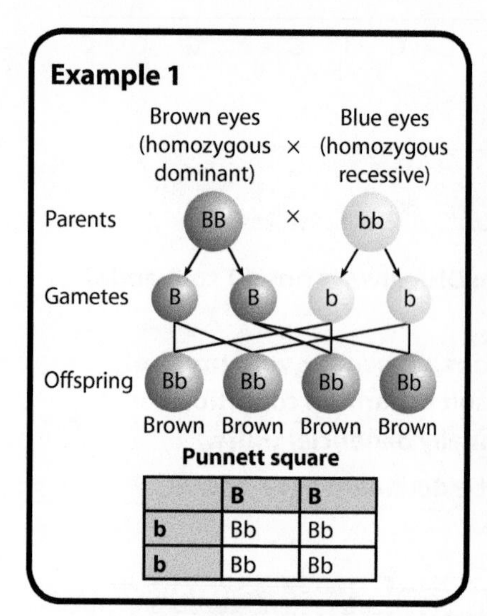

	B	B
b	Bb	Bb
b	Bb	Bb

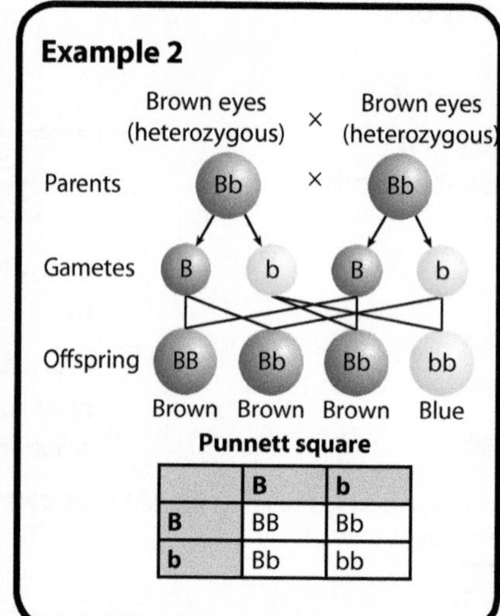

	B	b
B	BB	Bb
b	Bb	bb

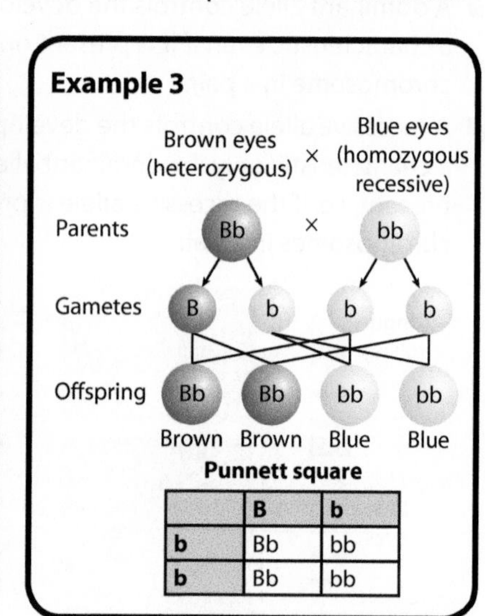

	B	b
b	Bb	bb
b	Bb	bb

WS You need to be able to interpret genetic diagrams and work out ratios of offspring.

- In Example 2 above, the phenotypes of the offspring are 'brown eyes' and 'blue eyes'. As there are potentially three times as many brown-eyed children as blue-eyed, the phenotypes are said to be in a 3:1 ratio.
- In Example 3 above, the ratio would be 1:1 because half of the theoretical offspring are brown-eyed and half blue-eyed. Another way of saying this is that the probability of parents producing a brown-eyed child is 50%, or ½, or 0.5.

Most traits result not from one pair of alleles but from multiple genes interacting, e.g. inheritance of blood groups in the **ABO** system.

HT In exams, you may be asked to construct your own punnett squares to solve genetic cross problems like the ones above.

Family trees

Family trees are another way of showing how genetic traits can be passed on. Here is an example.

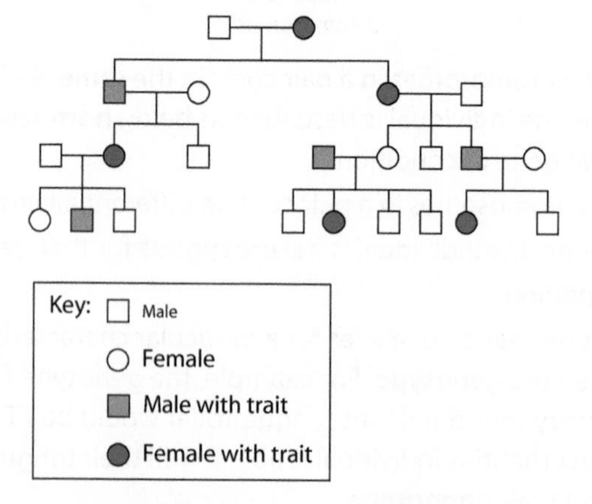

Inheritance of sex

Sex in humans/mammals is determined by whole chromosomes. These are the 23rd pair and are called sex chromosomes. There is an 'X' chromosome and a smaller 'Y' chromosome. The other 22 chromosome pairs carry the remainder of genes coding for the rest of the body's characteristics.

All egg cells carry X chromosomes. Half the sperm carry X chromosomes and half carry Y chromosomes. The sex of an individual depends on whether the egg is fertilised by an X-carrying sperm or a Y-carrying sperm.

If an X sperm fertilises the egg it will become a girl. If a Y sperm fertilises the egg it will become a boy. The chances of these events are equal, which results in approximately equal numbers of male and female offspring.

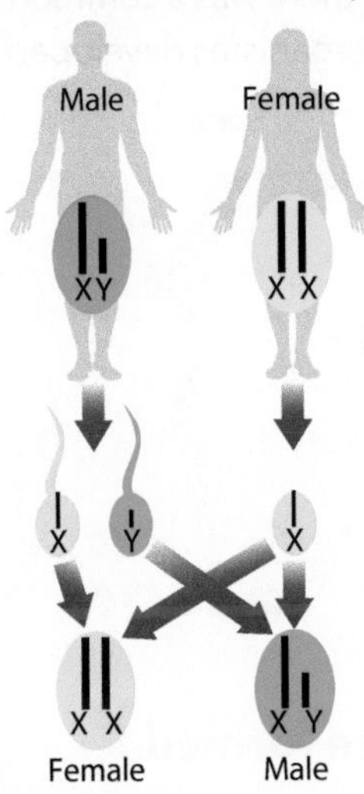

Inherited diseases

Some disorders are caused by a 'faulty' gene, which means they can be **inherited**. One example is **polydactyly**, which is caused by a dominant allele and results in extra fingers or toes. The condition is not life-threatening.

Cystic fibrosis, on the other hand, can limit life expectancy. It causes the mucus in respiratory passages and the gut lining to be very thick, leading to build-up of phlegm and difficulty in producing correct digestive enzymes.

Cystic fibrosis is caused by a recessive allele. This means that an individual will only exhibit symptoms if both recessive alleles are present in the genotype. Those carrying just one allele will not show symptoms, but could potentially pass the condition on to offspring. Such people are called **carriers**.

Conditions such as cystic fibrosis are mostly caused by **faulty alleles** that are **recessive**.

SUMMARY

- **Genetic diagrams show all possible combinations of alleles and outcomes for a gene. Family trees are another way of showing how genetic traits are passed on.**

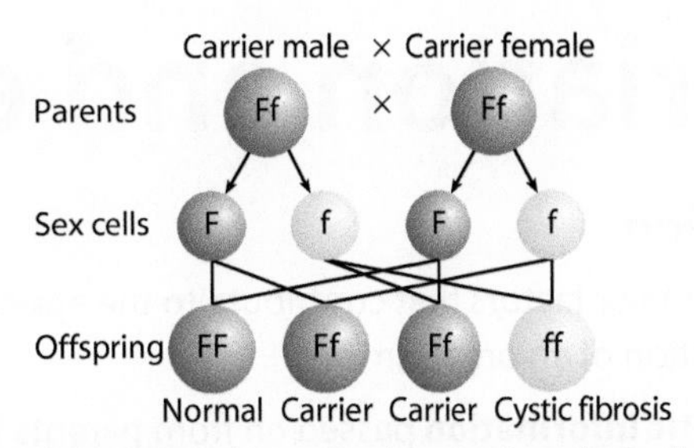

Knowing that there is a 1 in 4 chance that their child might have cystic fibrosis gives parents the opportunity to make decisions about whether to take the risk and have a child. This is a very difficult decision to make.

Technology has advanced and it is now possible to screen embryos for genetic disorders.

- If an embryo has a life-threatening condition it could be destroyed, or simply not be implanted if IVF was applied.
- Alternatively, new gene therapy techniques might be able to reverse the effects of the condition, resulting in a healthy baby.

As yet, these possibilities have to gain approval from ethics committees – some people think that this type of 'interference with nature' could have harmful consequences.

QUESTIONS

QUICK TEST

1. If a homozygous brown-eyed individual is crossed with a homozygous blue-eyed individual, what is the probability of them producing a blue-eyed child?

EXAM PRACTICE

1. A man who knows he is a carrier of cystic fibrosis is thinking of having children with his partner. His partner does not suffer from cystic fibrosis, nor is she a carrier.

   Suggest what a genetic counsellor might advise the couple.

   Use genetic diagrams to aid your explanation and state any probabilities of offspring produced. **[6 marks]**

DAY 6 35 Minutes

Variation and evolution

Variation

The two major factors that contribute to the appearance and function of an organism are:

- **genetic information** passed on from parents to offspring
- **environment** – the conditions that affect that organism during its lifetime, e.g. climate, diet, etc.

These two factors account for the large variation we see **within** and **between** species. In most cases, both of these factors play a part.

Evolution

Put simply, evolution is the theory that all organisms have arisen from simpler life forms over billions of years. It is driven by the **mechanism** of **natural selection**. For natural selection to occur, there must be genetic variation between individuals of the same species. This is caused by mutation or new combinations of genes resulting from sexual reproduction.

Most mutations have no effect on the phenotype of an organism. Where they do, and if the environment is changing, this can lead to relatively rapid change.

Evidence for evolution

Evidence for evolution comes from many sources. It includes:

- comparing genomes of different organisms
- studying embryos and their similarities
- looking at changes in species during modern times, e.g. antibiotic resistance in bacteria
- comparing the anatomy of different forms
- the fossil record.

One of the earliest sources of evidence for evolution was the discovery of fossils.

Further evidence for evolution – the pentadactyl limb

If you compare the forelimbs of a variety of vertebrates, you can see that the bone structures are all variations on a five-digit form, whether it be a leg, a wing or a flipper. This suggests that there was a common ancestral form from which these organisms developed.

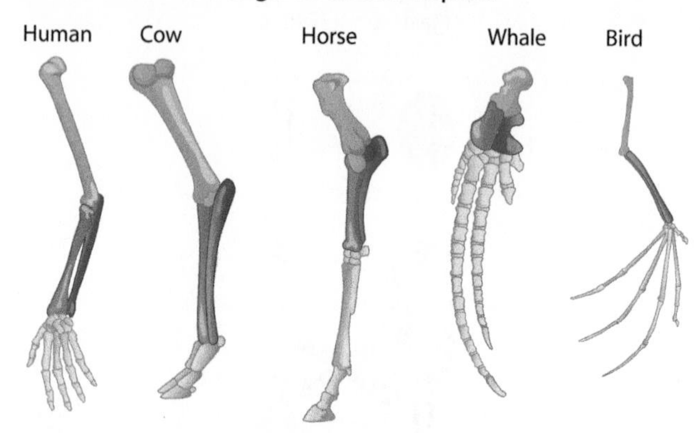

How fossils are formed

1. When an animal or plant dies, the processes of decay usually cause all the body tissues to break down. In rare circumstances, the organism's body is rapidly covered and oxygen is prevented from reaching it. Instead of decay, fossilisation occurs.

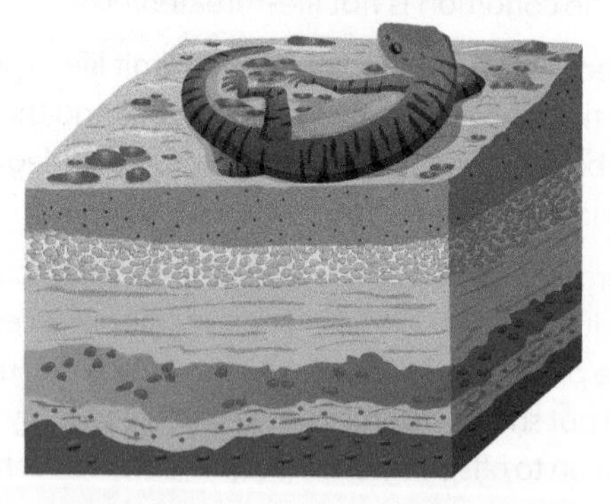

2. Over hundreds of thousands of years, further sediments are laid down and compress the organism's remains.

3. Parts of the organism, such as bones and teeth, are **replaced by minerals** from solutions in the rock.

4. Earth upheavals, e.g. **tectonic plate movement**, bring sediments containing the fossils nearer the Earth's surface.

5. Erosion of the rock by wind, rain and rivers exposes the fossil. At this stage, the remains might be found and excavated by **paleontologists**.

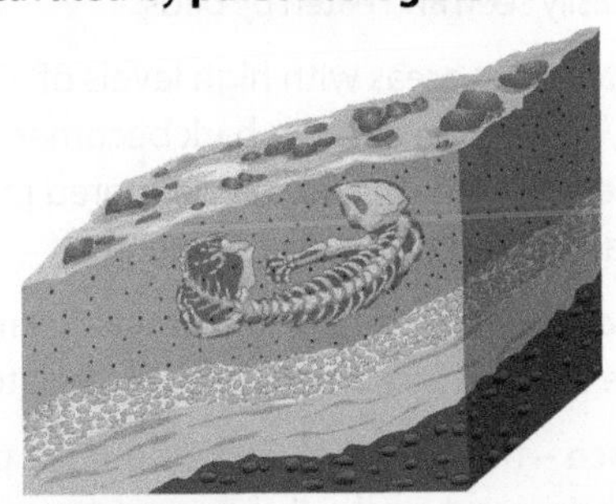

SUMMARY

- **Genetic information, and environment are the two major factors that contribute to variation.**
- **Evolution is the theory that all organisms have developed from simple life forms over billions of years.**
- **The fossil record tells us that most organisms that once existed have become extinct.**
- **Speciation occurs where members of an original population can no longer interbreed.**

Fossils can also be formed from footprints, burrows and traces of tree roots.

By comparing different fossils and where they are found in the rock layers, paleontologists can gain insights into how one form may have developed into another.

Difficulties occur with earlier life forms because many were **soft-bodied** and therefore not as well-preserved as organisms with bones or shells. Any that are formed are easily destroyed by Earth movements. As a result of this, scientists cannot be certain about exactly how life began.

Extinction

The fossil record provides evidence that most organisms that once existed have become **extinct**. In fact, there have been at least five **mass** extinctions in geological history where most organisms died out. One of these coincides with the disappearance of dinosaurs.

Causes of extinction include:

- **catastrophic events**, e.g. volcanic eruptions, asteroid collisions
- changes to the environment over geological time
- new **predators**
- new **diseases**
- new, more successful **competitors**.

QUESTIONS

QUICK TEST

1. Suggest two pieces of evidence that support the theory of evolution through natural selection.
2. Suggest two causes of extinction.

EXAM PRACTICE

1. The fore-limbs found in human, cow, horse, whale and bird are all based on the pentadactyl limb which means a five-digited limb.

   Scientists believe that the whale may have evolved from a horse-like ancestor that lived in swampy regions millions of years ago.

   Suggest how whales could have evolved from a horse-like mammal. In your answer, use Darwin's theory of natural selection. **[5 marks]**

Darwin and evolution

Human evolution

Modern man has evolved from a common ancestor that gave rise to all the primates: gorillas, chimpanzees and orangutans. DNA comparisons have shown we are most closely related to the chimpanzee.

The evolution of man can be traced back over the last four to five million years. Over this period of time, the **human form** (hominid) has developed:

- a more upright, bipedal stance
- less body hair
- a smaller and less 'domed' forehead
- greater intelligence and use of tools, initially from stone. These tools can be dated using scientific techniques, e.g. radiometric dating and carbon dating.

There have been significant finds of fossils that give clues to human evolution.

1. **Ardi** – at 4.4 million years old, this is the oldest, most complete hominid skeleton.
2. **Lucy** – from 3.2 million years ago, this is one of the first fossils to show an upright walking stance.
3. **Proconsul skull** – discovered by **Mary Leakey** and her husband; the hominid is thought to be from about 1.6 million years ago.

Darwin's theory of evolution through natural selection

Within a population of organisms there is a range of variation among individuals. This is caused by genes. Some differences will be beneficial; some will not.

Beneficial characteristics make an organism more likely to survive and pass on their genes to the next generation. This is especially true if the environment is changing. This ability to be successful is called **survival of the fittest**.

Species that are not well adapted to their environment may become extinct. This process of change is summed up in the theory of evolution through **natural selection**, put forward by **Charles Darwin** in the nineteenth century.

Many theories have tried to explain how life might have come about in its present form.

However, Darwin's theory is accepted by most scientists today. This is because it explains a wide range of observations and has been discussed and tested by many scientists.

Darwin's theory can be reduced to five ideas.

They are:

- variation
- competition
- survival of the fittest
- inheritance
- extinction.

Darwin's ideas are illustrated in the following two examples.

Example 1: peppered moths

Variation – most peppered moths are pale and speckled. They are easily camouflaged amongst the lichens on silver birch tree bark. There are some rare, dark-coloured varieties (that originally arose from genetic mutation). They are easily seen and eaten by birds.

Competition – in areas with high levels of air pollution, lichens die and the bark becomes discoloured by soot. The lighter peppered moths are now put at a competitive disadvantage.

Survival of the fittest – the dark (melanic) moths are now more likely to avoid detection by pedators.

Inheritance – the genes for dark colour are passed on to offspring and gradually become more common in the general population.

Extinction – if the environment remains polluted, the lighter form is more likely to become extinct.

Example 2: methicillin-resistant bacteria

The resistance of some bacteria to antibiotics is an increasing problem. MRSA bacteria have become more common in hospital wards and are difficult to eradicate.

Variation – bacteria mutate by chance, giving them a resistance to antibiotics.

Competition – the non-resistant bacteria are more likely to be killed by the antibiotic and become less competitive.

Survival of the fittest – the antibiotic-resistant bacteria survive and reproduce more often.

Inheritance – resistant bacteria pass on their genes to a new generation; the gene becomes more common in the general population.

Extinction – non-resistant bacteria are replaced by the newer, resistant strain.

To slow down the rate at which new, resistant strains of bacteria can develop:

- doctors are urged not to prescribe antibiotics for obvious viral infections or for mild bacterial infections
- patients should complete the full course of antibiotics to ensure that **all** bacteria are destroyed (so that none will survive to mutate into resistant strains).

Species become more and more specialised as they evolve and adapt to their environmental conditions.

The point at which a new species is formed occurs when the original population can no longer interbreed with the newer, 'mutant' population. For this to occur, **isolation** needs to happen.

Speciation

- Groups of the same species that are separated from each other by physical boundaries (like mountains or seas) will not be able to breed and share their genes. This is called **geographical isolation**.
- Over long periods of time, separate groups may specialise so much that they cannot successfully breed any longer and so two new species are formed – this is **reproductive isolation**.

SUMMARY

- **The ability to be successful is called survival of the fittest. Species that are not well adapted to their environment may become extinct.**
- **This process is the basis of the theory of evolution by natural selection, put forward by Charles Darwin.**

QUESTIONS

QUICK TEST

1. Define evolution.
2. What is the name of the oldest most complete hominid?
3. State two examples where natural selection has been observed by scientists in recent times.

EXAM PRACTICE

1. Alfred Wallace lived at the same time as Charles Darwin and also wrote papers on the idea of natural selection. In 1858 he wrote the following:

   "I ask myself: how and why do species change, and why do they change into new and well-defined species. Why and how do they become so exactly adapted to distinct modes of life; and why do all the intermediate grades die out?"

   Wallace had no knowledge of genes or DNA structure.

   Knowing what we know now, how would you answer his question about how organisms become 'exactly adapted?' **[2 marks]**

Evolution – the modern synthesis

The pioneers of evolution

Charles Darwin published his findings and theories in a book called *On the Origin of Species* (1859). At the time, the reaction to Darwin's theory, particularly from religious authorities, was hostile. They felt he was saying that 'people were descended from monkeys' (although he wasn't) and that he denied God's role in the creation of man. This meant that Darwin's theory was only slowly and reluctantly accepted by many people in spite of his many eminent supporters. In Darwin's time there wasn't the wealth of evidence we have today from the study of genetics and DNA. (The structure of DNA wasn't made plain until 1953.)

Jean-Baptiste Lamarck had published ideas about the gradual change in organisms before Darwin but he put forward a different approach (now known to be inaccurate). He believed that evolution was driven by the inheritance of acquired characteristics. However, he thought that:

- organisms changed during their lifetime as they struggled to survive
- these changes were passed on to the offspring.

For example, he said that when giraffes stretched their necks to reach leaves higher up, this extra neck length was passed on to their offspring.

Lamarck's theory was rejected because there was no evidence that the changes occurring in an individual's lifetime could alter their genes and so be passed on to their offspring. The difference between Darwin and Lamarck's approach is shown in the picture.

Lamarck believed that the necks of giraffes stretched during their lifetime to reach food in trees. They then passed this characteristic on to the next generation.

Darwin believed giraffes that had longer necks could reach more food in trees, so they were more likely to survive and reproduce (survival of the fittest).

Alfred Russel Wallace also put forward a theory of evolution by natural selection. In fact, he and Darwin jointly published writings a year before Darwin's *On the Origin of Species.*

Wallace gathered evidence from around the world to back up the theory of natural selection. In particular, he looked at how speciation could occur and wrote extensively about warning colouration in animals.

Gregor Mendel carried out breeding programmes for plants in the middle of the nineteenth century. He discovered that characteristics could be passed down as 'units' (which we now know as genes) in a predictable way.

HT WS The significance of Mendel's work was only realised after his death. This shows the importance of a scientist's work being published so that others can build on it. It also shows that, sometimes, apparently trivial phenomena and their study can lead to important theories. In this way, our understanding of the world we live in is enhanced.

In your course, you will need to give examples of **how a model can be tested** by **observation** and **experiment**. Mendel's work is a classic example of this because he could show that the ratios of offspring in his pea plant experiments could be accurately predicted.

Other evidence to support the theory of natural selection

Since the nineteenth century, a sequence of findings and techniques has further reinforced our ideas about evolution by natural selection.

- In the late nineteenth century, chromosomes were observed during cell division under the microscope.
- In the early twentieth century, the link between Mendel's 'units' and genes was made. The idea that genes were found on chromosomes was put forward.
- In the mid-twentieth century, the structure of DNA was worked out by Watson and Crick.

Cloning

Studies of genetics have led to many technological advancements. One of these is the ability to clone cells, organs or even whole organisms.

Cuttings

When a gardener has a plant with all the desired characteristics, he/she may choose to produce lots of them by taking stem, leaf or root cuttings. The cuttings are grown in a damp atmosphere until roots develop.

Cloning by tissue culture

To mass-produce plants that are genetically identical to the parent plant and to each other, horticulturalists follow this method.

1. Select a parent plant with the characteristics that you want.
2. Scrape off lots of small pieces of tissue into several beakers containing nutrients and hormones.
3. A week or two later there will be lots of genetically identical plantlets growing.
4. Repeat the process.

Embryo transplantation

Embryo transplantation is now commonly used to breed farm animals.

1. Sperm is collected from an adult male with desirable characteristics.
2. A selected female is artificially inseminated with the male animal's sperm.
3. The fertilised egg develops into an embryo that is removed from the mother at an early stage.
4. In the laboratory, the embryo is split to form several clones.
5. Each clone is transplanted into a female who will be the surrogate mother to the newborn.

Adult cell cloning

1. The diploid nucleus is taken from a mature cell (ordinary body cell) of the donor organism.
2. The diploid nucleus, containing all of the donor's genetic information, is inserted into an empty egg cell (i.e. an egg cell with the nucleus removed or enucleated). This is nuclear transfer. A tiny electric shock is applied at this stage to trigger cell division in the embryo.
3. The egg cell, containing the diploid nucleus, is stimulated so that it begins to divide by mitosis.
4. The resulting embryo is implanted in the uterus of a 'surrogate mother'.
5. The embryo develops into a foetus and is born as normal.

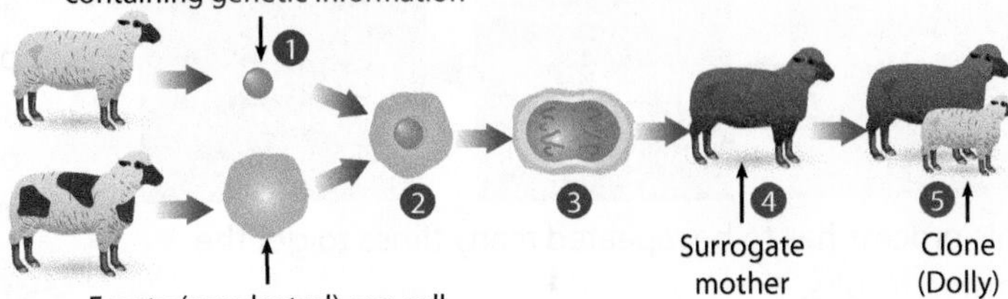

SUMMARY

- **Charles Darwin, Jean-Baptiste Lamarck, Alfred Russel Wallace and Gregor Mendel were pioneers of evolution.**
- **Advances in technology have led to the ability to clone cells, organs and whole organisms through cuttings, tissue culture, embryo transplantation and adult cell cloning.**

QUESTIONS

QUICK TEST

1. How did Darwin's mechanism of natural selection differ from Lamarck's?
2. Describe the advantage of cloning plants or animals.

QUESTIONS

EXAM PRACTICE

1. Describe the main sequence involved in mass producing a desirable plant through tissue culture. The first step has been done for you: **[3 marks]**

   – Select parent with desired characters …

2. Dolly the sheep was the first animal to be produced by adult cell cloning, but she only lived until the age of 8 years and it took over 100 attempts to produce a viable embryo. Some say, this is too high a price to pay for this technique.

   State the advantages of adult cell cloning to counter this argument. **[2 marks]**

Selective breeding and genetic engineering

Selective breeding

Farmers and dog breeders have used the principles of selective breeding for thousands of years by keeping the best animals and plants for breeding.

For example, to breed Dalmatian dogs, the spottiest dogs have been bred through the generations to eventually get Dalmatians. The factor most affected by selective breeding in dogs is probably temperament. Most breeds are either naturally obedient to humans or are trained to be so.

This is the process of selective breeding.

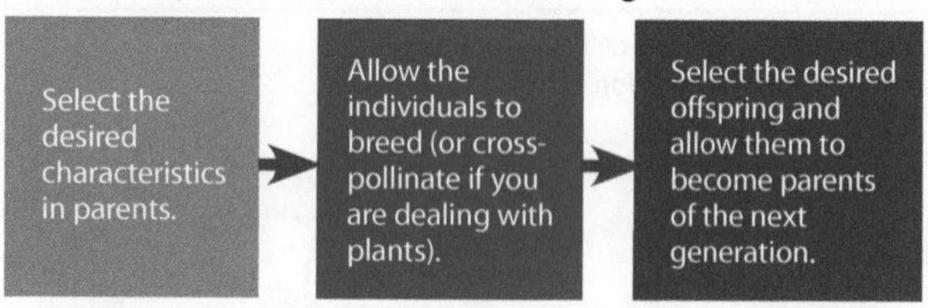

This process has to be repeated many times to get the desired results.

Advantages of selective breeding
• It results in an organism with the 'right' characteristics for a particular function. • In farming and horticulture, it is a more efficient and economically viable process than natural selection.
Disadvantages of selective breeding
• Intensive selective breeding reduces the gene pool – the range of alleles in the population decreases so there is **less variation**. • Lower variation reduces a species' ability to respond to environmental change. • It can lead to an accumulation of harmful recessive characteristics (in-breeding), e.g. restriction of respiratory pathways and dislocatable joints in bulldogs.

Examples of selective breeding

Modern food plants

Three of our modern vegetables have come from a single ancestor by selective breeding. (Remember, it can take many, many generations to get the desired results.)

Selective breeding in plants has also been undertaken to produce:

- disease resistance in crops
- large, unusual flowers in garden plants.

Modern cattle

Selective breeding can contribute to improved yields in cattle. Here are some examples.

- **Quantity of milk** – years of selecting and breeding cattle that produce larger than average quantities of milk has produced herds of cows that produce high daily volumes of milk.
- **Quality of milk** – as a result of selective breeding, Jersey cows produce milk that is rich and creamy, and can therefore be sold at a higher price.
- **Beef production** – the characteristics of the Hereford and Angus varieties have been selected for beef production over the past 200 years or more. They include hardiness, early maturity, high numbers of offspring and the swift, efficient conversion of grass into body mass (meat).

Genetic engineering

All living organisms use the same basic genetic code (DNA). So genes can be transferred from one organism to another in order to deliberately change the recipient's characteristics. This process is called genetic engineering or genetic modification (GM).

Altering the genetic make-up of an organism can be done for many reasons.

- **To improve resistance to herbicides**: for example, soya plants are genetically modified by inserting a gene that makes them resistant to a herbicide. When the crop fields are sprayed with the herbicide only the weeds die, leaving the soya plants without competition so they can grow better. Resistance to frost or disease can also be genetically engineered. Bigger yields result.
- **To improve the quality of food**: for example, bigger and more tasty fruit.
- **To produce a substance you require**: for example, the gene for human insulin can be inserted into bacteria or fungi, to make human insulin on a large scale to treat diabetes.
- **Disease resistance**: crop plants receive genes that give them resistance to the bacterium *Bacillus thuringiensis*.

Advantages of genetic engineering
● It allows organisms with new features to be produced rapidly. ● It can be used to make biochemical processes cheaper and more efficient. ● In the future, it may be possible to use genetic engineering to change a person's genes and cure certain disorders, e.g. cystic fibrosis. This is an area of research called gene therapy.
Disadvantages of genetic engineering
● Transplanted genes may have unexpected harmful effects on human health. ● Some people are worried that GM plants may cross-breed with wild plants and release their new genes into the environment.

HT Producing insulin

The following method is used to produce insulin.

1. The human gene for insulin production is identified and removed using a restriction enzyme, which cuts through the DNA strands in precise places.
2. The same restriction enzyme is used to cut open a ring of bacterial vector DNA (a plasmid).
3. Other enzymes called ligases are then used to insert the section of human DNA into the plasmid. The DNA can be 'spliced' in this way because the ends of the strands are 'sticky'.
4. The plasmid is reinserted into a bacterium, which starts to divide rapidly. As it divides, it replicates the plasmid.
5. The bacteria are cultivated on a large scale in fermenters. Each bacterium carries the instructions to make insulin. When the bacteria make the protein, commercial quantities of insulin are produced.

Sometimes, other vectors are used to introduce human DNA into organisms, e.g. viruses. It is important that the hybrid genes are transferred to the host organism at an early stage of its development, e.g. the egg or the embryo stage. As the cells are quite undifferentiated, the desired characteristics from the inserted DNA are more likely to develop.

SUMMARY

- **Selective breeding is done to ensure desirable characteristics are passed on.**
- **Genetic modification is when genes are transferred from one organism to another to change the recipient's characteristics.**

QUESTIONS

QUICK TEST

1. What advantages does genetic engineering have over selective breeding?
2. (HT) What is a plasmid?

EXAM PRACTICE

1. State one advantage and one disadvantage of selective breeding. **[2 marks]**
2. (HT) Describe the process where artificial insulin is produced via genetic engineering.

   State any enzymes involved and how the final product is obtained. The first stage is completed for you. **[5 marks]**

   Stage 1: Human gene for insulin identified

   Stage 2: ....

   Stage 3: ....

   Stage 4: ....

   Stage 5: ....

   Stage 6: ....

Classification

The origins of classification

In the past, observable characteristics were used to place organisms into categories.

In the eighteenth century, **Carl Linnaeus** produced the first classification system. He developed a hierarchical arrangement where larger groups were subdivided into smaller ones.

Kingdom	Largest group
Phylum	
Class	
Order	
Family	
Genus	
Species	Smallest group

Linnaeus also developed a binomial system for naming organisms according to their genus and species. For example, the common domestic cat is *Felis catus*. Its full classification would be:

- Kingdom: *Animalia*
- Phylum: *Chordata*
- Class: *Mammalia*
- Order: *Carnivora*
- Family: *Felidae*
- Genus: *Felis*
- Species: *Catus*.

Linnaeus' system was built on and resulted in a **five-kingdom system**. Developments that contributed to the introduction of this system included improvements in microscopes and a more thorough understanding of the biochemical processes that occur in all living things. For example, the presence of particular chemical pathways in a range of organisms indicated that they probably had a common ancestor and so were more closely related than organisms that didn't share these pathways.

Kingdom	Features	Examples
Plants	Cellulose cell wall Use light energy to produce food	Flowering plants Trees Ferns Mosses
Animals	Multicellular Feed on other organisms	Vertebrates Invertebrates
Fungi	Cell wall of chitin Produce spores	Toadstools Mushrooms Yeasts Moulds
Protoctista Protozoa	Mostly single-celled organisms	Amoeba Paramecium
Prokaryotes	No nucleus	Bacteria Blue–green algae

The classification diagram below illustrates how different lines of evidence can be used. The classes of vertebrates share a common ancestor and so are quite closely related. Evidence for this lies in comparative anatomy (e.g. the pentadactyl limb) and similarities in biochemical pathways.

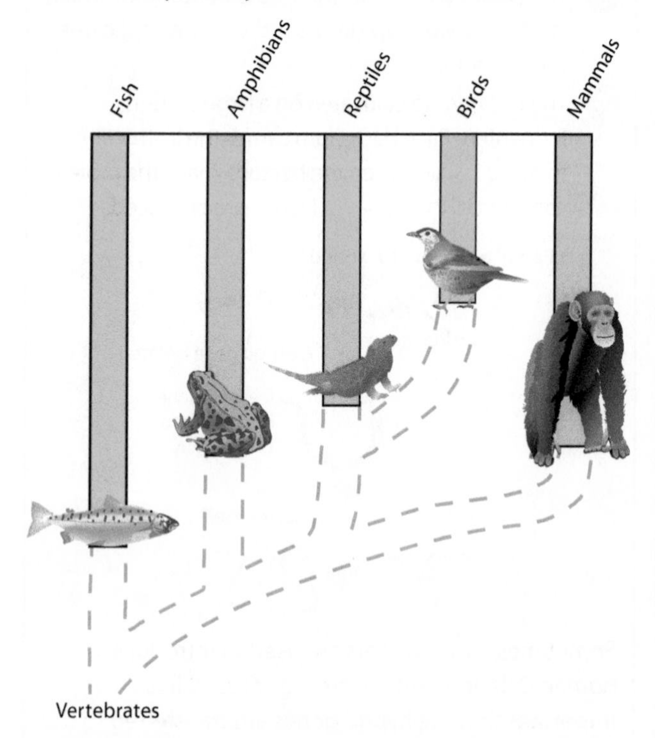

In more recent times, improvements in science have led to a **three-domain system** developed by **Carl Woese**. In this system organisms are split into:

- **archaea** (primitive bacteria)

- **bacteria** (true bacteria)
- **eukaryota** (including Protista, fungi, plants and animals).

Further improvements in science include chemical analysis and further refinements in comparisons between non-coding sections of DNA.

Evolutionary trees

Tree diagrams are useful for depicting relationships between similar groups of organisms and determining how they may have developed from common ancestors. Fossil evidence can be invaluable in establishing these relationships.

Here, two species are shown to have evolved from a common ancestor.

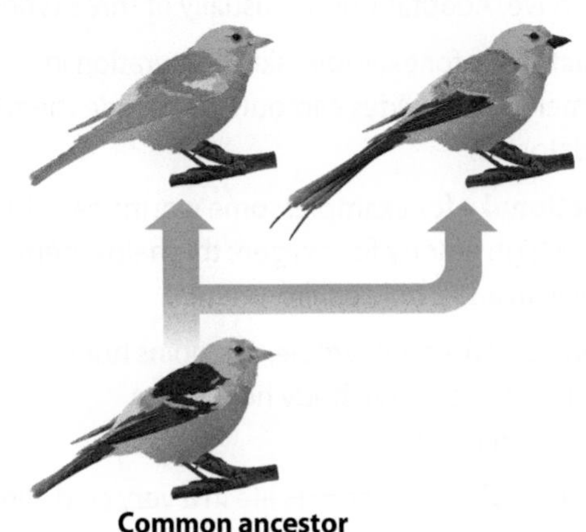

Common ancestor

WS You need to understand how new evidence and data leads to changes in models and theories.

In the case of the three-domain system, a more accurate and cohesive classification structure was proposed as a result of improvements in microscopy and increasing knowledge of organisms' internal structures.

These apes share a common ancestor

SUMMARY

- **There is a huge variety of living organisms. Scientists group or classify them using shared characteristics.**
- **This is important because it helps to: work out how organisms evolved on Earth; understand how organisms coexist in ecological communities and identify and monitor rare organisms that are at risk from extinction.**

QUESTIONS

QUICK TEST

1. What do the first and second words in the binomial name of an organism mean?
2. Who developed the recent three-domain system?

EXAM PRACTICE

1. In the Linnean system of classification, organisms are placed within a hierarchy of organisation.

   a) The following table shows this hierarchy with some missing levels. Complete the missing parts.

Kingdom
....
Class
....
Family
....
Species

[3 marks]

   b) Linnaeus also developed the bionomial system of classification. The domestic dog has the following classification:

   Kingdom – *Animalia*; Class – *Mammalia*; Family – *Canidae*; Genus – *Canis*; Species – *familiaris*

   Use this information to suggest the binomial name of the domestic dog. **[1 mark]**

Organisms and ecosystems

Communities

Ecosystems are physical environments with a particular set of conditions (**abiotic** factors), plus all the organisms that live in them. The organisms interact through competition and predation. An ecosystem can support itself without any influx of other factors or materials. Its energy source (usually the Sun) is the only external factor.

Other terms help to describe aspects of the environment.

- The **habitat** of an animal or plant is the part of the physical environment where it lives. There are many types of habitat, each with particular characteristics, e.g. pond, hedgerow, coral reef.
- A **population** is the number of individuals of a species in a defined area.
- A **community** is the total number of individuals of all the different populations of organisms that live together in a habitat at any one time.

An organism must be well-suited to its habitat to be able to compete with other species for limited environmental resources. Even organisms within the same species may compete in order to survive and breed. Organisms that are specialised in this way are restricted to that type of habitat because their adaptations are unsuitable elsewhere.

Resources that plants compete over include:

- light
- space
- water
- minerals.

Animals compete over:

- food
- mates.
- territory.

Interdependence

In communities, each species may depend on other species for food, shelter, pollination and seed dispersal. If one species is removed, it may have knock-on effects for other species.

Stable communities contain species whose numbers fluctuate very little over time. The populations are in balance with the physical factors that exist in that habitat. Stable communities include **tropical rainforests** and ancient oak **woodlands**.

Adaptations

Adaptations:

- are special features or behaviours that make an organism particularly well-suited to its environment and better able to compete with other organisms for limited resources
- can be thought of as a biological solution to an environmental challenge – evolution provides the solution and makes species fit their environment.

Animals have developed in many different ways to become well adapted to their environment and to help them survive. Adaptations are usually of three types:

- **Structural** – for example, skin colouration in chameleons provides camouflage to hide them from predators.
- **Functional** – for example, some worms have blood with a high affinity for oxygen; this helps them to survive in anaerobic environments.
- **Behavioural** – for example, penguins huddle together to conserve body heat in the Antarctic habitat.

Look at the **polar bear** and its life in a very cold climate. It has:

- small ears and large bulk to reduce its surface area to volume ratio and so reduce heat loss
- a large amount of insulating fat (blubber)
- thick white fur for insulation and camouflage
- large feet to spread its weight on snow and ice
- fur on the soles of its paws for insulation and grip
- powerful legs so it is a good swimmer and runner, which enables it to catch its food
- sharp claws and teeth to capture prey.

The **cactus** is well adapted to living in a desert habitat. It:

- has a rounded shape, which gives a small surface area to volume ratio and therefore reduces water loss
- has a thick waxy cuticle to reduce water loss
- stores water in a spongy layer inside its stem to resist drought
- has sunken stomata, meaning that air movement is reduced, minimising loss of water vapour through them
- has leaves that take the form of spines to reduce water loss and to protect the cactus from predators.

Some organisms have biochemical adaptations. **Extremophiles** can survive extreme environmental conditions. For example:

- bacteria living in deep sea vents have optimum temperatures for enzymes that are much higher than 37°C
- icefish have antifreeze chemicals in their bodies, which lower the freezing point of body fluids
- some organisms can resist high salt concentrations or pressure.

WS During your course you will be asked to suggest explanations for observations made in the field or laboratory. These include:

- suggesting factors for which organisms are competing in a certain habitat
- giving possible adaptations for organisms in a habitat.

For example, low-lying plants in forest ecosystems often have specific adaptations for maximising light absorption as they are shaded by taller plants. Adaptations might include leaves with a large surface area and higher concentrations of photosynthetic pigments to absorb the correct wavelengths and lower intensities of light.

SUMMARY

- **In a community, each species may depend on other species for food, shelter, pollination and seed dispersal. They are interdependent.**
- **Organisms need to be well adapted within the ecosystem in order to survive.**
- **Adaptations may be structural, functional or behavioural.**
- **Extremophiles can survive in extreme environments, e.g. very low temperatures.**

QUESTIONS

QUICK TEST

1. What is an ecosystem?
2. What three resources do animals compete over?
3. Give two examples of extremophiles.

EXAM PRACTICE

1. Amazonian tree frogs have adaptations suited to living in equatorial rainforests.

   Explain how each of the adaptations listed below helps the frog to survive.

   i) Sticky pads on its 'fingers' and toes.

   ii) Hides its bright colours when asleep by closing its eyes and tucking its feet under its body.

   iii) Long sticky tongue. **[3 marks]**

2. What is an extremophile?

   Give one example of how these organisms are well adapted to their habitats. **[2 marks]**

Studying ecosystems 1

Taking measurements in ecosystems

Ecosystems involve the interaction between **non-living** (**abiotic**) and **living** (**biotic**) parts of the environment. So it is important to identify which factors need to be measured in a particular habitat.

Abiotic factors include:

- light intensity
- temperature
- moisture levels
- soil pH and mineral content
- wind intensity and direction
- carbon dioxide levels for plants
- oxygen levels for aquatic animals.

Biotic factors include:

- availability of food
- new predators arriving
- new pathogens
- one species out-competing another.

Measuring biotic factors – sampling methods

It is usually impossible to count all the species living in a particular area, so a sample is taken.

When sampling, make sure you:

- **take a big enough sample** to make the estimate good and reliable – the larger the sample, the more accurate the results.
- **sample randomly** – the more random the sample, the more likely it is to be representative of the population.

Quadrats

Quadrats are square frames that typically have sides of length 0.5 m. They provide excellent results as long as they are placed randomly. The population of a certain species can then be estimated.

For example, if an average of 4 dandelion plants are found in a 0.25 m^2 quadrat, a scientist would estimate that 16 dandelion plants would be found in each 1 m^2 and 16 000 dandelion plants in a 1000 m^2 field.

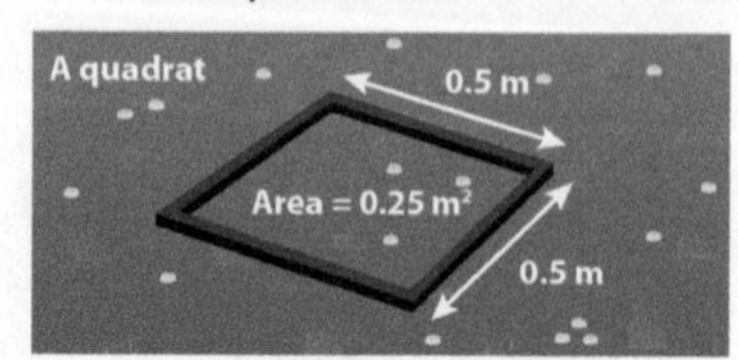

Transects

Sometimes an environmental scientist may want to look at how species change across a habitat, or the boundary between two different habitats – for example, the plants found in a field as you move away from a hedgerow.

This needs a different approach that is systematic rather than random.

1. Lay down a line such as a tape measure. Mark regular intervals on it.
2. Next to the line, lay down a small quadrat. Estimate or count the number of plants of the different species. This can sometimes be done by estimating the percentage cover.
3. Move the quadrat along at regular intervals. Estimate and record the plant populations at each point until the end of the line.

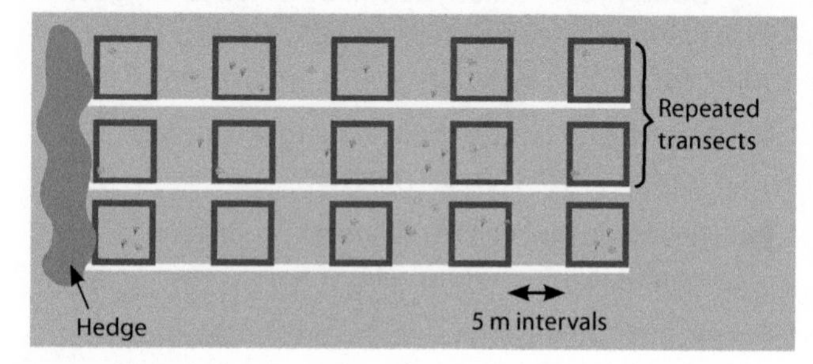

Sampling methods

Sampling animal populations is more problematic as they are mobile and well adapted to evade capture. Here are four of the main techniques used.

Pooters

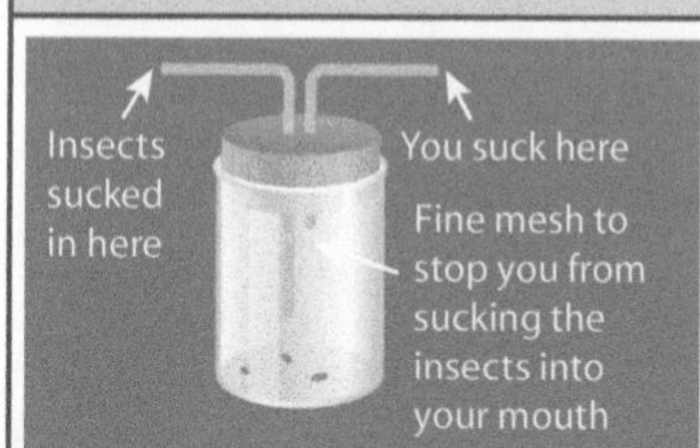

This is a simple technique in which insects are gathered up easily without harm. With this method, you get to find out which species are actually present, although you have to be systematic about your sampling in order to get representative results and it is difficult to get ideas of numbers.

Sweepnets

Sweepnets are used in long grass or moderately dense woodland where there are lots of shrubs. Again, it is difficult to get truly representative samples, particularly in terms of the relative numbers of organisms.

Pitfall traps

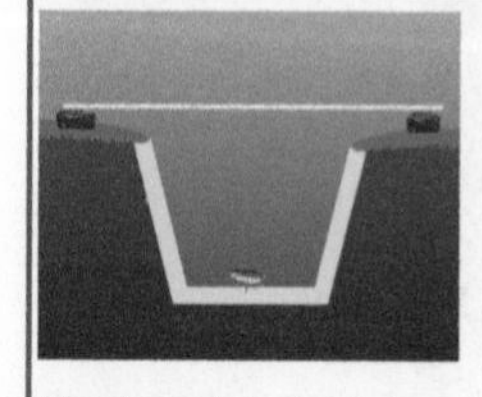

Pitfall traps are set into the ground and used to catch small insects, e.g. beetles. Sometimes a mixture of ethanol or detergent and water is placed in the bottom of the trap to kill the samples, and prevent them from escaping. This method can give an indication of the relative numbers of organisms in a given area if enough traps are used to give a representative sample.

Capture/recapture

The capture/recapture method is sometimes called the Lincoln index.

1. Animals are caught humanely – for example, woodlice are caught in traps overnight. Their number is counted and recorded.
2. The animals are marked in some way – for example, water boatmen (a type of insect) can be marked with a drop of waterproof paint on their upper surface.
3. The marked animals are then released back into the population for a suitable amount of time.
4. A second sample is obtained, which will contain some marked animals and some unmarked. The numbers in each group are again counted.

Certain assumptions are made when using capture/recapture data. These include:

- no death, immigration or emigration
- each sample being collected in exactly the same way without bias
- the marks given to the animals not affecting their survival rate, e.g. using paint on invertebrates requires care because if too much is added it can enter their respiratory passages and kill them.

The following formula can then be used to estimate the total population size in the habitat.

$$\text{population size} = \frac{\text{number in first sample (all marked)} \times \text{number in second sample (marked and unmarked)}}{\text{number in second sample that were previously marked}}$$

Example

Fifty-six mice are caught in woodland and a small section of fur removed from their tails with clippers. The mice are released. The next evening, a further sample of sixty-two mice are caught. Twenty-five of these mice have shaved tails. What is the total population size?

$$\text{Population size} = \frac{56 \times 62}{25} = 139 \text{ mice}$$

Keys

Correctly identifying species in a sample can be difficult. Using keys like this one can help to identify organisms.

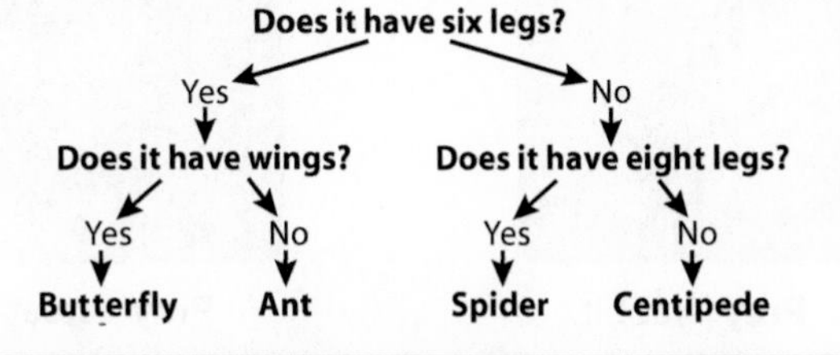

SUMMARY

- **Ecosystems involve interaction between abiotic and biotic parts of the environment.**
- **Biotic factors can be measured by sampling methods such as quadrats and transects.**
- **Animal numbers can be sampled using trading techniques.**

QUESTIONS

QUICK TEST

1. What information does a transect give you?
2. What is the difference between biotic and abiotic factors?

EXAM PRACTICE

1. A group of students were carrying out a quadrat survey to determine the population of daisies in a park. The park measured 60m x 90m.

   Their results are shown in the table.

Quadrat	1	2	3	4	5	6	7	8	9	10
No. daisies	5	2	1	0	4	5	2	0	6	3

   a) They placed the quadrats randomly.

   Why was this? **[1 mark]**

   b) Calculate the median number of daisies per quadrat. **[3 marks]**

   c) Another student calculated the mean number of daisies per quadrat to be 2.3.

   If the quadrats measured $1m^2 \times 1m^2$, calculate the estimated number of daisies in the park.

   Show your working. **[2 marks]**

Studying ecosystems 2

Measuring abiotic factors

Many measurements connected with climate and the weather can be measured using meteorological equipment. Measuring chemical pollutant levels is particularly useful as they give information about concentrations of pollutants. This can be done by **direct chemical tests**.

For example, water can be tested for pH and samples can be assessed for metal ion content, e.g. mercury.

Indicator species

The occurrence of certain indicator species can be observed by sampling an area and noting whether the species is present or not, together with the overall numbers of the species.

For example, insect larvae and other invertebrates can act as an indicator of water pollution. When sewage works outflow into a stream, this pollutes the water by altering the levels of nitrogen compounds in the stream and reducing oxygen levels. This has an impact on the organisms that can survive in the stream.

Organisms that can cope with pollution include the rat-tailed maggot, the bloodworm, the water louse and the sludge worm. However, some organisms are very sensitive to this type of water pollution so they are not found in areas where the levels are high.

Organisms such as mayfly and stonefly larvae are killed by high levels of water pollution (they cannot tolerate low oxygen levels), so they are indicators of clean water.

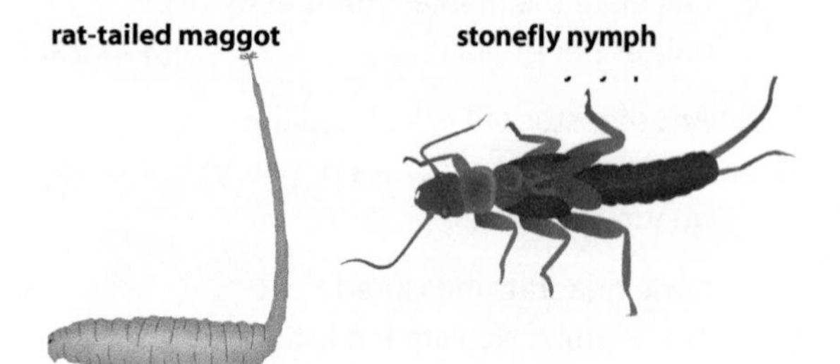

Air quality can be measured using lichens as **indicator species**.

Predator–prey relationships

Animals that kill and eat other animals are called **predators** (e.g. foxes, lynx). The animals that are eaten are called **prey** (e.g. rabbits, snowshoe hares).

Many animals can be both predator and prey. For instance, a stoat is a predator when it hunts rabbits and it is the prey when it is hunted by a fox.

Predator – stoat

Prey – rabbit

Predator – fox

Prey – stoat

In nature there is a delicate balance between the population of a predator (e.g. lynx) and its prey (e.g. snowshoe hare). However, the prey will always outnumber the predators.

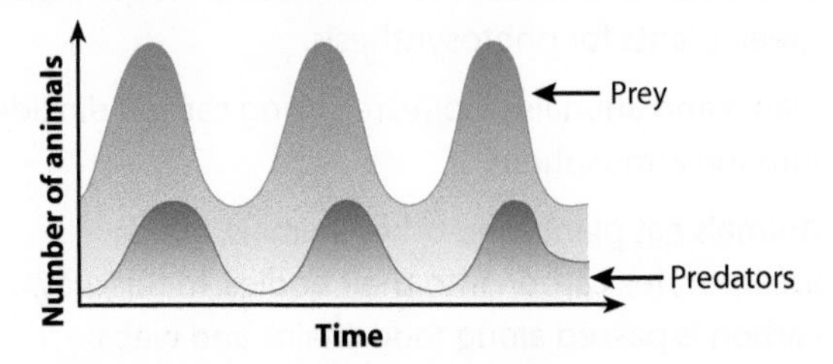

The number of predators and prey follow a classic population cycle. There will always be more hares than lynx and the population peak for the lynx will always come after the population peak for the hare. As the population cycle is cause and effect, they will always be out of phase.

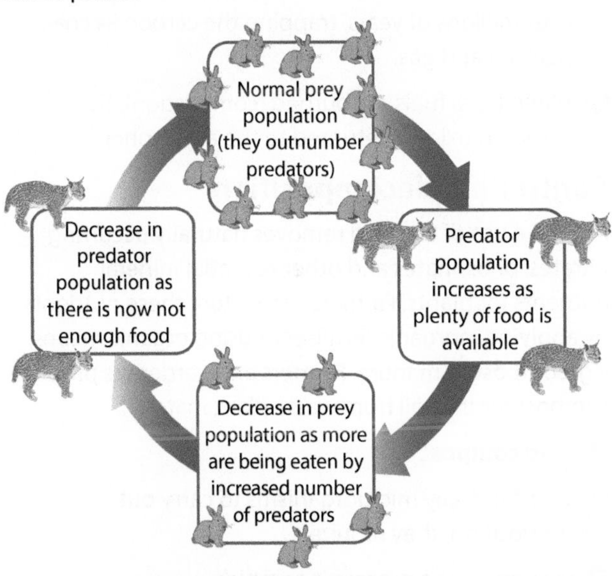

SUMMARY

- **Indicator species can be observed by sampling an area to see whether the species is present or not.**
- **Predators kill and eat prey. The prey will always outnumber predators. Numbers of predators and prey vary in cycles.**

QUESTIONS

QUICK TEST

1. What is an indicator species?
2. What is a predator?

EXAM PRACTICE

1. Jenna has been using a net to capture organisms from a fast-flowing stream. After half an hour, she counts and identifies the organisms she has found.

   Her results are displayed in the table.

Species	Number
Stonefly nymphs	1
Mayfly nymphs	5
Caddis fly larvae	2
Water fleas	75

   a) Jenna says that because of the indicator species she has found, the water is fairly unpolluted.

   Do you agree?

   Give a reason for your answer. **[2 marks]**

   b) Apart from further samples, what other data would she need to collect in order to verify this conclusion? **[1 mark]**

2. Rabbits and stoat numbers tend to follow a cyclical pattern as their numbers change over the years.

   Explain why this is so. **[3 marks]**

Recycling

The water cycle

Water is a vital part of the biosphere. Most organisms consist of over 50% water.

The two key processes that drive the water cycle are **evaporation** and **condensation**.

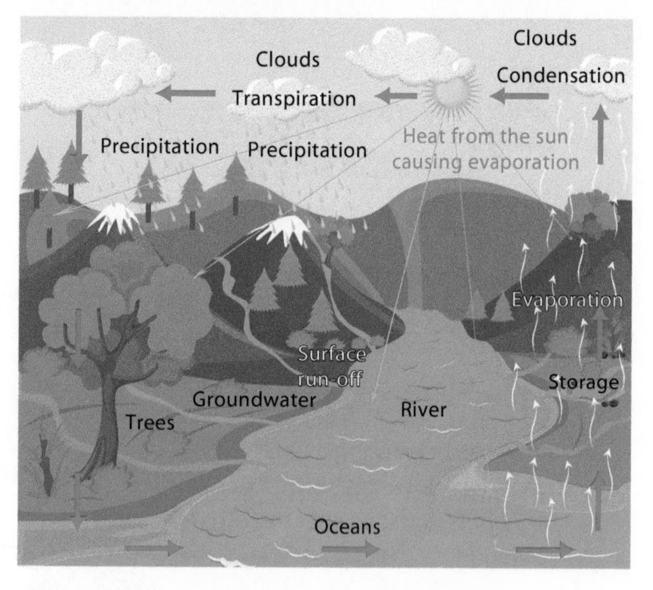

The carbon cycle

The constant recycling of carbon is called the carbon cycle.

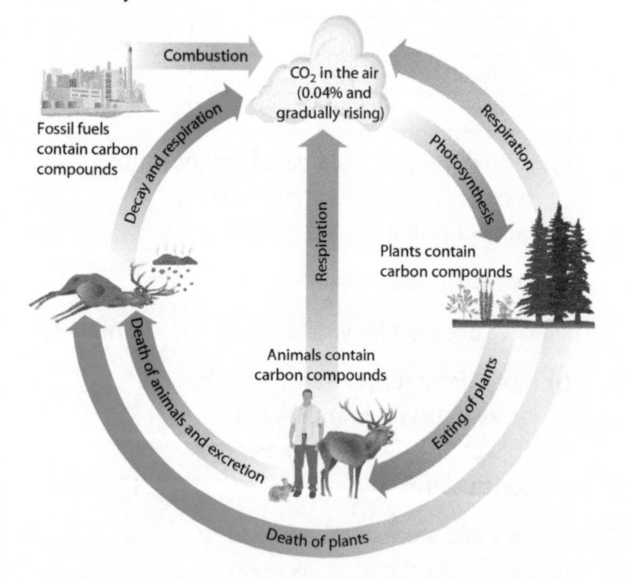

- Carbon dioxide is removed from the atmosphere by green plants for photosynthesis.
- Plants and animals respire, releasing carbon dioxide into the atmosphere.
- Animals eat plants and other animals, which incorporates carbon into their bodies. In this way, carbon is passed along food chains and webs.
- Microorganisms such as fungi and bacteria feed on dead plants and animals, causing them to decay. The microorganisms respire and release carbon dioxide gas into the air. Mineral ions are returned to the soil through decay.
- Some organisms' bodies are turned into fossil fuels over millions of years, trapping the carbon as coal, peat, oil and gas.
- When fossil fuels are burned (combustion), the carbon dioxide is returned to the atmosphere.

Controlling decomposition

Intensive use of the land removes naturally occurring nitrates, phosphates and other essential mineral nutrients for plants. Farmers can restore these nutrients by applying inorganic fertiliser or using organic means, e.g. compost or manure. Farmers and gardeners produce compost for the soil from waste plant material.

Making compost

In order for decay microorganisms to carry out decomposition, they require:

- oxygen (as the process is aerobic)
- moisture
- a warm temperature.

The presence of any of these factors will increase the rate of decomposition.

One of your required practicals will be to investigate the effect of one factor on the rate of decay.

Biogas digesters

A simple biogas generator

Gas trapped beneath metal gas holder

Gas release tap

Waste material

Residual 'digested' sludge

Anaerobic decay produces methane gas (sometimes called **biogas**). Biogas can be made on a large scale using a continuous-flow method in a digester. Organic material is added daily and the biogas is siphoned off and stored. The remaining solid sludge can be used as fertiliser for crops.

The biogas produced by the digesters can be:

- burned to generate electricity
- burned to produce hot water for steam central heating systems
- used as a fuel for buses.

SUMMARY

- **Materials within ecosystems are constantly being recycled and used to provide the substances that make up future organisms.**
- **Two of these substances are water and carbon.**
- **Biogas digestors produce a gas that can be used as a fuel.**

QUESTIONS

QUICK TEST

1. Which two processes drive the water cycle?
2. What is biogas?
3. What three factors are needed for microorganisms to carry out decomposition?

EXAM PRACTICE

1. Carbon is recycled in the environment in a process called the carbon cycle.

   The main processes of the carbon cycle are shown below:

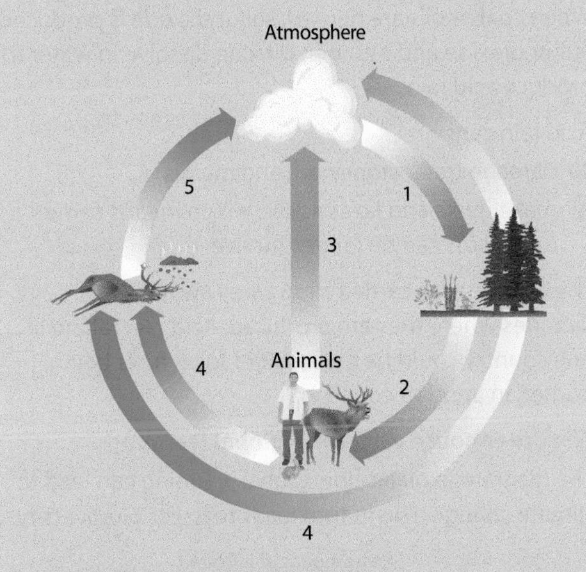

   a) Name the process that occurs at stage 1 in the diagram. **[1 mark]**

   b) If organic material is allowed to decompose anaerobically it produces a different product.

   Which gas is formed from this type of decay? **[1 mark]**

   c) Name one commercial use of the gas produced in this way. **[1 mark]**

Environmental change & biodiversity

Environmental change

Waste management

The human population is increasing exponentially (i.e. at a rapidly increasing rate). This is because birth rates exceed death rates by a large margin.

So the use of finite resources like fossil fuels and minerals is accelerating. In addition, waste production is going up:

- **on land**, from domestic waste in landfill, toxic chemical waste, pesticides and herbicides
- **in water**, from sewage fertiliser and toxic chemicals
- **in the air**, from smoke, carbon dioxide and sulfur dioxide.

Acid rain

When coal or oils are burned, sulfur dioxide is produced. Sulfur dioxide and nitrogen dioxide dissolve in water to produce acid rain.

Acid rain can:

- damage trees, stonework and metals
- make rivers and lakes acidic, which means some organisms can no longer survive.

The acids can be carried a long way away from the factories where they are produced. Acid rain falling in one country could be the result of fossil fuels being burned in another country.

The greenhouse effect and global warming

The diagram explains how global warming can lead to climate change. This in turn leads to lower biodiversity.

Small amount of infra-red radiation transmitted into space

Infra-red radiation re-radiated into atmosphere

Carbon dioxide and methane trap the infra-red radiation (heat) as the longer wavelengths are not transmitted as easily.

UV rays from the sun reach Earth and are absorbed

The consequences of global warming are:

• a rise in sea levels leading to flooding in low-lying areas and loss of habitat
• the migration of species and changes in their distribution due to more extreme temperature and rainfall patterns; some organisms won't survive being displaced into new habitats, or newly migrated species may outcompete native species. The overall effect is a loss of biodiversity.

WS You may be asked to evaluate methods used to address problems caused by human impact on the environment.

For example, here are some figures relating to quotas and numbers of haddock in the North Sea in two successive years.

	2009	2010
Haddock quota (tonnes)	27 507	23 381
Estimated population (thousands)	102	101

What conclusions could you draw from this data? What additional information would you need to give a more accurate picture?

Biodiversity

Biodiversity is a measure of the number and variety of species within an ecosystem. A healthy ecosystem:

- has a large biodiversity
- has a large degree of interdependence between species
- is stable.

Species depend on each other for food, shelter and keeping the external physical environment maintained. Humans have had a negative impact on biodiversity due to:

- pollution killing plants and animals
- degrading the environment through deforestation and removing resources such as minerals and fossil fuels
- over-exploiting habitats and organisms.

Only recently have humans made efforts to reduce their impact on the environment. It is recognised that maintaining biodiversity is important to ensure the continued survival of the human race.

Impact of land use

As humans increase their economic activity, they use land that would otherwise be inhabited by living organisms. Examples of habitat destruction include:

- farming
- quarrying
- dumping waste in landfill.

Peat bogs

Peat bogs are important habitats. They support a wide variety of organisms and act as carbon sinks.

If peat is burned it releases carbon dioxide into the atmosphere and contributes to global warming. Removing peat for use as compost in gardens takes away the habitat for specialised animals and plants that aren't found in other habitats.

Deforestation

Deforestation is a particular problem in tropical regions. Tropical rainforests are removed to:

- **release land for cattle and rice fields** – these are needed to feed the world's growing population and for increasingly Western-style diets
- **grow crops for biofuel** – the crops are converted to **ethanol-based** fuels for use in petrol and diesel engines. Some specialised engines can run off pure ethanol.

The consequences of deforestation are:

There are fewer plants, particularly trees, to absorb carbon dioxide. This leads to increased carbon dioxide in the atmosphere and accelerated global warming.
Combustion and decay of the wood from deforestation releases more carbon dioxide into the atmosphere.
There is reduced biodiversity as animals lose their habitats and food sources.

Maintaining biodiversity

To prevent further losses in biodiversity and to improve the balance of ecosystems, scientists, the government and environmental organisations can take action.

- Scientists establish **breeding programmes** for **endangered species**. These may be captive methods where animals are enclosed, or protection schemes that allow rare species to breed without being poached or killed illegally.
- The government sets **limits** on **deforestation** and **greenhouse gas emissions**.

Environmental organisations:

- **protect and regenerate** shrinking habitats such as mangrove swamps, heathlands and coral reefs
- **conserve and replant** hedgerows around the margins of fields used for crop growth
- introduce and encourage **recycling** initiatives that reduce the volume of landfill.

SUMMARY

- **The human population is increasing exponentially. Waste production is increasing.**
- **Burning of fossil fuels contributes to acid rain and global warming.**
- **Biodiversity is the number and variety of species in an ecosystem. Maintaining biodiversity is vital.**

QUESTIONS

QUICK TEST

1. Why are tropical rainforests removed?
2. Why is high biodiversity seen as a good thing?

EXAM PRACTICE

1. In Ireland, four species of bumble bee are now endangered. Scientists are worried that numbers may become so low that they are inadequate to provide pollination to certain plants.

   a) Explain how the disappearance of these four species of bumble bee might affect biodiversity as a whole. **[3 marks]**

   b) There are many threats to bumble bees, such as the possible effect of pesticides like *neonicotinoids*; loss of suitable habitat, disappearance of wildflowers and new diseases.

   Suggest two conservation measures that might help increase numbers of bumble bees in the wild. **[2 marks]**

DAY 7 40 Minutes

Energy and biomass in ecosystems

Trophic levels

Communities of organisms are organised in an ecosystem according to their feeding habits.

Food chains show:

- the organisms that consume other organisms
- the transfer of energy and materials from organism to organism.

Energy from the Sun enters most food chains when green plants absorb sunlight to photosynthesise. Photosynthetic and chemosynthetic organisms are the producers of biomass for the Earth. Feeding passes this energy and biomass from one organism to the next along the food chain.

A food chain

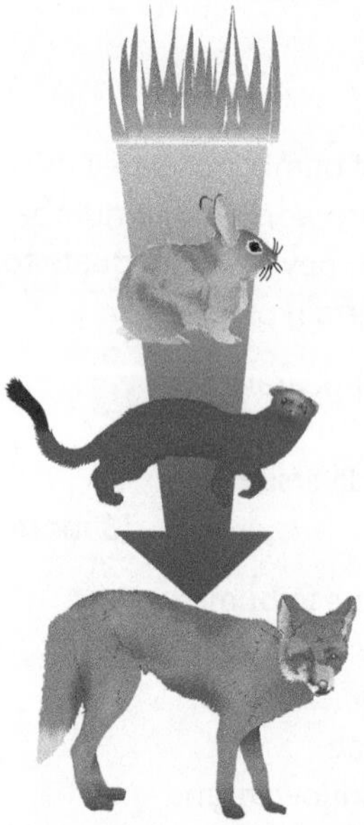

Green plant:
producer
Rabbit:
primary consumer
Stoat:
secondary consumer
Fox:
tertiary consumer

The arrow shows the flow of energy and biomass along the food chain.

- All food chains start with a producer.
- The rabbit is a herbivore (plant eater), also known as the primary consumer.
- The stoat is a carnivore (meat eater), also known as the secondary consumer.
- The fox is the top carnivore in this food chain, the tertiary consumer.

Each consumer or producer occupies a **trophic level** (feeding level).

- Level 1 are producers.
- Level 2 are primary consumers.
- Level 3 are secondary consumers.
- Level 4 are tertiary consumers.

Excretory products and uneaten parts of organisms can be the starting points for other food chains, especially those involving decomposers.

Decomposers

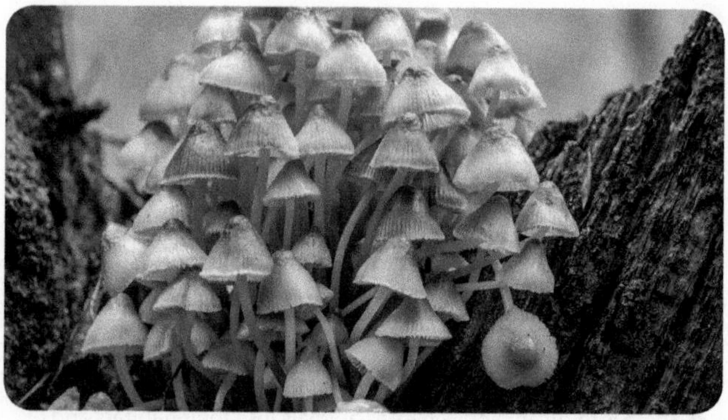

Pyramids of biomass

Pyramids of biomass deal with the dry mass of living material in the chain. The width of each trophic level can be calculated by multiplying the number of organisms by their dry mass.

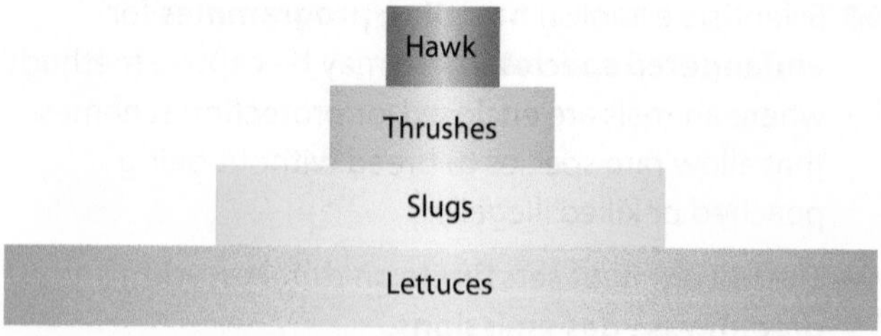

Hawk
Thrushes
Slugs
Lettuces

Efficiency of energy transfer

If you know how much energy is stored in the living organisms at each level of a food chain, the efficiency of energy transfer can be calculated.

To do this, divide the amount of energy used usefully (e.g. for growth) by the total amount of energy taken in.

$$\textbf{energy efficiency (\%)} = \frac{\textbf{energy used usefully}}{\textbf{total energy taken in}} \times 100$$

Example

A sheep eats 100 kJ of energy in the form of grass but only 9 kJ becomes new body tissue; the rest is lost as faeces, urine or heat. Calculate the efficiency of energy transfer in the sheep.

$$\text{energy efficiency} = \frac{9}{100} \times 100$$
$$= 9\%$$

Similar efficiency calculations can be performed for biomass.

WS During your course you will be expected to use formulae to answer questions and to re-arrange the subject of an equation. Calculating energy efficiency is one example of when you may need to do this.

For example, you might be given the energy efficiency for a particular trophic level and asked to find out the total energy taken in. In this case:

$$\text{energy efficiency (\%)} = \frac{\text{energy used usefully}}{\text{total energy taken in}} \times 100$$

would be re-arranged to form:

$$\text{total energy taken in} = \frac{\text{energy used usefully}}{\text{energy efficiency}} \times 100$$

Energy

The fox gets the last tiny bit of energy left after all the others have had a share. This explains why food chains rarely have fourth degree or fifth degree consumers – they would not get enough energy to survive.

The stoats run around, mate, excrete, keep warm, etc. They pass on about a tenth of all the energy they get from the rabbits.

The rabbits run around, mate, excrete, keep warm, etc. They pass on about a tenth of all the energy they get from the grass.

The Sun is the energy source for all organisms. However, only about 10% of the Sun's energy is captured in photosynthesis.

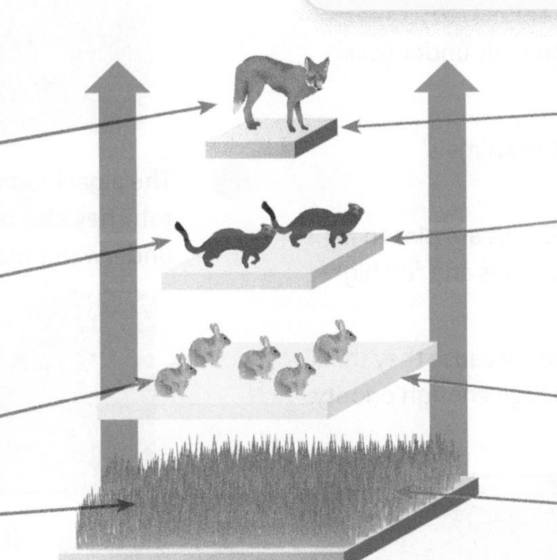

Biomass

The fox gets the remaining biomass.

The stoats lose quite a lot of biomass in faeces (egestion), urine (excretion), carbon dioxide and water (respiration).

The rabbits lose quite a lot of biomass in faeces (egestion), urine (excretion), carbon dioxide and water (respiration).

A lot of the biomass remains in the ground as the root system.

SUMMARY

- **Food chains show the transfer of energy and materials from organism to organism.**
- **Pyramids of biomass show the overall dry mass of organisms in different trophic levels.**
- **Efficiency of energy transfer can be calculated by working out how much energy is used usefully.**

QUESTIONS

QUICK TEST

1. What is a trophic level?
2. List three ways in which energy is lost in a food chain.

QUESTIONS

EXAM PRACTICE

1. a) The following organisms are found in a rock pool on a beach:

   Seaweed (producer/plant)

   Sea snails (herbivores)

   Starfish (carnivore)

   Using this information, sketch a likely pyramid of biomass for these organisms. **[2 marks]**

   b) Explain why the pyramid has the shape that it does. **[3 marks]**

Farming and food security

Farming techniques

Intensive farming is characterised by:

- high yields
- use of mechanised planting and harvesting
- extensive use of inorganic fertilisers, pesticides and herbicides.

In order to maximise profits, food production efficiency is the priority. This is achieved by:

- restricting energy transfer from animals to the environment, e.g. keeping animals under cover
- limiting animals' movement
- controlling the temperature of the animals' surroundings.

For example, battery chickens and calves are placed in cages or pens. Fish can be grown in cages and fed high-protein diets.

Many people object to these practices because they think animal welfare has not been given a high enough priority.

Eutrophication

Overusing fertilisers in intensive farming can lead to eutrophication.

1. Fertilisers or sewage can run into the water and pollute it. As a result, there are a lot of nitrates and phosphates, which leads to rapid growth of algae.

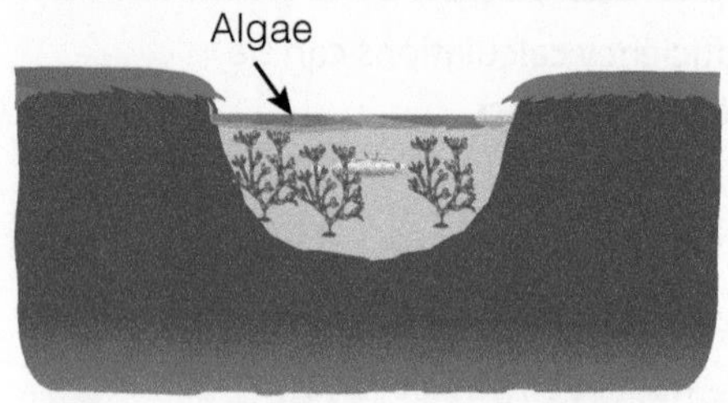

2. The algal blooms reproduce quickly, then die and rot. They also block off sunlight, which causes underwater plants to die and rot.

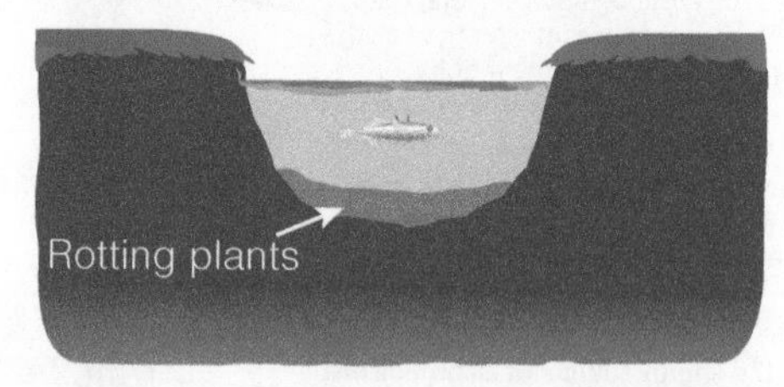

3. The number of aerobic bacteria increase. As they feed on the dead organisms they use up oxygen. This causes larger organisms and plants to die because they are unable to respire.

Food security

Food security is a global issue. It recognises the importance of allowing the world's growing population to have access to an adequate diet. Greater industrialisation and higher standards of living are leading to changing diets in countries such as China. So-called 'Western diets' have a higher red meat content. This increases the demand for more land devoted to raising livestock.

Sustainable methods of food production are constantly sought, but many factors need to be taken into account when trying to achieve food security.

- Many developed countries have a demand for scarce food resources, meaning that some foods are transported many thousands of miles from their point of origin.
- Due to the massive use of pesticides, new pests develop that are resistant to existing chemicals. Pathogenic microorganisms also evolve to become resistant to methods of control such as antibiotics.
- Climate change means that some countries are deprived of rainfall, resulting in a failure of harvests. This leads to widespread famine and armed conflicts as populations migrate to find work and food.
- Cost of farming methods continues to rise.

In the future, agricultural solutions to the increasing problem of food security might include:

- the increased use of hydroponics – a way of growing plants without soil under very controlled conditions
- biological control – where natural invertebrate predators are used to control pests
- gene technology – including pest-resistant GM crops
- fertilisers and pesticides.

Sustainable fishing

Fish stocks are declining across the globe. The problem is so big that unless methods are used to halt the decline, some species (such as the northwest Atlantic cod) might disappear.

Methods to make fishing sustainable include:

- governments imposing **quotas** that limit the weight of fish that can be taken from the oceans on a yearly basis – this practice is not always followed and illegal fishing is difficult to combat
- increasing the mesh size of nets to allow smaller fish to escape and reach adulthood, so that they can breed.

Biotechnology

Biotechnology, particularly the use of genetic engineering, has allowed new food substances to be produced relatively cheaply. Microorganisms can be grown in large industrial vats.

For instance, the fungus, Fusarium, can be used to produce **mycoprotein**, which is protein-rich and suitable for vegetarians. The fungus is grown on glucose syrup in aerobic conditions in a matter of days. Once grown, the biomass is harvested and purified.

GM crops can increase yields and add nutritional value by increasing the vitamin content of crops such as golden rice.

SUMMARY

- **Intensive farming produces high yields but also has drawbacks, including questions around animal welfare, and the overuse of fertilisers causing eutrophication.**
- **Sustainable methods of food production, such as sustainable fishing and biotechnology, are vital in ensuring the world's population has enough to eat.**

QUESTIONS

QUICK TEST

1. What is eutrophication?
2. What are the advantages and disadvantages of keeping battery hens for food production?

EXAM PRACTICE

1. A farmer breeds cattle for beef. The animals are kept in enclosures inside a barn so that the temperature can be regulated and they cannot move around too much.

   a) In terms of food production, explain why the farmer would want to…

   i) regulate the temperature of the environment in which the cows are kept.

   ii) restrict how much the cattle can move around. **[2 marks]**

   b) Some people object to livestock being raised in this way.

   Suggest a reason for this. **[1 mark]**

Answers

Page 5
QUICK TEST
1. False. Eukaryotic cells are more complex than prokaryotic cells.
2. False – most cells do have a nucleus, but others, e.g. red blood cells, bacterial cells, do not.
3. Ribosomes

EXAM PRACTICE
1. a) Plasmids **[1]**
   b) They possess a cell wall. **[1]**
   c) Muscle cells release a lot of energy **[1]** mitochondria release this energy (for contraction) **[1]**
   d) 10/500 = 0.002 × 1000
   = 2 µm **[2 marks for correct answer. If incorrect, 1 mark for correct working.]**

Page 7
QUICK TEST
1. Ciliated epithelial cells
2. Xylem
3. Meristems

EXAM PRACTICE
1. Differentiation **[1]**
2. a) **One from:** Long, slender axons for carrying impulses long distances; Many dendrites for connecting with other nerve cells **[1]**
   b) **Two from:** Therapeutic cloning; Treating paralysis; Repairing nerve damage; Cancer research; Growing new organs for transplantation **[2]**
   c) **Two from:** Stem cells are sometimes obtained from human embryos; They believe it is wrong to use embryos (which might be deemed living) for these purposes; Risk of viral infections **[2]**

Page 9
QUICK TEST
1. Light microscopes and electron microscopes
2. Food/nutrients, warmth and moisture/water

EXAM PRACTICE
1. 82.61 mm^2 **[3] [If incorrect, 1 mark each for correct working:]** Mean radius = 5.13 mm; area = 3.14 × (5.13)^2
2. a) Objective lenses **[1]**
   b) Mitochondria are very small **[1]** Resolving power of light microscope is not great enough **[1]**
   c) Scanning electron microscope (SEM) **[1]**
   Light microscopes (and transmission electron microscopes) cannot produce 3D images **[1]**

Page 11
QUICK TEST
1. Gametes
2. Chromosomes
3. 23

EXAM PRACTICE
1. a) Meiosis **[1]**
   b) Four cells produced (in second meiotic division) **[1]**
   Each cell is haploid (not diploid) **[1]**

Page 13
QUICK TEST
1. Aerobic
2. A reaction that gives out heat
3. Lactic acid
4. A catabolic reaction

EXAM PRACTICE
1. a) Glucose → Lactic acid + energy released **[1]**
   Glucose → Carbon dioxide + Ethanol + Energy released **[1]**
   b) **1 mark must be one from:** Ethanol excreted/removed from cell; But cannot be removed from surroundings; Humans sometimes extract it for the alcoholic drinks industry **[1]**
   Two from: Lactic acid causes oxygen debt; Increased breathing and heart rates deliver more blood to muscles; Oxygen removes lactic acid; Lactic acid oxidised **[2]**

Page 15
QUICK TEST
1. pH and temperature
2. Lipases
3. Amino acids

EXAM PRACTICE
1. a) **Three from:** High temperature denatures enzyme; Active site shape changed; Starch no longer fits active site; No maltose formed **[3]**
   b) **Three from:** Acidic pH keeps active site in correct shape; Protein substrate fits active site; Enzyme puts a strain on protein molecule/breaks it down; To products/peptides/amino acids **[3]**

Page 17

QUICK TEST

1. To calculate rates of diffusion
2. The plant cells have a higher water potential/lower solute concentration. The water moves **down** an osmotic gradient from inside the cells into the salt solution by osmosis.

EXAM PRACTICE

1. **Two from:** Ions enter by active transport; requiring release of energy; from respiration; via protein carriers in the cell membrane **[2]**
2. **Four from:** The cell has become plasmolysed; The outside solution has lower water potential/higher solute concentration than cell contents; Water leaves cell; By osmosis; Down concentration/water potential/osmotic gradient **[4]**

Page 19

QUICK TEST

1. The leaf
2. Lignin

EXAM PRACTICE

1. a) **One from:** Xylem consist of hollow tubes while phloem have sieve end-plates; Xylem vessels do not possess companion cells. **[1]**
   b) Phloem carries sugar/sucrose **[1]**
   Aphids need sugar for diet/have well-adapted mouthparts to reach source of sugar. **[1]**

Page 21

QUICK TEST

1. It would increase the (evapo)transpiration rate.
2. Guard cells
3. Potometer

EXAM PRACTICE

1. a) 1st 60 mins: Bubble moves 1.2 cm (2.3 – 2.1) **[1]**
   therefore rate is 1.2 / 60 = 0.02 cm per min. **[1]**
   2nd 60 mins: Bubble moves 3.5 cm **[1]**
   Therefore rate is 3.4 / 60 = 0.06 cm per min **[1]**
   [Correct answer in each case gains full 2 marks automatically.]
   b) **Three from:** Moving air increases rate of water absorption; Water vapour lost more rapidly from leaves/faster transpiration; Moving air increases diffusion gradient between leaf interior and atmosphere; Evaporation/diffusion is therefore more rapid **[3]**

Page 23

QUICK TEST

1. Veins
2. True
3. The pulmonary vein

EXAM PRACTICE

1. a) Artery has to withstand/recoil with higher pressure, elasticity allows smoother blood flow/second boost to blood when recoils **[1]**
   b) Vein has valves to prevent backflow of blood/compensate for low blood pressure **[1]**
2. a) The blood passes through the heart twice for every complete circuit of the body. **[1]**
   b) **One from:** Allows a higher pressure to be maintaine[d] around the body; Allows maximum efficiency for delivering oxygen/materials to body cells. **[1]**

Page 25

QUICK TEST

1. Red blood cells, white blood cells, plasma and platele[ts]
2. Oxyhaemoglobin

EXAM PRACTICE

1. **Two from:** Involved in defence against infection/pa[rt] of immune system; Some engulf foreign cells; Other[s] produce antibodies **[2]**
2. **Three from:** Tar layer increases diffusion path for oxygen; Less oxygen is absorbed into the bloodstream; Less oxygen is delivered to respiring cells; Breathing rate is increased to compensate for this **[3]**

Page 27

QUICK TEST

1. Photosynthesis takes **in** carbon dioxide and water, and requires an energy input. It produces glucose and oxyge[n]. Respiration **produces** carbon dioxide and water, and releases energy. It absorbs glucose and oxygen.
2. It is needed for cell walls.

EXAM PRACTICE

1. a) Increased light intensity increases rate of photosynthesis **[1]** More photosynthesis means more carbohydrate produced in a given time **[1]** More carbohydrate means bigger/more tomato[es] **[1]** She should install more powerful lighting/artificial lighting at night **[1]**
   b) **Two from:** Rate of photosynthesis does not increase after a certain light intensity; Therefore any extra investment in lighting will not result in any more yield; Money will be wasted **[2]**

Answers

Page 29

QUICK TEST

1. Body mass index
2. A microorganism that causes disease

EXAM PRACTICE

1. **Two from:** Emphysema; Heart disease; Stroke; **Any other reasonable answer. [2]**
2. a) 26.7 **[2] [If incorrect, allow 1 mark for correct working:]** $64/1.55^2$
   b) She needs to lose some weight **[1]** Her BMI category is 'overweight.' **[1]**

Page 31

QUICK TEST

1. Cholera bacteria are found in human faeces, which contaminate water supplies.
2. **Any reasonable answer, e.g.:** athlete's foot

EXAM PRACTICE

1. a) Pathogens **[1]**
   b) **Two from:** Bacteria and viruses reproduce rapidly in the body; Viruses cause cell damage; Toxins are produced that damage tissues **[2]**
2. Warm temperatures are ideal for mosquito to thrive **[1]** Stagnant water is ideal habitat for mosquito eggs to be laid/larvae to survive. **[1]**

Page 33

QUICK TEST

1. Phagocyte engulfs a pathogen then digests it using the cell's enzymes.
2. An antigen is a molecular marker on a pathogen cell membrane that acts as a recognition point for antibodies.
3. Lysozyme

EXAM PRACTICE

1. a) Antibodies **[1]**
   b) i) Lymphocyte **[1]**
      ii) Antigen is identified by the immune system **[1]** Memory cells are activated **[1]** A large amount of antibodies is produced rapidly to combat the fresh invasion of pathogens. **[1]**

Page 35

QUICK TEST

1. An antibiotic is a drug that kills bacteria.
2. Active and passive
3. Do not prescribe antibiotics for viral/non-serious infections; Ensure that the full course of antibiotics is completed.

EXAM PRACTICE

1. Dead/inactive form of virus is injected into person **[1]** this triggers production of antibodies **[1]** and formation of memory cells **[1]** If actual infection occurs, antibodies produced rapidly to deal with the pathogen **[1]**
2. a) Inhibit cell processes in bacteria **[1]**
   b) **Two from:** Antibiotics do not work against viruses/only work against bacteria; Flu caused by a virus; Over prescription risks antibiotic resistance **[2]**
3. MRSA is resistant to most modern antibiotics **[1]** caused by doctors over-prescribing antibiotics **[1]** new antibiotics need to be developed, which is a slow process/takes many years. **[1]**

Page 37

QUICK TEST

1. A double blind trial is when both volunteers and doctors don't know whether the volunteers have been given the new drug or a placebo.
2. True
3. **Two from:** Cancer treatment; Pregnancy testing kits; Measuring hormone levels; Research

EXAM PRACTICE

1. a) **Two from:** To ensure that the drug is actually effective/more effective than placebo; To work out the most effective dose/method of application; To comply with legislation **[2]**
   b) **One from:** Computer modelling; Experimenting on cell cultures **[1]**
   c) Antibodies sensitised to the hormone HCG are bound to a test strip **[1]** Strip changes colour when exposed to urine from a pregnant woman **[1]**

Page 39

QUICK TEST

1. **Three from:** Stunted growth; Leaf spots; Areas of decay; Growths/tumours; Malformed stems and leaves; Discolouration; Presence of pests
2. Fungal disease
3. To form chlorophyll, for photosynthesis

EXAM PRACTICE

1. **Three from:** Make up 'acid rain' solution (e.g. a dilute sulfuric or nitric acid solution, or a solution with a pH in the range 2-6); Minimum of 2 groups of infected roses – spray one group with acid solution; Control variable e.g. same place in garden; Measurement/assessment technique e.g. number of bushes with black spot infection **[3]**
2. Mechanical: **One from:** Thorns/hairs; Leaves that droop/curl on contact; Mimicry **[1]**
   Physical: **One from:** Tough, waxy leaves; Cellulose cell walls; Layers of dead cells on stems **[1]**

Page 41

QUICK TEST

1. 37°C
2. **Two from:** Osmoregulation/water balance; Balancing blood sugar levels; Maintaining a constant body temperature; Controlling metabolic rate

EXAM PRACTICE

1. Pancreas – Produces insulin
   Skin receptor – Detects pressure
   Pituitary gland – Releases TSH
   Hypothalamus – Releases TRH
   [3 marks for 4 correct, 2 marks for 3 correct, 1 mark for 2 correct]

Page 43

QUICK TEST

1. Synapses
2. They ensure a rapid response to a threatening/harmful stimulus, e.g. picking up a hot plate. As a result, they reduce harm to the human body.

EXAM PRACTICE

1. **a)** X is the stimulus **[1]**
   b) It ensures a rapid response to a threatening/harmful stimulus **[1]** It is an unconscious reaction **[1]**
   c) **Four from:** Pain/pressure receptor in skin stimulated; Sensory neurone sends impulse; To spine; Intermediate neurone relays impulse to motor neurone; Motor neurone sends impulse to arm muscle; Arm muscle contracts **[4]**

Page 45

QUICK TEST

1. Central nervous system
2. Vasodilation is when blood vessels widen causing blood to flow closer to the skin's surface.

EXAM PRACTICE

1. **Six from:** Thermoregulatory centre/hypothalamus detects a rise in core temperature; Nervous signals/impulses are sent from this centre; Sweat glands release more sweat; Sweat evaporates; Endothermic change; Absorbs heat from skin; Vasodilation occurs/arterioles or blood vessels near surface of skin increase in diameter; More blood flows near surface of skin; More heat lost by radiation; When core temperature falls negative feedback reverses these changes **[6]**

Page 47

QUICK TEST

1. The retina
2. An eyeball that is too long, or weak suspensory ligaments

EXAM PRACTICE

1. **a)** Pupil decreases in diameter/size **[1]** due to contraction of the iris muscle **[1]** to prevent light damaging the retina **[1]**
   b) **Three from:** Ciliary body/muscle relaxes; Stretching lens more thinly/lens adopts less convex shape; Light rays refracted less; In order to converge/focus at a sharp point on the retina **[3]**

Page 49

QUICK TEST

1. Thyroxine
2. Pancreas
3. Reduced ability of cells to absorb insulin and therefore high levels of blood glucose. This leads to tiredness, frequent urination, poor circulation, eye problems, etc.

EXAM PRACTICE

1. **a)** **Two from:** Causes all body cells to absorb glucose; Causes liver to convert glucose to glycogen; Blood glucose level is reduced **[2]**
   b) Insulin injections and monitoring of diet **[1]**
   c) The body's cells often no longer respond to insulin **[1]** It can be managed by adjusting carbohydrate intake **[1]**

Answers

Page 51

QUICK TEST

1. **Three from:** Glucose; Amino acids; Fatty acids; Glycerol; Some water
2. The pituitary gland

EXAM PRACTICE

1. **a)** Excretion **[1]**
   b) Amino acids **[1]**

Page 53

QUICK TEST

1. FSH, LH, progesterone, oestrogen
2. **One from:** Production of sperm in testes; Development of muscles and penis; Deepening of the voice; Growth of pubic, facial and body hair

EXAM PRACTICE

1. **a)** Oestrogen – promotes repair of uterus wall and stimulates egg release **[1]**
   Progesterone – maintains lining of uterus **[1]**
   b) Release of egg/ovulation **[1]**
2. Breast development **[1]** Sperm production **[1]** Menstrual cycle **[1]**

Page 55

QUICK TEST

1. Spermicidal agent. It is cheap.
2. They provide a barrier/prevent transferral of the virus during sexual intercourse.

EXAM PRACTICE

1. Advantage – very effective against STIs **[1]**
   Disadvantage – **One from:** Can only be used once; May interrupt sexual activity; Can break; May be allergic to latex **[1]**
2. Woman is given FSH and LH to stimulate the production of several eggs **[1]** Sperm and eggs are then introduced together outside the body in a petri dish **[1]** Successfully growing embryos can then be transplanted into the woman's uterus **[1]**

Page 57

QUICK TEST

1. Positive phototropism
2. Auxin
3. Gibberellins

EXAM PRACTICE

1. Plant hormones **[1]**
2. Light causes auxin to build up on lower side of shoot/side of shoot away from light **[1]** This causes greater cell elongation on this side of the shoot **[1]**

Page 59

QUICK TEST

1. Fertilisation
2. Via runners
3. By spores

EXAM PRACTICE

1. Advantage – **One from:** Produces variation; Survival advantage when the environment is changing; Can be used by humans for selective breeding purposes **[1]**
   Disadvantage: **One from:** Slow process; Disadvantage in stable environments; More resources required; Results of selective breeding are unpredictable **[1]**
2. **a)** Not as dependent on one species/if one species is low in numbers or becomes extinct there is another organism available to act as a 'host.' **[1]**
   b) **One from:** Fungi; Mosses; Ferns; Protista **[1]**

Page 61

QUICK TEST

1. It has allowed the production of linkage maps that can be used for tracking inherited traits from generation to generation. This has led to targeted treatments for these conditions.
2. A, T, C and G

EXAM PRACTICE

1. By collecting and analysing DNA samples from different groups of peoples **[1]** human migration patterns can be deduced **[1]**
2. Backbone consisting of a sugar molecule **[1]** attached to a phosphate molecule **[1]** nitrogenous base attached **[1]**

Page 63

QUICK TEST

1. Dominant and recessive
2. messenger RNA (involved in transcription) and transfer RNA (involved in translation)

EXAM PRACTICE

1. A recessive allele controls the development of a characteristic only if a dominant allele is not present. **[1]** So, the recessive allele must be present on both chromosomes in a pair. **[1]**
2. Messenger RNA – Constructed in the nucleus **[1]** Forms a complementary strand to DNA **[1]**
   Transfer RNA – Found in the cytoplasm (or temporarily in the ribosome) **[1]** Carry amino acids to add to a forming protein **[1]**

Page 65

QUICK TEST

1. 0%

EXAM PRACTICE

1. Parents: Ff x FF **[1]**
   Gametes: F f x F **[1]**
   Offspring FF Ff **[1]**
   Phenotypes Normal Carrier **[1]**
   50 : 50 **[1]**
   No children will suffer from cystic fibrosis; 50% chance of a child being a carrier **[1]**

Page 67

QUICK TEST

1. **Two from:** The fossil record; Comparative anatomy (pentadactyl limb); Looking at changes in species during modern times; Studying embryos and their similarities; Comparing genomes of different organisms
2. **Two from:** Catastrophic events, e.g. volcanic eruption; Changes to the environment; New predators; New diseases; New competitors

EXAM PRACTICE

1. Variation – horse-like ancestor adapted to environment/had different characteristics, named examples of different characteristics, **[1]** e.g. some horse-like mammals had more flipper-like limbs (as whales have flippers), mutation in genes allowed some individuals to develop these advantageous characteristics **[1]**

   Idea of competition for limited resources; examples of different types of competition e.g. obtaining food in increasingly water-logged environment **[1]**

   Idea of survival of the fittest – named examples of different adaptations, e.g. some horse-like mammals had more flipper-like limbs that allowed them to swim well in water **[1]**

   Idea of successful characteristics being inherited – genes passed on to next generation, extinction – variety which is least successful does not survive and becomes extinct **[1]**

Page 69

QUICK TEST

1. The biological changes seen in species over millions of years
2. Ardi
3. **Any two examples**, e.g. Peppered moths; Antibiotic resistance in bacteria

EXAM PRACTICE

1. **Two from:** Changes in DNA/genes; Mutation; Production of new proteins; Proteins determine structural or behavioural adaptations **[2]**

Page 71

QUICK TEST

1. In Darwin's theory, change in the inherited material occurs first. This leads to change in the phenotype/external appearance. Lamarck's theory stated that th physical change occurs first.
2. Cloning allows mass production of individuals with beneficial traits in a short space of time.

EXAM PRACTICE

1. – Scrape off small pieces of tissue into vessels containing nutrient and hormones **[1]**
   – Remove plantlets/clones **[1]**
   – Repeat process **[1]**
2. It is possible to mass produce organisms with specif characteristics beneficial to humans **[1]** Stated example: e.g. High meat quality in cattle/pigs/sheep Greater milk yields in dairy cattle; High quality wool sheep, etc. **[maximum 2]**

Answers

Page 73

QUICK TEST

1. Genetic engineering is more precise and it takes less time to see results.
2. A plasmid is a ring of DNA found in bacteria.

EXAM PRACTICE

1. Advantage – **One from:** It results in an organism with the 'right' characteristics for a particular function; In farming and horticulture, it is a more efficient and economically viable process than natural selection **[1]**
   Disadvantage – **One from:** Intensive selective breeding reduces the gene pool – the range of alleles in the population decreases so there is less variation; Lower variation reduces a species' ability to respond to environmental change; It can lead to an accumulation of harmful recessive characteristics (in-breeding) **[1]**
2. Stage 2: Gene removed using restriction enzyme **[1]**
   Stage 3: Bacterial plasmid 'cut open' using restriction enzyme **[1]**
   Stage 4: Ligase enzyme used to insert human gene into bacterial plasmid **[1]**
   Stage 5: Bacterial cells containing plasmids rapidly reproduce **[1]**
   Stage 6: Insulin purified from fermenter culture and produced in commercial quantities **[1]**

Page 75

QUICK TEST

1. Genus and species
2. Carl Woese

EXAM PRACTICE

1. Phylum **[1]** Order **[1]** Genus **[1]**
2. Canis familiaris **[1]**

Page 77

QUICK TEST

1. An ecosystem includes the habitat, its communities of animals and plants, together with the physical factors that influence them.
2. Food, mates, territory
3. **Any two suitable examples**, e.g. bacteria living in deep sea vents and icefish (exist in waters less than 0°C in temperature).

EXAM PRACTICE

1. **i)** Allow it to grip/stick to trees **[1]**
   ii) Camouflage so it won't be seen **[1]**
   iii) To catch insects/prey **[1]**
2. Extremophiles are organisms that can survive in extreme environmental conditions **[1]**
   One from: Bacteria living in deep sea vents have optimum temperatures for enzymes that are much higher than 37°C; Icefish have antifreeze chemicals in their bodies which lower the freezing point of body fluids; Some organisms resist high salt concentrations or pressure **[1]**

Page 79

QUICK TEST

1. It would tell you how numbers of different species vary along the line of the transect.
2. Biotic factors are living; abiotic factors are non-living

EXAM PRACTICE

1. **a)** To ensure they obtained representative samples **[1]**
   b) Working – numbers arranged in order of magnitude: 0,0,1,2,2,3,4,5,5,6
   The middle 2 numbers are 2 and 3, therefore the median is 2.5 **[3 marks for correct answer. If incorrect, 1 mark for each correct working]**
   c) Area of the field is $60 \times 90 = 5400\ m^2$
   Therefore, number of daisies = $5400 \times 2.3 = 12,420$ **[2 marks for correct answer. If incorrect, 1 mark for correct working]**

Page 81

QUICK TEST

1. An organism used as an indicator of pollution
2. An animal that kills and eats other animals

EXAM PRACTICE

1. a) Yes **[0 marks]** There are no species present that you might find in very polluted water e.g. rat tailed maggot, bloodworm **[1]** There are many 'clean-water' indicator species e.g. stonefly nymphs **[1]**
   b) Direct chemical tests e.g. pH of the water, oxygen concentrations **[1]**
2. As the rabbit population decreases, there is less food available for the stoats **[1]** so the stoat population decreases **[1]** fewer rabbits are eaten, so the rabbit population increases **[1]**

Page 83

QUICK TEST

1. Condensation and evaporation
2. Methane gas
3. Oxygen, warmth, moisture

EXAM PRACTICE

1. a) Photosynthesis **[1]**
   b) Methane **[1]**
   c) **One from:** Electricity generation; Generation of steam for central heating systems; Used as bus fuel **[1]**

Page 85

QUICK TEST

1. **One from:** To provide land for cattle and rice fields; To grow crops for biofuel
2. With more species, there are more relationships between different organisms and therefore more resistance to disruption from outside influences.

EXAM PRACTICE

1. a) **Three from:** Certain plant species/crops become scarce; Due to lack of pollinators; Other species which depend on these plants will be endangered/reduced in number; Biodiversity falls **[3]**
   b) **Two from:** Reduce usage of pesticides such as *neonicotinoids*; Plant more species which attract bumble bees (flowers, etc.); Farm bees in hives and release into the wild; Disease prevention measures to act against foul-brood disease/isolate colonies/ remove diseased colonies **[2]**

Page 87

QUICK TEST

1. The level that an organism feeds at, e.g. producer, primary consumer level. It refers to layers in a pyramid of biomass.
2. **Three from:** Respiration; Movement; Reproduction; Maintaining constant body temperature; Urine; Faeces; Not all parts of the animal/plant are eaten.

EXAM PRACTICE

1. a)

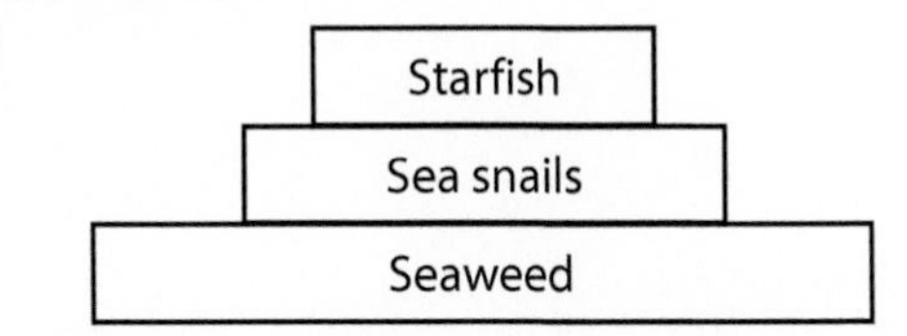

   [1 mark for correct sequence of organisms, 1 mark for correct shape of pyramid]

   b) Energy lost between trophic levels **[1]** through respiration, movement of organism, body heat, undigested food **[1]** so less energy available for next trophic level **[1]**

Page 89

QUICK TEST

1. A process where nitrates and phosphates enrich waterways, causing massive growth of algae and loss of oxygen.
2. **Advantages:** cheap cost of production; rapid growth of stock **Disadvantage:** reduced animal welfare

EXAM PRACTICE

1. a) i) So energy is not wasted by the cows generating heat to keep warm **[1]**
   ii) So energy is not lost through movement of cows **[1]**
   b) They believe it is cruel because the cows cannot behave in a natural way. **[1]**

Notes

Notes

Notes

Notes

Notes

Glossary

Active ingredient – Chemical in a drug that has a therapeutic effect (other chemicals in the drug simply enhance flavour or act as bulking agents)

Active site – The place on an enzyme molecule into which a substrate molecule fits

Alleles – Alternative forms of a gene on a homologous pair of chromosomes

HT **Ammonia** – Nitrogenous waste product

Anatomy – The study of structures within the bodies of organisms

Anthropologists – Scientists who study the human race and its evolution

Antibodies – Proteins produced by white blood cells (particularly lymphocytes). They lock on to antigens and neutralise them

Antigen – Molecular marker on a pathogen cell membrane that acts as a recognition point for antibodies

Aphid – A type of sap-sucking insect

Binary fission – Process where bacterial cells divide into two

Binomial system – 'Two name' system of naming species using Latin

Biomass – Mass of organisms calculated by multiplying their individual mass by the number that exist

Biosphere – Area on the Earth's crust that is inhabited by living things

Carbon sinks – Resources that lock up carbon in their structure rather than allowing them to form carbon dioxide, e.g. peat bogs, oceans, limestone deposits

Catalyst – A substance that controls the rate of a chemical reaction without being chemically changed itself

Cellulose – Large carbohydrate molecule found in all plants; an essential constituent of cell walls

Chlorophyll – A molecule that gives plants their green colour and absorbs light energy

Chlorosis – Where leaves lose their colour as a result of mineral deficiency

Cilia – Microscopic hairs found on the surface of epithelial cells; they 'waft' from side to side in a rhythmic manner

Clone – A genetically identical cell, tissue, organ or organism

Colony – A large number of microorganisms (of one type) growing in a location, e.g. a circular colony on the surface of agar

Common ancestor – Organism that gave rise to two different branches of organisms in an evolutionary tree

Competition – When two individuals or populations seek to exploit a resource, e.g. food. One individual/population will eventually replace the other

HT **Complementary strand** – A sequence of bases that can bond with an equivalent sequence in DNA. Complementary strands can be DNA or mRNA

Compost – Fertiliser produced from the decay of organic plant material

Compress – Squash or squeeze. In geology, this is usually due to Earth movements or laying down sediments

Concave – Lens that curves inwards on one or both sides

Contraception – Literally means 'against conception'; any method that reduces the likelihood of a sperm meeting an egg

Converge – Come together at a point

Convex – Lens that curves outwards on one or both sides

Coronary artery – The blood vessel delivering blood to the heart muscle

Daughter cells – Multiple cells arising from mitosis and meiosis

HT **Deamination** – Process in the liver in which nitrogen is removed from an amino acid molecule

Decomposers – Microorganisms that break down dead plant and animal material by secreting enzymes into the environment. Small, soluble molecules can then be absorbed back into the microorganism by the process of diffusion

Denaturation – When a protein molecule such as an enzyme changes shape and makes it unable to function

Dermis – The layer of tissue in the skin that lies beneath the epidermis. It contains many receptors and structures involved in temperature control

HT **Diagnostic** – A process used to identify or recognise a disease or pathogen

Differentiation – A process by which cells become specialised, i.e. have a particular structure to carry out their function

Glossary

Diploid – A full set of chromosomes in a cell (twice the haploid number)

Diverge – Spread outwards from a point

Droplet infection – Transmission of microorganisms through the aerosol (water droplets) produced through coughing and sneezing

Ductless gland – A gland that does not secrete its chemicals through a tube. The pancreas is an exception to this rule as it contains a duct for delivering enzymes, but its hormones are released directly into the bloodstream

Emissions – Gaseous products usually connected with pollution, e.g. carbon dioxide emissions from exhausts

Endangered – Category of risk attached to rare species of plants and animals. This usually triggers efforts to preserve the species' numbers

Endocrine system – System of ductless organs that release hormones

Endothermic – A change that requires the input of energy

Energy demand – Energy required by tissues (particularly muscle) to carry out their functions

Environmental resources – Materials or factors that organisms need to survive, e.g. high oxygen concentration, living space or a particular food supply

Epidermis – Outer layer of tissue in the skin

Epithelial – A single layer of cells often found lining respiratory and digestive structures

Eutrophication – A process where nitrates and phosphates enrich waterways, causing massive growth of algae and loss of oxygen

(Evapo)transpiration – Evaporation of water from stomata in the leaf

Exothermic reaction – A reaction that gives out heat

External fertilisation – Gametes join outside the body of the female

Extinction – When there are no more members of a species left living

Fittest – The most adapted individual or species

Follicle stimulating hormone – A hormone produced by the pituitary gland that controls oestrogen production by the ovaries

Fusion – Joining together; in biology the term is used to describe fertilisation

Gamete – A sex cell, i.e. sperm or egg

Glucagon – Hormone released by the pancreas that stimulates the conversion of glycogen to glucose

Glycogen – Storage carbohydrate found in animals

Gravitropism (geotropism) – Growth response in plants against or with the force of gravity

Haemoglobin – Iron-containing molecule that binds to oxygen molecules in red blood cells

Haploid – A half set of chromosomes in a cell; haploid cells are either eggs or sperm

Haploid nucleus – Nucleus with a half chromosome set

Herbicide – A chemical applied to crops to kill weeds

Herd immunity – Vaccination of a significant portion of a population (or herd) makes it hard for a disease to spread because there are so few people left to infect. This gives protection to susceptible individuals such as the elderly or infants

Homologous chromosomes – A pair of chromosomes carrying alleles that code for the same characteristics

HT **Hybridoma cells** – Cells produced by the fusion of lymphocytes and a tumour cell

Hydrogen bond – A bond formed between hydrogen and oxygen atoms on different molecules close to each other

Immune system – A system of cells and other components that protect the body from pathogens and other foreign substances

Immunological memory – The system of cells and cell products whose function is to prevent and destroy microbial infection

Implantation – Process in which an embryo embeds itself in the uterine wall

Impulses – Electrical signals sent down neurones that trigger responses in the nervous system

Indicator species – Organism used as an indicator of pollution

Inhibition – The effect of one agent against another in order to slow down or stop activity, e.g. chemical reactions can be slowed down using inhibitors. Some hormones are inhibitors

Internal fertilisation – Gametes join inside the body of the female

Intrauterine – Inside the uterus

Kinetic energy – Energy possessed by moving objects, e.g. reactant molecules such as enzyme and substrate molecules

Lignin – Strengthening, waterproof material found in walls of xylem cells

Limiting factor – A variable that, if changed, will influence the rate of reaction most

Malnutrition – A diet lacking in one or more food groups

Menstruation – Loss of blood and muscle tissue from the uterus wall during the monthly cycle

Meristems – Growth regions in plants where stem cells are produced, e.g. apical bud

Mucus – Thick fluid produced in the lining of respiratory and digestive passages

Neuroscience – Study of the nervous system

Non-communicable – Disease or condition that cannot be spread from person to person via pathogen transfer

Oral contraceptive – Hormonal contraceptive taken in tablet form

Organelle – A membrane-bound structure within a cell that carries out a particular function

Organic – Material obtained from living things; either their dead bodies or their waste. Organic can also refer to a type of farming that avoids overuse of intensive practices such as pesticides and inorganic fertilisers

Oxygen debt – The oxygen needed to remove lactic acid after exercise

Partially permeable membrane – A membrane with microscopic holes that allows small particles through (e.g. water) but not large ones (e.g. sugar)

Partially permeable membrane – Artificial or organic layer that only allows small molecules through

Pathogen – A harmful microorganism

Pesticide – Chemical sprayed on crops to kill invertebrate pests

Pharmaceutical drug – Chemicals that are developed artificially and taken by a patient to relieve symptoms of a disease or treat a condition

Photosynthetic organism – Able to absorb light energy and manufacture carbohydrate from carbon dioxide and water

Phototropism – Growth response in plants, towards or away from the direction of incoming light

Placebo – A substitute for the medication that does not contain the active ingredient

Plaque – Fatty deposits that can build up in arteries

HT **Plasmid** – A ring of DNA found in bacteria

Polymer – A long chain molecule made up of individual units called monomers

HT **Polypeptide** – A sequence of peptides, which in turn are smaller chains of amino acids

Receptor molecule – Protein on the outer membrane of a cell that binds to a specific molecule, such as transmitter substance

Regeneration – Rebuilding or regrowth of a habitat, e.g. flooding peatland to encourage regrowth of mosses and other plants

Resolution – The smallest distance between two points on a specimen that can still be told apart

Sample – A small area or population that is measured to give an indication of numbers within a larger area or population

Selective re-absorption – Occurs in the kidney tubules and allows useful substances to pass from the kidneys back into the blood

Sensitisation – Cells in the immune system are able to 'recognise' antigens or foreign cells and respond by attacking them or producing antibodies

Spinal cord – Nervous tissue running down the centre of the vertebral column; millions of nerves branch out from it

Stem cells – Undifferentiated cells that can become any type of cell

Stimuli – Changes in the internal or external environment that affect receptors

Substrate – The molecule acted on by an enzyme

Surface area to volume ratio – A number calculated by dividing the total surface area of an object by its volume. When the ratio is high, the efficiency of diffusion and other processes is greater

Glossary

Surrogate mother – An adult female animal that has had an embryo implanted into her uterus. The surrogate gives birth to the offspring but has no genetic relationship to it

Sustainability – Carrying out human activity, e.g. farming, fishing and extraction of resources from the ground, so that damage to the environment is minimised or removed

Symptoms – Physical or mental features that indicate a condition or disease, e.g. rash, high temperature, vomiting

Therapeutic cloning – A process where embryos are grown to produce cells with the same genes as a particular patient

HT **Transcription** – Process by which a complementary strand of mRNA is made in the nucleus

HT **Translation** – Process occurring in ribosomes where polypeptides are produced from the code in mRNA

Translocation – Process in which sugars move through the phloem

Transpiration – Flow of water through the plant ending in evaporation from leaves

Turgidity – Where plant cells fill with water and swell as a result of osmosis

Urea – Nitrogenous waste product

Vectors – Small organisms (such as mosquitoes or ticks) that pass on pathogens between people or places

Ventilation – Process of drawing air into and out of the lungs. It involves the ribs, intercostal muscles and diaphragm

Vertebrate – Animal with a backbone

Water potential/diffusion gradient – A higher concentration of particle numbers in one area than another; in living systems, these areas are often separated by a membrane or cell wall

Yield – The weight of living material harvested in farming and fishing

Index

Acknowledgements

The author and publisher are grateful to the copyright holders for permission to use quoted materials and images.

All images are © Shutterstock and © HarperCollins Publishers.

Every effort has been made to trace copyright holders and obtain their permission for the use of copyright material. The author and publisher will gladly receive information enabling them to rectify any error or omission in subsequent editions. All facts are correct at time of going to press.

Published by Collins
An imprint of HarperCollins*Publishers*
1 London Bridge Street
London SE1 9GF

ISBN: 9780008276041

First published 2018

This edition published 2020

Previously published as Letts

10 9 8 7 6 5 4 3 2

British Library Cataloguing in Publication Data.

A CIP record of this book is available from the British Library.

Author: Tom Adams
Commissioning Editors: Clare Souza and Kerry Ferguson
Editor/Project Manager: Katie Galloway
Cover Design: Kevin Robbins, Gordon MacGilp and Sarah Duxbury
Inside Concept Design: Ian Wrigley
Text Design and Layout: Nicola Lancashire at Rose & Thorn Creative Services, and Ian Wrigley
Production: Natalia Rebow
Printed by CPI Group (UK) Ltd, Croydon CR0 4YY

MIX
Paper from responsible source
FSC™ C007454

This book is produced from independently certified FSC™ paper to ensure responsible forest management.

For more information visit: www.harpercollins.co.uk/green

Introducción

Las energías renovables son energías alternativas disponibles en la naturaleza y que pueden ayudarnos a reducir la dependencia de las energías de origen fósil, como son el carbón, petróleo y gas natural y otros. La realidad en este momento es, que todas juntas, no llegan a cubrir el 20% la necesidad de energía que el mundo actual tiene. Sin embargo, su aprovechamiento y utilización es muy importante en muchas aplicaciones.

Son energías renovables las energías que aunque se consuman, se vuelven a reponer, tal es el caso del calor y la luz del Sol, la fuerza del viento, el agua en el cauce de los ríos, la fuerza del mar, el gradiente térmico de la tierra, la biomasa, etc.

Hay energías alternativas y renovables cuya aplicación resulta rentable, como es el caso de la energía hidráulica, en menor cuantía la energía eólica. Otras energías necesitan inversiones importantes, que en el momento actual y a pesar de los precios de las energías de origen fósil, no son competitivas aparentemente, si sólo se tiene en cuenta el precio del kilovatiohora obtenido.

Las energías alternativas renovables tienen otras ventajas, especialmente para los países que no son productores de energías de origen fósil, ya que permite reducir dependencia y aprovechar sus recursos naturales.

Otro aspecto que se busca en este tipo de energías es que sean poco contaminantes. Pueden ser menos contaminantes de la atmósfera y la tierra que las de origen fósil, pero tienen el problema de impacto medio ambiental (generación fotovoltaica, eólica, hidráulica).

En esta obra se tratan todas las energías renovables y alternativas más importantes, mostrando sus ventajas e inconvenientes, y los procedimientos de recuperación de su energía y transformación en otras formas de energía para su aprovechamiento.

Se muestran abundantes tablas para que el lector o estudioso de esta materia se sitúe en lo que significan estas energías.

Lo real es que necesitamos energía, y cada vez más. Mucha de la energía que consumimos tiene fecha de caducidad. No aparecen en el horizonte a corto y medio plazo, alternativas energéticas a las que actualmente consumimos.

Uno de los objetivos de esta obra es la de presentar estas energías alternativas de una forma clara, con datos suficientes para que sean entendidos sus principios, mostrando las posibilidades reales que tienen como alternativa al suministro de las energías convencionales, buscando además que puedan favorecer una reducción de los niveles de contaminación atmosférica. El texto está acompañado de fotografías, esquemas y fórmulas de cálculo.

Índice

1. Necesidad de energía en el mundo actual

2. Fuentes de energía

3. Introducción a las energías renovables

4. El Sol, fuente de energía y de vida

5. Código Técnico de la Edificación (CTE)

6. Energía Solar Térmica

7. Energía Fotovoltaica

8. Energía eólica

10. Biomasa

11. Biocombustibles

12. Otras energías alternativas

13. Contaminación

1 Necesidad de energía en el mundo actual

La demanda de energía se estima que crezca a razón de 1,8% anual, hasta el año 2030, lo que supone un crecimiento del 55% sobre las necesidades energéticas globales que se tienen en la actualidad. La demanda será más fuerte en países emergentes (China, India y Brasil), mientras que en los más industrializados el incremento será menor.
Las energías renovables se incrementarán de forma muy apreciable, pasando del 7%, hasta el 20% en el año 2020.
La energía lo mueve todo, nuestro mundo actual se mueve consumiendo energía, y una parte importante de la misma no tiene reposición.

La energía lo mueve todo, en este caso es el petróleo y la fuerza del viento

1.1. Reseña histórica

Nota:

El gran inventor Nicola Tesla (1857-1943) fue uno de los hombres más importantes que ha tenido la humanidad, y que con sus descubrimientos y desarrollos en electricidad, permitieron dar un gran impulso a la tecnología eléctrica, la industrialización y al confort de los hogares.

El despegue industrial (industrialización) y empleo masivo de energías comenzó con la llamada Revolución Industrial que se inició a mediados del siglo XVIII. Este período se divide en dos épocas:

- ***Primera Revolución Industrial*** (1760-1870), que se inicia con la llegada a la industria y los transportes de la máquina de vapor.
- ***Segunda Revolución Industrial*** (1870 hasta 1914), período de desarrollo de la energía eléctrica (motores eléctricos y alumbrado), así como el inicio de los automóviles como medio de transporte.

La tabla 1.1 recoge los inventos más importantes que impulsaron la industrialización, y por tanto, ayudaron al desarrollo de la humanidad. La lista es mucho más extensa, pero los que se citan tuvieron una importancia muy relevante.

Tabla 1.1. Grandes inventos que colaboraron a la Revolución Industrial.

Fecha	Inventor	Innovación
1690	Denis Papín	Primera máquina que utilizó vapor de agua.
1712	Newcomen	Máquina de vapor Newcomen.
1769	James Watt	Máquina de vapor.
1770	Cugnot	Carro de vapor.
1784	James Vatt	Máquina de doble acción de Watt.
1785	Cartwright	Telar mecánico.
1792	Murdoch	Lámpara de gas.
1796	Trevithich	Locomotora. Primera máquina de vapor de alta presión.
1800	Volta	Pila eléctrica.
1814	Stephenson	Máquina a vapor industrial.
1825	Inglaterra	Primera línea de pasajeros en tren.
1826	Josef Ressel	Hélice para barcos.
1834	Richard Roberts	Telar y máquina de hilar.
1837	Davenport	Motor de corriente continua (CC).
1837	Morse	Telégrafo.
1849	Bourding	Turbina de gas.
1849	Francis	Turbina hidráulica.
1868	Gramme	Dinamo eléctrica (CC)
1876	Otto	Motor de combustión de 4 ciclos.
1876	Bell	Teléfono.
1879	Edison	Lámpara de incandescencia.
1885	Benz	Automóvil.
1882	Tesla	El motor eléctrico de corriente alterna.
1885	Stanley/Tesla	El transformador eléctrico que permitía el transporte de energía eléctrica a grandes distancias.
1887	Tesla	Motor de inducción.
1889	Daimier	Motor de gasolina.
1890	Tesla	Motor de corriente alterna (CA).
1894	Tesla	Generador eléctrico de corriente alterna (CA).
1895	Diesel	Motor diesel.

En la tabla 1.2 se resumen los cambios tecnológicos y sociales acaecidos en las dos épocas de la Revolución Industrial, así como las fuentes de energía utilizadas.

Tabla 1.2. La Revolución Industrial

	Primera Revolución Industrial (1760-1870)	Segunda Revolución Industrial (A partir de 1870 hasta 1914)
Energías utilizadas	• Fuerza del hombre • Fuerza animal • Leña • Carbón • Hidráulica • Viento (eólica)	• Carbón • Hidráulica • Electricidad • Hidrocarburos
Máquinas utilizadas	• Máquina de vapor • Turbinas. Saltos de agua (Hidráulica)	• Máquina de vapor • Turbinas. Saltos de agua (Hidráulica) • Motores eléctricos • Motores de gasolina y diesel
Industrias desarrolladas	• Textil • Minería • Industria metalúrgica • Transporte (ferrocarriles)	• Textil • Minería • Siderurgia • Industria metalúrgica • Industria eléctrica • Industria del automóvil
Consecuencias	• Éxodo rural. • Migraciones internacionales. • Desarrollo del capitalismo. • Burguesía industrial. • Nace el proletariado. • Aparición de la sociedad moderna y desaparición de la sociedad rural hasta entonces controlada por la nobleza. • Incipiente mecanización del campo. • Modificación de las clases sociales, movimientos demográficos, mentalidad, costumbres, etc. • Se inicia la industrialización. • Desarrollo industrial y minero. • Aumento de la producción. • Crecimiento de ciudades. • Mejora del comercio. • Cambio social (caída de la mortalidad, elevada tasa de natalidad, aumento de la esperanza de vida). • Cambios de los métodos de trabajo, incorporación de la electricidad, montaje en cadena y la automatización. • Mejora y desarrollo de las comunicaciones. • Nuevos medios de transportes (automóvil, ferrocarril, navegación y aviación).	

Algunos ejemplos de la evolución industrial
Sólo un pequeño muestrario de la evolución industrial en todos los campos.

El transporte

Las comunicaciones

La agricultura

Figura 1.1. Tres ejemplos de evolución tecnológica.

1.2. Evolución del consumo de energía en los últimos siglos

Ha sido el hombre el que ha hecho uso de las energías para ayudarse en el desarrollo de su bienestar. La primera fuente de energía fue el Sol, a la que se le fueron añadiendo la biomasa, la energía hidráulica, la fuerza del viento, el carbón, luego el petróleo y la electricidad, más tarde llegó la energía nuclear y el gas natural, y más recientemente las energías renovables, tales como la eólica, solar térmica, solar fotovoltaica, el aprovechamiento de residuos urbanos, los biocombustibles y otras energías alternativas y renovables.

Ha sido en los siglos XIX y XX cuando se ha dado un consumo acelerado de energía, y se han buscado otras fuentes de energía para cubrir la demanda creciente que se viene produciendo en todo este tiempo, especialmente a partir del inicio de la Revolución Industrial.

Todos los niveles de nuestra sociedad (hogar, servicios, industria y sector terciario) necesitan energía para poder funcionar y proporcionarnos confort.

Doce energías básicas:
Petróleo
Gas natural
Carbón
Nuclear
Biomasa
Biocombustibles
Hidráulica
Eólica
Incineración de residuos urbanos
Solar térmica
Solar fotovoltaica
Aprovechamiento de los mares

Hay otras formas de energía, pero son de menor identidad o derivadas de estas.

1.2.1. Energía primaria

Se denomina energía primaria a toda forma de energía disponible en la naturaleza, antes de ser convertida o transformada en otra energía.

1.2.2. Nuevas energías

A corto plazo, no se ven alternativas a las energías que estamos empleando en la actualidad. Sí que hay previstas mejoras de los sistemas de transformación de energía, pero que no son suficientes para cubrir las necesidades futuras.

Se pretende suprimir en lo posible, la utilización de energías contaminantes (energías de origen fósil).

En la actualidad se está experimentando sobre la fusión nuclear mediante el proyecto ITER (Reactor Internacional Termonuclear Experimental). La energía básica es barata y los efectos de la reacción no parecen que tengan riesgo, como sucede con la fisión nuclear (energía atómica).

El reactor experimental está instalado en Cadarache a 70 km de Marsella y en este proyecto participan: UE, EE.UU., Rusia, Corea del Sur, India, Japón y China.

La primera fusión se pretende realizar en 2019 y si los ensayos son satisfactorios, no antes de 2050 habrá reactores comerciales para producir vapor con el que se accionen turbinas que muevan alternadores que generen electricidad barata.

1.3. Dependencia energética en el mundo actual

Muchos países no tienen en su territorio fuentes de energía suficientes para cubrir sus necesidades, como el caso de España, que tiene que importar en torno al 80% de la energía que consumimos.

La energía es cara, con muchas alteraciones de precio, e incluso, con problemas de suministro que no siempre está asegurado.

Aprovechar las energías alternativas, aunque sean más caras que las convencionales, tienen la ventaja de que están en el propio país.

Todos los países intentan reducir la dependencia, por dos razones fundamentales. El suministro de la energía supone un gasto muy importante para las economías nacionales, y en segundo lugar, el interés normal para asegurar un suministro continuado, sin sobresaltos.

1.4. Principales fuentes de energía utilizadas en la actualidad

La entrada en la época de la industrialización, ha llevado a un incremento progresivo de consumo de las energías, de acuerdo con este orden:

- Petróleo (gasolina, gasóleo y derivados).
- Carbón. Sigue siendo un producto energético muy utilizado.
- El gas natural y otros gases (butano y propano).
- La electricidad que es una energía que se obtiene por transformación de otras energías. Las energías son fundamentales para el funcionamiento del mundo actual, ya que son necesarias en todas las actividades que realiza el hombre.
- Las energías provienen de fuentes que son: renovables y no renovables (agotables).
- Las energías pueden ser: contaminantes y no contaminantes (limpias).
- Las energías alternativas se vienen desarrollando en los últimos años, pero a pesar de las elevadas inversiones que se hacen, no suponen una alternativa real al problema energético. Las energías renovables se obtienen de fuentes naturales virtualmente inagotables, unas porque es inmensa su cantidad, y otras, porque se renuevan por medios naturales.

La energía alternativa que mayor rendimiento tiene es la eólica, seguida de la hidráulica. La energía fotovoltaica tiene muchas y buenas aplicaciones, sin embargo, no es una alternativa de cara al suministro industrial de energía eléctrica.

La energía solar térmica tiene muchas posibilidades de aplicación para calentar agua para usos domésticos, calefacción, generación de electricidad e instalaciones frigoríficas.

- Energías verdes son aquellas que son respetuosas con el medio ambiente, es decir, que no provocan efecto invernadero, no contaminan y no contribuyen al calentamiento global, o lo hacen a pequeña escala.

Tabla 1.3. Procedencia de las energías consumidas. (Porcentaje a nivel mundial)

Tipo de energía primaria	1973	2000	2005
Carbón	24,4%	23,5%	25,3%
Petróleo	46,2%	34,9%	35,0%
Gas natural	16,0%	21,1%	20,7%
Nuclear	0,9%	6,8%	6,3%
Hidráulica	1,8%	2,3%	2,2%
Renovables y biomasa	10,6%	11,0%	10,0%
Otros [1]	0,1%	0,5%	0,5%
Total	100%	100%	100%
Total (Mtep)	**6.128**	**9.963**	**11.435**

Mtep – Millones de toneladas equivalentes a petróleo.

[1] Corresponde a energías solar, eólica y geotérmica.

Tabla 1.4. Sistema Eléctrico Español en 2009.

Origen	Porcentaje
Renovable [1]	27,9%
Cogeneración de alta eficiencia [2]	2,3%
Cogeneración [3]	9,3%
CC Gas natural [4]	27,3%
Carbón [5]	12,1%
Fuel/Gas [6]	0,7%
Nuclear [7]	19,3%
Otras	1,3%

[1] Electricidad que se obtiene a partir de energías que son renovables, como la hidráulica, la eólica, la fotovoltaica.

[2] La cogeneración para que sea de alta eficiencia (según Directiva 8/2004/CE), deberá cumplir los criterios siguientes:

- Deberá aportar un ahorro de energía primaria de al menos el 10%, en relación a que se destinara dicha energía sólo a producir electricidad o calor (produce electricidad y calor al mismo tiempo).
- La producción de las unidades de cogeneración a pequeña escala (< 1 MW) y de microgeneración (< 50 kW) que aporten un ahorro de energía primaria, podrá considerarse cogeneración de alta eficiencia.

[3] La cogeneración es una técnica que permite producir calor y electricidad. Se trata de pequeñas centrales situadas en el lugar donde se aprovecha el calor. La electricidad sobrante se revierte en la red general eléctrica.

[4] CC gas natural. Ciclo combinado de gas natural.
Las centrales de gas natural aprovechan la energía dos veces, la primera para alimentar el motor o turbina de gas para accionar un generador y la segunda, para con los humos de la combustión (800 °C), producir vapor de agua con el que se acciona una turbina que mueve otro alternador.

[5] Energía eléctrica procedente de centrales térmicas que queman carbón para producir vapor de agua y generar electricidad.

[6] Energía eléctrica procedente de centrales nucleares.

[7] Otras formas de generar electricidad a partir de energías varias.

1.5. Aprovisionamiento futuro de energías

Como se ha dicho, cada vez se necesita más y más energía. Los países emergentes están aumentando de forma considerable la demanda de materias primas y de energía.

Las energías principales que ahora se emplean (energías de origen fósil: petróleo, carbón y gas natural) tienen un plazo de caducidad, que será más o menos largo, pero al final se terminarán, porque no tienen reposición.

Por otro lado, estas energías están en manos de muy pocos países, algunos con regímenes totalitarios, lo que hace que el mercado esté a expensas de los vaivenes políticos, que da lugar a inseguridad en el suministro de los productos energéticas.

Mucho se habla de las energías alternativas, pero al menos a corto plazo no son una solución a la fuerte demanda.

La energía nuclear para generar electricidad tiene muchos detractores, y con razón, cuando es una buena alternativa, ya que se ha mejorado mucho en la seguridad de las instalaciones y se ha reducido la vida de sus residuos.

La energía hidráulica está al límite de sus posibilidades, poco más se puede hacer.

El paso de los vehículos automóviles a consumir electricidad en lugar de derivados del petróleo, todavía tiene un camino muy largo para que pueda significar una alternativa real para reducir el consumo de energías de origen fósil, ya que para generar electricidad hay que seguir consumiendo carbón y gas natural que son combustibles de origen fósil.

1.6. Qué pueden significar las energías alternativas

Las energías alternativas aprovechan productos energéticos que se generan o producen en una determinada zona, como es el caso del viento, luz y calor procedente del Sol, saltos de agua, mareas, calor de la tierra, producción agrícola (biocarburantes), bosques y cultivos varios (biomasa), etc.

Hay lugares que son más ventajosos que otros para aprovechar unas determinadas energías, razón por la cual, no en todas partes se pueden obtener todo tipo de energías alternativas.

Como puede apreciarse por las tablas que se insertan en esta obra, las energías alternativas pueden significar hasta un 25% del suministro de energía eléctrica.

Por ejemplo, la energía fotovoltaica permite suministros puntuales de energía eléctrica para aplicaciones diversas y en lugares remotos.

Hay que buscar también que las energías alternativas sean competitivas en precio, con las tradicionales (energías de origen fósil y nuclear). El fuerte desarrollo que han tenido algunas energías ha sido por las elevadas subvenciones (fotovoltaica, solar, eólica y otras) por parte de las diferentes administraciones, en las que se ha buscado más el rendimiento económico respecto a la inversión, que la propia generación de energía alternativa, para reducir dependencia energética y contaminar menos.

1.7. La dependencia energética española

Como es sabido por todos, España es un país que no tiene petróleo ni gas natural. Las minas de carbón son escasas, de difícil extracción y de regular calidad energética.

Dependemos del suministro exterior y de la situación del mercado mundial, en cada momento, sufriendo de lleno las crisis que periódicamente se producen en el mundo.

La Administración española está muy interesada en conseguir el máximo aprovechamiento de las energías alternativas, para reducir en lo posible la dependencia exterior. Sin embargo, no se está por la energía nuclear que es otra alternativa competitiva, como la tienen nuestros vecinos del norte (Francia). Ante la proximidad de sus centrales, en caso de siniestros o accidentes, no nos veríamos libres de las consecuencias de sus efectos. Tenemos el riesgo pero no las ventajas de una energía más barata.

1.8. Origen de la electricidad en España

Varias son las fuentes de energía primaria que se transforman en electricidad, tal como indica la tabla 1.4.

1.9. Fuentes de la energía primaria en España

La industria, los servicios, el transporte, las viviendas, el sector agrario y otros, necesitan mucha energía. Todavía son las energías tradicionales y contaminantes las más utilizadas, mientras que las energías renovables no llegan a ser una alternativa real.

Las previsiones para el año 2020 son de duplicar la utilización de energías renovables, pero si contamos que se va a consumir más energía en esa fecha, los incrementos no van a suponer una reducción importante de la dependencia exterior.

1.10. Previsión de demanda mundial de energía primaria

La tabla 1.9, recoge la previsión de demanda mundial de energía para el año 2030 comparada con la que consumía España el año 2000. Los porcentajes nos dan una idea de por dónde van a ir los consumos en el futuro.

El incremento de la demanda de energía casi se duplica en treinta años. Las energías de origen fósil siguen sin variación y suponen, casi el 90% del total.

Los cambios de tendencia no son significativos. A nivel mundial, los combustibles fósiles seguirán siendo los más consumidos.

Tabla 1.5. Balance de la energía eléctrica en España el año 2008.

Energías base	Porcentaje sobre el total
Gas natural	38,9%
Renovables [1]	19,7%
Nuclear	18,6%
Carbón	15,8%
Productos petrolíferos	6,0%
Producción por bombeo	0,9%

[1] Del 19,7% de la energía eléctrica que se obtiene de energías renovables, en la tabla 1.6 se indican los porcentajes correspondientes a las diferentes energías renovables.

Tabla 1.6. Porcentaje de las diferentes energías renovables en el balance de la generación eléctrica en España.

Energías renovables	Porcentaje
Eólica	10,0%
Hidráulica	7,3%
Biomasa	0,8%
Fotovoltaica	0,8%
R.S.U. [1]	0,6%
Biogás	0,2%
Termoeléctrica	0,005%

[1] R.S.U. – Residuos Sólidos Urbanos.

Nota:

Si se exceptúan las energías eólica e hidráulica, el resto de las energías alternativas son muy limitadas en su aportación energética respecto a la generación eléctrica

Tabla 1.7. Balance de la energía primaria consumida en España el año 2008.

Energía base	Porcentaje sobre el total
Petróleo	47,9%
Gas natural	24,5%
Nuclear	10,8%
Carbón	9,8%
Renovables	7,6%

Tabla 1.8. Balance de la energía renovable en el conjunto de la energía consumida en España el año 2008.

Energías renovables	Porcentaje
Biomasa	2,9%
Eólica	1,9%
Hidráulica	1,4%
R.S.U.	0,5%
Biocarburantes	0,4%
Biogás	0,2%
Fotovoltaica	0,2%
Solar térmica	0,1%
Geotérmica	0,01%
Solar termoeléctrica	0,004%

Tabla 1.9. Previsión de demanda mundial de energía primaria por combustibles.

Energía primaria	A nivel mundial	
	2000	2030
Carbón	26%	24%
Petróleo	39%	37%
Gas natural	23%	28%
Nuclear	7%	5%
Hidroeléctrica	3%	2%
Renovables	2%	4%
Total	**9.179 Mtep**	**15.267 Mtep**

Mtep : Millones de toneladas equivalentes en petróleo.

1.11. Situación actual de las energías renovables en España

Como puede apreciarse tras el análisis de las tablas anteriores en lo que se refiere a las energías alternativas, tan sólo las energías eólica e hidráulica son significativas, el resto de las energías renovables sólo se pueden considerar como testimoniales.

Difícilmente van a ser una solución al problema energético que tenemos, si los porcentajes sobre el resto de energías que consumimos son tan pequeños.

Esta obra que estudia las energías alternativas, pretende que conozcamos sus principios básicos y veamos las ventajas e inconvenientes que tienen, así como las dificultades que suponen algunos de sus procesos de generación.

Al no ser competitivo este tipo de energías, y si realmente interesa su obtención por los beneficios que puedan aportar (estratégicos, reducción de la contaminación, etc.) deberemos pagarlas más caras o que estén subvencionadas por las Administraciones, lo que supone más impuestos, cosa que hay que justificar muy bien para que se entienda.

España ha tenido una evolución positiva en la generación de energías alternativas, como lo demuestra la tabla 1.10, sin embargo, dada nuestra dependencia exterior, los avances no son suficientes ni a corto, ni a medio plazo. Hay que tener en cuenta, que aumentar la generación de energías renovables, supone un gran esfuerzo económico para todos, ya que muchas de ellas están subvencionadas y en porcentajes demasiado elevados. Los inversores han buscado el interés económico, sobre el interés estratégico.

Tabla 1.10. Evolución y previsión de energías primarias hasta 2012.

Energía primaria	2000		2006		2012	
	ktep	%	ktep	%	ktep	%
Carbón	21.635	17,3	17.999	12	14.113	7,8
Petróleo	64.663	51,7	75.315	50,3	84.820	46,9
Gas natural	15.223	12,2	26.905	18	42.535	23,5
Nuclear	16.211	13	16.570	11,1	16.602	9,2
Energías renovables	7.061	5,6	12.464	8,3	22.218	12,3
Saldo eléctrico (Imp. - Exp.)	382	0,3	385	0,3	385	0,2

De acuerdo con esta previsión, baja significativamente el carbón, un poco el petróleo y energía nuclear, y sube el gas natural, y las energías renovables triplican su aportación energética.

1.12. Previsión de Plan de Energía Renovables 2011-2020

España prevé que en 2020 la participación de las energías renovables en nuestro país serán del 22,7% sobre la energía final y un 42,3% de la generación eléctrica. Estas estimaciones han sido comunicadas a la Comisión Europea en cumplimiento de la Directiva de Energías renovables recientemente aprobada. Objetivo fijado por la UE ha sido de 40%. Los datos están contenidos en el anticipo del Plan de Renovables 2011-2020, enviado por el Ministerio de Industria, Turismo y Comercio a la Comisión Europea en cumplimiento de la propia Directiva comunitaria sobre la materia (2009/28/CE), que contempla objetivos obligatorios de energías renovables para la UE y para cada uno de los Estados miembros en el año 2020, y la elaboración por parte de éstos de planes de acción nacionales para alcanzar dichos objetivos.

1.12.1. Plan de Energías Renovables

Plan de Energías Renovables (PER).

El Plan de Acción Nacional de Energías Renovables (PANER) 2010/2020, comprende las siguientes energías:

- Solar
- Eólicas
- Biomasa
- Biocombustibles
- Hidroeléctrica
- Energías del mar
- Geotermia

La tabla 1.11 muestra el consumo español de energías renovables y su aportación en la energía final (Metodología Comisión Europea).

Tabla 1.11. Consumo español de Energías Renovables y su aportación en la Energía Final (Metodología Comisión Europea).

Consumo final de energías renovables (ktep)	2008	2012	2016	2020
Energías renovables para generación eléctrica	5.342	8.477	10.682	13.495
Energías renovables para calefacción/refrigeración	3.633	3.955	4.740	5.618
Energías renovables en transporte	601	2.073	2.786	3.500
Total en renovables, en ktep (según Directiva)	**9.576**	**14.504**	**18.208**	**22.613**
Total en Renovables según Directiva	10.687	14.505	17.983	22.382

Consumo de energía final de energía y porcentaje de las renovables sobre el total				
Consumo de energía final	2008	2012	2016	2020
Consumo de energía bruta final (ktep)	101.918	93.321	95.826	98.677
% Energías renovables/energía final	10,5%	15,5%	18,8%	22,7%

Análisis de la tabla 1.11:

a) Vemos que el consumo de energía bruta final se mantiene con muy poca variación a la baja hasta el año 2020.

b) Las energías renovables experimentan un incremento importante, pero no llegan a ser una alternativa real al problema energético sobre dependencia exterior.

Nota:

Las tablas de previsiones difieren bastante según quien sea el organismo que las confecciona. En algunas, se aprecia un optimismo exagerado que parece no ir con la realidad, porque tiene muchas variables que a veces no se analizan o cambian con el paso del tiempo.

1.12.2. Previsiones futuras sobre producción energética, a partir de las fuentes de energías alternativas

La Asociación de Productores de Energías Renovables (APPA) ha realizado las previsiones de generación para el año 2020. Estima que el 54% de la energía eléctrica producida en España podría tener su origen en energías renovables.

La Directiva 2009/28/CE recoge los objetivos mínimos vinculantes para España que se deberán alcanzar en 2020 en energías renovables, respecto al total de energía consumida, marcada en 20%. Por su parte, el gobierno español fija en 22,7% el objetivo mínimo a alcanzar en 2020.

Con un escenario del 23,4% proveniente de energías renovables respecto a la energía total para el año 2020, se podría llegar a generar 43% de la energía eléctrica en España, lo que significaría una reducción de la dependencia energética exterior.

1.12.3. Mix energético para España

La dependencia energética es un lastre para la economía española. No existe una receta que nos proporcione el mejor mix energético ya que está sujeto a muchas variables de tipo económico y político. Se habla mucho del ***Mix Energético*** pero poco se hace y planifica, y menos, se lleva a la práctica.

Uno de los objetivos es la reducción porcentual de los combustibles fósiles, incluido el carbón, a no ser que, se desarrollen y apliquen soluciones tecnológicas viables en lo económico, para reducir significativamente los actuales niveles de contaminación que origina su utilización.

Las energías renovables van a ser protagonistas forzosos del sistema energético de cualquier país.

Respecto a la energía nuclear, falta tomar una decisión concreta.

1.13. Qué futuro nos espera

El futuro es bastante incierto. Llevamos muchos años conociendo el problema de falta de recursos que tiene nuestro país y muchos otros, pero poco se hace, o lo que se hace es insuficiente y muy caro.

Lo que necesitamos es encontrar una fuente de energía que sea abundante, sin barreras de suministro y especialmente, que no sea contaminante.

Se llevan muchos años señalando el problema, pero no se encuentran alternativas para cubrir las necesidades crecientes de nuestra civilización.

La energía es imprescindible para nuestro mundo actual, todo se mueve con energía. Si nos falta energía, se para toda la actividad.

Imaginemos que nos faltara el suministro de energía eléctrica en casa: no funciona el ascensor, no se bombea el agua a nuestro piso, no tendremos luz artificial, no funcionaría el frigorífico, ni la vitrocerámica para cocinar, ni el microondas, ni la televisión, ni el ordenador, ni el teléfono, ni los juegos, ni el despertador, ni el aire acondicionado, ni la calefacción tanto que sea eléctrica como a gas o gasóleo, ni... nada.

1.14. Moderación en el consumo de energía

La energía que gastamos es finita, y la que no lo es, es escasa. Tenemos que concienciarnos todos (grandes y pequeños consumidores) en la obligación de ahorrar energía, consumir lo necesario y ahorrar lo que no necesitamos. Utilizar en lo posible procedimientos que economicen energía, como son los transportes colectivos, usando las cosas con criterios de economía, buscando siempre el ahorro, aunque lo podamos pagar.

Sabemos que estamos en una época de cambios climáticos, debidos a causas inciertas que no vamos a tratar ni cuestionar, pero ahí están. Debemos hacer todo lo que esté de nuestra parte para reducir el grado de contaminación que hoy día padecemos en todos los niveles; doméstico, terciario, industrial, transporte, agrícola, etc.

Figura 1.2. Dos ejemplos de evolución tecnológica.

1.15. Mejora de la eficiencia de los procesos

Es muy importante que se mejoren los rendimientos o eficiencia en los procesos de transformación de la energía, ya que los hay con eficiencia muy baja, por lo que se da un desaprovechamiento de los recursos. A veces, se gasta mucho, para conseguir poco, esto no se puede permitir por dos razones principales; consumimos energía que se agota y contaminamos más de lo que se debería, si consumiéramos menos para conseguir lo mismo.

Respecto a su utilización de las energías:

En los procesos de transformación de la energía se pierde una parte de la misma respecto a su finalidad, tal como vemos en la tabla 1.12.

Tabla 1.12. Rendimientos en la conversión de energía de algunos procesos.

Proceso	% [1]	Conversión
Generador eléctrico	94 a 99	Mecánica/Electricidad
Motor eléctrico	80 a 95	Electricidad/Mecánica
Caldera de vapor grande	88	Química/Térmica
Estufa de gas doméstica	85	Química/Térmica
Estufa doméstica de petróleo	66	Química/Térmica
Turbina de vapor	49	Térmica/mecánica
Turbina hidráulica	80 a 95	Hidráulica/Mecánica
Motor diesel	30 a 45	Química/Térmica/mecánica
Turbina de gas	25 a 33	Química/Térmica/mecánica
Motor de gasolina	25 a 30	Química/Térmica/mecánica
Panel fotovoltaico	14 a 20	Radiante/eléctrica

[1] Rendimientos aproximados.

Figura 1.3. Motores de gasolina. El rendimiento es muy bajo.

Figura 1.4. Motor eléctrico. El rendimiento es bueno, especialmente los de gran potencia.

Comentario:

Las tablas 1.13 y 1.14 nos muestran la realidad de lo que se puede obtener con las energías renovables si exceptuamos la energía eólica que casi será igual a la suma de las energías nuclear y térmica por carbón para el año 2020.

Hay procesos que tienen varias transformaciones que dan lugar a que las perdidas de energía sean muy elevadas, y por tanto, que los rendimientos sean muy bajos. Queda mucho por conseguir en este campo, si nos fijamos por ejemplo en el rendimiento.

El rendimiento de un motor o una transformación de energía viene dado por la siguiente fórmula:

$$\text{Rendimiento: } \eta = \frac{\textit{Energía útil}}{\textit{Energía absorbida}} = \frac{E_u}{E_a}$$

Los motores eléctricos, especialmente los de gran potencia tienen un buen rendimiento, en torno al 90%. En media potencia el rendimiento es superior al 80%.

Los motores de gasolina tienen un rendimiento en torno al 30%, que es muy bajo.

1.16. Previsiones de producción de energía eléctrica para el año 2020, con energías renovables

Contribución de las energías renovables (ER) a la generación de energía eléctrica en España, según el PANER en el año 2020.
PANER – Plan de Acción Nacional de Energías Renovables.

Tabla 1.13. Previsión de producción neta de electricidad para el año 2020, en España. Cálculo de la Directiva 2009/28/CE.

Forma de energía	GWh	Porcentaje (%)
Gas natural con cogeneración	141.741	35,4
Nuclear	55.600	13,9
Carbón	33.500	8,4
P. Petrolíferos + RSU (no recuperables)	8.721	2,2
Hidroeléctrica por bombeo	8.023	2,0
Energías renovables [1]	**152.835**	**38,1**
TOTAL	**400.420**	**100**

RSU – Residuos sólidos urbanos.
[1] Conjunto de energías renovables que se desglosan en la tabla.

Tabla 1.14. Previsión de producción neta de electricidad a partir de energías renovables para el año 2020, en España.

Forma de energía	GWh	Porcentaje (%)
Eólica terrestre y marina	79.489	19,8
Hidráulica sin bombeo	33.140	8,3
Solar termoeléctrica	15.353	3,8
Solar fotoeléctrica	14.316	3,6
Solar biomasa, biogás y RSU	10.017	2,5
Energías del mar y geotérmicas	520	0,1
TOTAL	**152.835**	**38.1**

1.17. Preguntas con respuesta

En este apartado se recogen a modo de resumen, algunos de los temas que se tratan a lo largo de esta obra y que conviene tenerlos claros desde su inicio.

- ***¿Qué son las energías alternativas?***
  Energías alternativas son las energías obtenidas de fuentes distintas a las clásicas (petróleo, carbón y gas natural) y que son: la solar (calor y luz), eólica, hidráulica, geotérmica, mareomotriz y la biomasa.
- ***¿Qué son energías renovables?***
  Energías renovables son las energías que se obtienen de fuentes naturales, y son virtualmente inagotables, porque se renuevan.
- ***¿Qué son las energías verdes?***
  Energías verdes son aquellas que están generadas a partir de energías primarias respetuosas con el medio ambiente. Son energías renovables no contaminantes.
- ***¿Qué es energía fotovoltaica?***
  La energía fotovoltaica se obtiene de la luz del Sol (fotones) transformando la energía de la luz, en energía eléctrica por medio de células fotoeléctricas.
- ***¿Qué es energía térmica solar?***
  Es la energía térmica que procede del Sol y que la podemos aprovechar con diferentes fines, además del que tiene sobre la superficie de la Tierra.
- ***¿Qué es energía cinética?***
  Se dice que un cuerpo o masa tiene energía cinética, cuando está en movimiento y es capaz de desarrollar un trabajo.
- ***¿Qué es energía potencial?***
  Se dice que un cuerpo o materia tiene energía potencial cuando la tiene almacenada y es capaz en un momento dado, de desarrollar un trabajo.
- ***¿Qué es energía hidráulica?***
  Energía hidráulica es la energía que desarrolla el agua cuando está en movimiento, por ejemplo: en una minicentral o en un pantano, para generar electricidad.
- ***¿Qué es energía geotérmica?***
  La energía geotérmica consiste en al aprovechamiento del calor del interior de la tierra, a muchos metros de profundidad con temperaturas elevadas.
- ***¿Qué es la energía maremotriz?***
  La energía mareomotriz está en el aprovechamiento del desnivel de las mareas (grandes masas de agua desplazadas que pueden mover turbinas para generar electricidad.
- ***¿Qué es la energía undimotriz?***
  La energía undimotriz es la energía producida por el movimiento y la fuerza de las olas.
- ***¿Qué es el ozono?***
  El ozono (O_3) es un gas de color azul y olor picante. En elevadas concentraciones (100%) es altamente tóxico para el ser humano.
- ***¿Qué es la capa de ozono?***
  El ozono atmosférico se encuentra entre 10 y 40 km sobre el nivel del mar, teniendo su máxima concentración a 25 km (estratosfera). El ozo-

no actúa en la atmósfera como un filtro de los rayos ultravioleta (UV) que provienen del Sol. Sin este filtrado, no sería posible la vida en la Tierra.

- ***¿Qué es el agujero de la capa de ozono?***
  El agujero de ozono corresponde a una parte de la capa de ozono donde se producen reducciones anormales de su densidad, circunstancia que se produce sobre las regiones polares en primavera y se recupera en verano.
- ***¿Qué es la biomasa?***
  La biomasa está contenida en la materia orgánica que se produce sobre la tierra. Conjunto de materia viva vegetal o animal destinada a la creación de energía.
- ***¿Qué son los biocombustibles?***
  Son combustibles de origen biológico que están obtenidos de materias renovables a partir de restos orgánicos que proceden habitualmente de la caña de azúcar, trigo, maíz, remolacha y semillas oleaginosas.
- ***¿Qué son los biocarburantes?***
  Los biocarburantes son biocombustibles utilizados como combustibles, para reemplazar combustibles de origen fósil (gasolina y gasoil preferentemente).
- ***¿Qué es la contaminación?***
  La contaminación es una degradación del medio ambiente debida a los residuos, materiales y gases emitidos por la actividad del hombre.
- ***¿Qué es la contaminación atmosférica?***
  Contaminación de la atmósfera (aire) como consecuencia de los gases y materias emitidas por actividad del hombre y por emisiones de la naturaleza y el propio planeta (volcanes).
- ***¿Qué es la lluvia ácida?***
  La lluvia ácida se forma cuando la humedad contenida en el aire se combina con óxidos de nitrógeno y dióxido de azufre emitido por la combustión de carbón y petróleo en fábricas y centrales térmicas, y que luego cae sobre las plantas y la tierra en forma de lluvia. La lluvia ácida supone una acidificación del medio ambiente.
- ***¿Qué es el CO_2?***
  El CO_2 es dióxido de carbono que es un gas incoloro, denso y reactivo y está generado por la combustión de combustibles de origen fósil y de materia orgánica. Se eleva hasta la troposfera donde participa en el efecto invernadero.
- ***¿Qué es el efecto invernadero?***
  El dióxido de carbono (CO_2) junto con el vapor de agua y otros que constituyen los llamados gases efecto invernadero (GEI). El exceso de CO_2 y gases invernadero en la atmósfera dificulta la salida de calor desde la Tierra hacia el exterior, haciendo que el calor se acumule sobre la superficie de la Tierra y suba su temperatura, con los problemas que origina.
  En los últimos años se ha pasado de 280 a 390 ppm (2009), que equivale a 0,039% del contenido atmosférico. Condiciona la radiación terrestre y contribuye al calentamiento global.
- ***¿Qué son los productos biodegradables?***
  La biodegradabilidad es la propiedad que tienen algunas sustancias químicas (materias) para que los microorganismos las conviertan en

sustrato, mediante energía y sustancias, como: aminoácidos, tejidos y organismos.
La degradación puede ser aerobia o anaerobia.
La diferencia entre productos biodegradables y no biodegradables está en el tiempo que tardan en desintegrarse y reintegrarse a la naturaleza. Debemos evitar en lo posible, la utilización de productos no biodegradables.

- *¿Qué es el ecologismo?*
  El ecologismo también llamado movimiento verde es un movimiento político, social y global que defiende la protección del medio ambiente para satisfacer la necesidad humana, incluyendo necesidades espirituales, sociales y otras.
  El movimiento ecologista se suele mostrar algunas veces violento en sus acciones, buscando llamar la atención, de unas formas que pueden parecer desproporcionadas. Son una corriente político-ideológica. Es contrario a los excesos en el uso de las energías contaminantes y de forma especial a ciertas energías.
- *¿Qué es la ecología?*
  No se debe confundir el ecologismo con ecología que es una ciencia, y no una teoría como es el movimiento ecológico.

2 Fuentes de energía

La producción de energía nacional el año 2009 fue de 29.971 ktep.

Energía	%	ktep
Nuclear	45,9	13.750
Carbón	12,6	3.778
Petróleo	0,4	107
Gas natural	0,0	12
Energías renovables	33,6	10.067
Hidráulica	7,5	2.258

ktep – kilotoneladas equivalentes de petróleo.
[1] Las tecnologías renovables suponen el 41,1% (12.325 ktep) de la energía producida en España.

La tabla muestra nuestra dependencia exterior. La producción nacional cubre en torno al 20% de las necesidades totales de energía.

El carbón y el butano representan a dos combustibles muy utilizados en nuestra sociedad.

Tabla 2.1. Principales energías renovables y no renovales.

Energías renovables
• Energía solar térmica • Energía solar luminosa (fotosíntesis) • Energía hidráulica • Energía fotovoltaica • Energía eólica • Energía geotérmica [(1)] • Mareomotriz • Energías procedentes de la biomasa
Energías no renovables
• Combustibles de origen fósil - Petróleo - Carbón - Gas natural • Energía nuclear

(1) Algunos autores incluyen la energía geotérmica en la columna de energías no renovables, por el hecho de que la Tierra se enfría con el paso del tiempo. El mismo criterio podríamos aplicar al Sol, pero creo que no procede porque en ambos casos es a muy largo plazo.

2.1. Introducción

El hombre desde la antigüedad ha necesitado energías. El Sol es la principal fuente de energía que proporciona al hombre y a la naturaleza luz y calor, energías fundamentales para que haya vida sobre la Tierra. En sus orígenes, el hombre tuvo que combatir el frío a base de fuego (leña) que siguió utilizando para preparar alimentos, luego para fundir metales y elaborar utensilios e imágenes (además de la leña, añadió carbón), también para navegar por el mar utilizó la fuerza del viento, para más recientemente pasar a otras fuentes de energía.

A medida que el hombre ha ido evolucionando y desarrollándose, ha necesitado de más y más energía, llegando al momento actual, en el que el aprovisionamiento de energías es un problema a nivel de todos los países de la Tierra.

Resolver el problema del abastecimiento de energía es un problema de difícil solución. Los científicos e investigadores buscan energías alternativas a las que venimos usando y que sean menos contaminantes y no se agoten como sucede con las principales energías que hoy día utilizamos, como son fundamentalmente el carbón, el petróleo, el gas natural y la energía nuclear.

2.2. Principales fuentes de energía

Las energías que utilizamos las podemos dividir en dos grandes grupos:

2.2.1. Clasificación de las energías

Dos tipos de energía:

b) Energías renovables

Las energías que una vez consumidas, se pueden volver a consumir porque se reponen, no se agotan.

Ejemplos de estas energías son: la luz y el calor que llega del Sol, el viento, el agua de un río, la fuerza del mar, la energía mareomotriz, la masa arbórea (biomasa).

a) Energías no renovables

Las energías que una vez consumidas, no tienen recuperación.

Ejemplos de estas energías son: el petróleo, el gas natural y otros gases, la energía nuclear y el carbón.

2.2.2. Principales fuentes de energía

La tabla 2.1 recoge las principales fuentes de energía actualmente utilizadas. El mayor porcentaje corresponde a las energías no renovables (80%). No se considera la electricidad como fuente de energía ya que se trata de una energía transformada desde otras fuentes de energía.

La energía eléctrica es básica para la humanidad, ya que es el motor de todo. En todos los procesos está presente la electricidad, por que es fácil de transportar hasta el usuario y limpia en su aplicación.

2.2. 3. Energías contaminantes

Las energías no renovables son más contaminantes que las energías renovables. En primer lugar está el carbón, seguido del petróleo y en tercer lugar el gas natural.

La energía nuclear tiene otros niveles de contaminación que pueden ser debidos a escapes radioactivos, accidentes durante su utilización y a la larga vida que tienen los desechos radioactivos antes de perder la radioactividad.

2.3. Energía solar

Del Sol obtenemos dos energías básicas para la vida en el planeta Tierra, que son el calor y la luz, ambas energías son imprescindibles para la vida de las plantas, los animales y la humanidad.

Ambas energías permiten su aprovechamiento para calentar agua mediante paneles térmicos y paneles fotovoltaicos para generar electricidad.

Las energías que proceden del Sol son las que producen los fenómenos atmosféricos del viento (energía eólica) y el ciclo del agua (energía hidráulica) y la fotosíntesis en las plantas y las mareas y olas en los mares.

2.3.1. Diversos aprovechamientos de las energías del Sol

En este apartado se presentan varios ejemplos sobre diversas formas de aprovechar la energía solar.

a) Bosques y agricultura (biomasa) que se benefician de la luz y el calor

De la biomasa se obtienen diversos productos, como son:

- Bioetanol (carburante).
- Biodiesel (carburante).
- Calor (por combustión de la biomasa).
- Aceites vegetales.
- Biogás.
- Biometanol.
- Biocarburantes sintéticos.

b) Paneles solares térmicos para calentar agua sanitaria (ACS)

Los paneles o captadores solares térmicos permiten aprovechar el calor que nos llega desde el Sol para calentar agua, bien sea para reforzar el fluido de las calefacciones por agua, como para proporcionarnos agua caliente sanitaria (ACS). Los captadores solares térmicos tienen un fluido que se calienta por la acción solar y que luego ceden este calor, en un intercambiador. El agua calentada queda almacenada en un acumulador en espera de su consumo.

c) El agua de los ríos

Son muchas las utilidades de estas masas de agua, tales como: suministro a domicilios, servicios e industria, regar los campos, generar electricidad, etc.

La fuerza del agua proporciona energía hidráulica. Esta energía se transforma en energía mecánica que antes se aprovechaba para mover las piedras de un molino, o para accionar las máquinas de una empresa textil u otra manufactura, y ahora, para generar electricidad.

d) Generación eléctrica mediante paneles fotovoltaicos

La luz del Sol (energía fotónica) se transforma en energía eléctrica (electricidad) por medio de pequeños generadores, como son, las células fotovoltaicas. Un panel o módulo fotovoltaico contiene un número elevado de estos pequeños generadores (células fotovoltaicas).

Figura 2.1. Bosque con pinos gigantes en el parque Yosemite (California), y finca con planta de pimiento del piquillo en Navarra.

Figura 2.2. Paneles solares térmicos para calentar agua sanitaria, instalados en la cubierta de un edificio.

Figura 2.3. Central eléctrica al pie de una presa.

Figura 2.4. Panel fotovoltaico que alimenta con corriente eléctrica un puesto de toma de recibos de la ORA en un aparcamiento, aislado de suministro eléctrico.

Figura 2.5. Central eólica en el desierto de Sonora (California).

Figura 2.6. La fuerza de las olas y las mareas son una fuente de energía.

Figura 2.7. Restos de una antigua mina de carbón.

e) Generación eléctrica mediante la fuerza del viento

La fuerza del viento dentro de unos límites de velocidad mediante "molinos" permite con su giro, mover generadores eléctricos (alternadores o dinamos) transformando la energía eólica del viento en energía eléctrica.

El aprovechamiento de la fuerza del viento ya se hacía desde la antigüedad para impulsar los barcos y mover las piedras en los molinos de granos.

f) Las mareas

El origen de las mareas que es el cambio periódico del nivel del mar se debe a las fuerzas gravitacionales que ejerce la Luna sobre la masa de agua de los mares y en menor fuerza, el Sol.

Las mareas son mayores en los grandes mares (grandes masas de agua) y menores en los mares pequeños.

2.4. Carbón

El carbono se encuentra libre en la naturaleza formando el diamante y el grafito.

El carbono arde formando anhídrido carbónico cuando la combustión es completa.

El carbón suministra casi el 25% de la energía mundial consumida.

El carbón se encuentra en la naturaleza en grandes concentraciones, con mayor o menor pureza y su extracción se hace en las minas, que pueden ser a cielo abierto o a diferentes profundidades.

El carbón inició su extracción industrial coincidiendo con la Primera Revolución Industrial y desde entonces sigue siendo un combustible prioritario, especialmente dedicado a la generación de energía eléctrica.

2.4.1. Carbones naturales y artificiales

Los carbones naturales que se encuentran en la naturaleza se clasifican en cuatro tipos: antracita, hulla, lignito y turba.

Tipos de carbón:

a) Carbón mineral

Carbón natural, sólido de consistencia pétrea o terrosa y combustible.

b) Carbón vegetal

Se obtiene de la madera quemada en combustión incompleta, para lo que se cubre con tierra para evitar la entrada o contacto con el aire.

c) Carbón de petróleo

Se obtiene por destilación del petróleo.

d) Carbón bituminoso

Se trata de una variedad de carbón cuyas características son intermedias entre la antracita y el lignito.

e) Carbón activado

Carbón que se utiliza como absorbente y que tiene una gran porosidad que le da esta propiedad.

f) Carbón de coque (cok)

La hulla esta impregnada de sustancias bituminosas de cuya destilación se obtienen hidrocarburos aromáticos y un tipo de carbón llamado coque (cok) que se utiliza como combustible en la siderurgia.

El carbón está muy repartido sobre la Tierra y se extrae de las minas, que pueden ser:

- Mina de pozo.
- Mina de talud.
- Mina de galerías.
- Mina a cielo abierto.

Existen variedades de carbón atendiendo a su poder calorífico, como son de más a menos: antracita, hulla, lignito y turba.

Tabla 2.4. Poder calorífico del carbón y otros combustibles sólidos.

Combustible	Densidad media kg/m³	PCI kJ/kg	PCS kJ/kg	Combustible	PCI kJ/kg	PCS kJ/kg
Turba	360	21.300	22500	Aglomerados de carbón	31.300	35.600
Lignito	1.050	28.400	29.600	Carbón de madera	31.400	33.700
Hulla	1.350	30.600	31.400	Coque	29.300	33.700
Antracita	875	34.300	34.700	Coque de petróleo	34.100	36.500

PCI – Poder calorífico inferior. **PCS** – Poder calorífico superior.

Equivalencias:

1 kcal = 1.000 cal = 4.184 J = 4,184 kJ.

1 kcal = 4,184 kJ

kcal/kg → kilocaloría por kg de materia.

kJ/kg → kilojulios por kilogramo de materia.

2.4.2. Óxido de carbono

El óxido de carbono (CO) se forma por la combustión incompleta del carbón, y es el gas producido en los gasógenos ordinarios, al pasar el aire a través de carbón incandescente, que incorpora nitrógeno contenido en el aire.

2.4.3. Principales países productores de carbón

El carbón se encuentra preferentemente en el hemisferio Norte.

En la actualidad, China es el principal productor mundial de carbón.

2.4.4. Aplicaciones del carbón

Del carbón se aprovecha preferentemente su poder calorífico, aunque también tiene otras aplicaciones.

El carbón suministra en torno al 25% de la energía mundial actualmente consumida, y una parte es transformada en electricidad.

a) Como combustible para calefacciones

En la actualidad el carbón ha perdido importancia al ser sustituido por gas natural, butano, propano, paneles solares y electricidad.

b) Como combustible para empresas siderúrgicas

En siderurgia (hornos altos), que representa el 12% del consumo mundial.

Carbón de coque (cok) o alquitrán de hulla obtenido por pirólisis del carbón en ausencia de aire.

Figura 2.8. Vagoneta cargada de carbón en la montaña de Palencia.

Figura 2.9. Montaña de carbón en una central térmica.

Tabla 2.2. Porcentaje de carbono de los diferentes tipos de carbono.

Tipo de carbón	% de carbono
Antracita	90 a 95
Hulla	75 a 92
Lignito	55 a 75
Turba	< 55

Tabla 2.3. Poder calorífico del carbón.

Productos combustibles	Poder calorífico (calorías)
Diamante	8.400
Antracita	8.300
Hulla	8.000
Coque	7.800
Lignito	4.700
Turba	3.000
Madera seca	3.700

Tabla 2.5. Principales países productores de carbón.

Principales productores mundiales de carbón	
• China • Alemania • Reino Unido • EE.UU. • Polonia	• Canadá • Rusia • India • Sudáfrica • Australia

Tabla 2.6. Resumen de ventajas e inconvenientes sobre la utilización del carbón.

CARBÓN
Ventajas
• Muy utilizado como materia energética. • Se obtienen diversos subproductos. • Energía todavía básica para generar electricidad y para la industria. • Su localización y producción se encuentra muy distribuida. • Se ha mejorado su rendimiento en el proceso de transformación a electricidad.
Inconvenientes
• No renovable. • Los humos de su combustión son muy contaminantes para la atmósfera y la naturaleza. • Impacto ambiental en los lugares de extracción. • Su extracción da lugar a la pérdida de muchas vidas (accidentes).

c) Para generación de energía eléctrica

El 40% de la producción total de carbón se utiliza para su transformación en energía eléctrica en centrales térmicas.

d) En la industria de transformación

Por ejemplo en las cementeras.

e) En carboquímica.

Proceso de gasificación con el que se obtiene gas de síntesis (hidrógeno y monóxido de carbono) que se transforma en diferentes productos químicos como: amoníaco, metanol, gasolina y gasóleo.

f) Petróleo sintético

Obtención por licuefacción directa. Fue un procedimiento que utilizó Alemania durante la II Guerra Mundial. Ahora en desuso.

g) Alquitrán de hulla

De su destilación se obtienen:

- Aceites ligeros. Un 5% de la cantidad destilada y contiene benceno y tolueno.
- Aceites medios. Un 12% de la cantidad destilada y contiene principalmente naftaleno y fenol.
- Aceites pesados. Un 9% de la cantidad destilada. Se emplea para quemar en calefacción y motores diesel.
- Aceite de antraceno. Representa el 19%.
- El resto no destilable, se emplea para el asfaltado de carreteras.

2.4.5. Ventajas e inconvenientes de esta forma de energía

(Ver Tabla 2.6.).

2.5. Petróleo

El petróleo sigue siendo el producto energético más importante de todos los que empleamos, no sólo como energía básica, sino como materia prima para la obtención de muchos subproductos básicos para la industria.

El petróleo ya lo utilizaban los chinos y después los árabes, antes que Edwin L. Drake lo redescubriera y extrajera a una profundidad de 20 m en Titusville (Pensilvania) en 1859.

2.5.1. Características principales del petróleo

Las características del petróleo extraído, varían unas de otras, en función de su procedencia, por lo que las características que aquí se exponen, lo son con carácter general.

- Se trata de mezclas complejas de hidrocarburos líquidos que llevan en disolución hidrocarburos sólidos.
- Estos hidrocarburos están constituidos básicamente por hidrógeno y carbono.
- Es un líquido oleoso, inflamable, color amarillo oscuro tirando a negro, de olor característico, compuesto por hidrocarburos.
- Densidad: entre 0,8 y 0,95 kg/dm^3.
- Principal fuente de energía utilizada en la actualidad.
- Asociado a los yacimientos de petróleo, y en otros niveles o capas, se encuentra el gas natural.
- Tiene elevado poder calorífico.

- La densidad es una de las características que más definen a los diferentes petróleos, como combustible.
- Principales componentes del petróleo: además de hidrógeno y carbono, contiene azufre, oxígeno y nitrógeno.
- Una de las principales características del petróleo es su **poder calorífico**, que lo hace utilizable como fuentes de energía. Este parámetro varía en función de su densidad, y por tanto, de su composición química concreta.

Figura 2.10. Silueta característica de un puesto de bombeo para la extracción de petróleo.

Tabla 2.7. Clasificación del petróleo por su densidad API

Aceite crudo	Densidad (g/cm^3)	Densidad en grados API
Extrapesado	< 1,0	10,0
Pesado	1,0 – 0,92	10,0 – 22,3
Mediano	0,92 – 0,87	22,3 – 31,11
Ligero	0,87 – 0,83	31,1 – 39
Superligero	> 0,87	> 31

API - Parámetro internacional establecido por Instituto Americano del Petróleo.

Figura 2.11. Depósitos que almacenan derivados del petróleo.

Comparación del poder calorífico entre los tres principales combustibles de procedencia sólida (carbón, petróleo y gas natural) para obtener la misma cantidad de calor.

1,5 t/carbón de hulla = 1 t/petróleo = 1.000 m³ de gas

2.5.2. Refinado del petróleo

El petróleo tal como se extrae de los yacimientos, no es utilizable, por lo que hay que pasarlo por el proceso de refino.

Para realizar el refino, se somete al petróleo a un proceso de destilación fraccionada por calentamiento y que pasa por las siguientes fases:

- Hasta 45 °C desprende gases combustibles.
- De 45 a 160 °C se obtiene gasolina, que corresponde aproximadamente el 40% del petróleo o crudo refinado.
- De 160 a 300 °C se obtiene el queroseno.
- De 300 a 350 °C se obtiene el gasoil, aceite pesado de densidad 0,89.
- A más de 350 °C se obtienen los aceites lubricantes grasos, desprovistos de acidez.
- De los residuos del refino queda el alquitrán.
- Enfriando a 0 °C se obtiene aceites pesados (gasoil) y separando la parte sólida de la líquida, después de filtrada, se obtiene la parafina, empleada en múltiples aplicaciones.
- Por destilación fraccionada de la gasolina se obtienen, entre otros, los siguientes productos: bencina, nafta, éter de petróleo, etc.

Tabla 2.8. Poder calorífico en función de la densidad:

Densidad API	Poder calorífico (cal/g)
0,7 a 0,8	11.700 a 11.100
0,8 a 0,9	11.100 a 11.675
0,9 a 0,95	10.675 a 10.500

Figura 2.12. Vista parcial de una refinería de petróleo.

Nota:

El petróleo como puede deducirse de este resumen, es una de las materias básicas de la que se extraen muchos subproductos de los que nos beneficiamos. En la actualidad, resulta imprescindible.

Tabla 2.9. Resumen de las fases del refinado.

Producto obtenido	Temperatura	Densidad (kg/dm^3)
Gasolina (40%) (1)	45 °C a 160 °C	0,7
Queroseno	160 °C a 300 °C	0,8
Aceites pesados o gasoil	300 °C a 350 °C	0,9
Aceites lubricantes	> de 350 °C	
Residuos		

(1) Por destilación fraccionada de la gasolina se obtienen: éter de petróleo, bencina, nafta, etc.

2.5.3. Productos derivados del petróleo

Resumen de los muchos subproductos que se obtienen del petróleo y entre los más importantes están:

a) Metano, etano, propano y butano.

En las refinerías se obtienen metano y etano que se queman, y propano y butano que se licuan, comprimen y almacenan en bombonas para su uso como combustibles de uso doméstico e industrial.

b) Gasolina

Destilada del petróleo crudo, se usa como combustible para motores de explosión (automóviles y aviones).

c) Queroseno

Se usa como combustible para motores (aviones).

d) Gasoil

Se usa como combustible para motores diesel.

e) Fuel-oil

Se usa como combustible en hornos, calefacciones, locomotoras, etc.

f) Parafinas y vaselina

Utilizadas en la fabricación de productos varios y farmacéuticos.

g) Asfalto

Utilizado en la pavimentación de carreteras y calles.

g) Otros muchos productos

Plásticos, pinturas y barnices, disolventes, fertilizantes e insecticidas, detergentes, cauchos artificiales, farmacia, negro de humo y muchos más.

Tabla 2.10. Principales países productores de petróleo.

Principales productores mundiales de petróleo (1)		
Rusia	Irak	Irán
EE.UU.	Nigeria	Kazanstán
Arabia Saudí	Argelia	Venezuela
Irán	Libia	México
China	Emiratos Árabes Unidos	Argentina
Reino Unido	Kuwait	Colombia
Noruega	Qatar	Bolivia
Canadá	Indonesia	Ecuador

(1) Dos tercios de las reservas mundiales de petróleo se encuentran en Oriente Próximo.

La tabla 2.11 reúne a los principales países exportadores de petróleo, según cifras de la Administración de Información de Energía de Estados Unidos (EIA, por sus siglas en inglés).

Tabla 2.11. Principales países exportadores de petróleo.

País	Exportaciones en 2008
Arabia Saudita	8.408
Rusia	6.882
Emiratos Árabes Unidos	2.584
Kuwait	2.416
Irán	2.394
Noruega	2.245
Angola	1.950
Argelia	1.929
Nigeria	1.883
Venezuela	1.863
Las exportaciones de petróleo están expresadas en millones de barriles por día e incluyen las de crudo, petróleo de esquisto bituminoso, arenas petroleras y líquidos gasificados, el líquido contenido en el gas natural que se obtiene separadamente.	

Tabla 2.12. Consumo de petróleo a nivel mundial. (Estimación en barriles/día).

Consumo por zona geográfica	Millones de barriles/día (m b/d
Estados Unidos	25,43
Europa	15,58
Pacífico	8,64
Ex Unión Soviética	3,80
Otros	0,75
China	6,59
Resto de Asia	8,78
América Latina	5,00
Oriente Medio	6,12
África	2,91
Estimación de consumo	83,60

2.5.4. Países que forman parte de la OPEP

OPEP – Organización de Países Exportadores de Petróleo.
Registrada en Naciones Unidas el 6 de noviembre de 1962.
Producen más del 35% del crudo mundial.

Nota:

*La magnitud técnica utilizada para medir la cantidad o volumen de petróleo extraído es el **barril**, que equivale a **159** litros.*

Tabla 2.13. Componentes de la Organización OPEP [1].

Componentes
Angola (2007) **Arabia Saudita** **Argelia** **Emiratos Árabes Unidos** **Indonesia** **Libia** **Nigeria** **Irán** **Irak** **Kuwait** **Qatar** **Venezuela** **Ecuador** (abandonó la OPEP) **Gabón** (abandonó la OPEP)
Adheridos
México Noruega Rusia Kazanstán Omán Egipto *Nota: estos países participan regularmente como observadores en las reuniones regulares de la OPEP.*

(1) La sede la OPEP está en Viena (Austria).

Tabla 2.14. Resumen de la clasificación de los mercados según API.

Crudo **liviano**	Tiene gravedad API superior a: 31,1 °API.
Crudo **mediano**	Tiene gravedad API comprendida entre: 22,3 y 31,1 °API.
Crudo **pesado**	Tiene gravedad API comprendida entre: 10 y 22,3 °API.
Crudo **extrapesado**	Tiene gravedad API inferior a: 10 °API.

API - Parámetro internacional establecido por Instituto Americano del Petróleo.
***API** - American Petroleum Institute*

Tabla 2.15. Clasificación de los crudos por su referencia.

Brent (Blend)	Denominación empleada por 15 crudos de campos de extracción que aplican el sistema Brent de los campos del Mar del Norte. Se aplica en Europa, África y Oriente Medio.
WTI (West Texas Intermediate)	Para el crudo producido en EE.UU.
Tapis (Malasia)	Referencia del crudo ligero del Lejano Oriente.
Minas (Indonesia)	Referencia del crudo pesado del Lejano Oriente.

2.5.5. Indicadores de precios de diferentes petróleos

Para un determinado precio, tomado en el mes de abril/2008 [1], vemos que hay diferencias entre los principales tipos de petróleo.

- West Texas .... 104,70 $
- Brent .... 103,70 $
- Cesta Venezolana .... 94,53 $
- Cesta OPEP .... 98,63 $

[1] En esta fecha nos encontrábamos en plena crisis de precios del petróleo. En marzo de 2011, también se han alcanzado y superado estos precios.

En el mercado pueden encontrarse otros precios que dependerán del tipo de petróleo y del contrato que haya realizado la petrolera con el propietario de la extracción.

2.5.6. Clasificación de los petróleos por su gravedad API

Una de las características que define al petróleo es el de su gravedad que determina su composición química. La densidad aumenta con el incremento de hidrocarburos y productos pesados (resinas y asfaltenos), y disminuye con la temperatura.

La densidad viene dada en g/ml (gramos/mililitro) o g/cm^3 (gramos/centímetro cúbico), o de la forma más común de denominarlo, que es, el grado API (°API).

1 g/ml = 10 °API (crudos pesados)

0,77 g/ml = 50 °API (crudos ligeros)

La tabla 2.16 considera un grupo de combustibles líquidos, además de derivados del petróleo.

Tabla 2.16. Poder calorífico de combustibles líquidos.

Combustible	PIC kJ/kg	PCS kJ/kg	Combustible	PCI kJ/kg	PCS kJ/kg
Aceite de esquistos	-	38830	Fuel-oil nº1	40600	42695
Alcohol comercial	23860	26750	Fuel-oil nº2	39765	41860
Alquitrán de hulla	-	37025	Gasóleo [1]	42275	43115
Alquitrán de madera	36420	-	Gasolina [2]	43950	46885
Etanol puro [4]	26790	29720	Petróleo bruto	40895	47970
Metanol [4]	19250	-	Queroseno [3]	43400	46500

[1] Densidad a 15 °C, 850 kg/m^3
[2] Densidad a 20 °C, 730 kg/m^3
[3] Densidad a 15 °C, 780 kg/m^3
[4] Densidad a 20 °C, 790 kg/m^3

2.5. 7. Reservas de petróleo estimadas en el 2005

Las reservas de petróleo a nivel mundial eran de 1.293.000 millones de barriles, siendo Venezuela con 315.000 m b/d la primera del mundo en reservas, seguida de Arabia Saudita con 261.000 m b/d, Canadá con 178.800 m b/d, continuando con Irán, Irak, EAU, Kuwait, Rusia, Libia, Nigeria, EE. UU., China, etc.

> **Nota:**
> *Las reservas pueden aumentar a medida que se descubran otros yacimientos que hasta ahora no se han descubierto. En el mes de abril de 2008, se ha descubierto una gran bolsa de petróleo en Brasil, y también otra en el 2009.*

2.5.8. Aplicaciones de los gases licuados del petróleo (GLP)

Muchas son las aplicaciones de los gases (GN y GLP), además de las domésticas, son muy importantes en los servicios y la industria, incluso en la generación de energía eléctrica.

El gas licuado del petróleo (GLP) es una mezcla de gases condensables que están presentes en el gas natural o disueltos en el petróleo.

Los GLP que a temperatura y presión de ambiente son gaseosos, son fáciles de condensar cuando se aumenta su presión o se disminuye la temperatura. Los gases más comunes son el propano y el butano.

Los recuadros siguientes recogen las principales aplicaciones de los GLP:

a) En el hogar

- Gasodomésticos
- Agua caliente
- Calefacción
- Aire acondicionado

b) Servicios

- Restaurantes
- Hoteles
- Hospitales
- Balnearios
- Universidades
- Ayuntamientos
- Polideportivos
- Camping
- Conventos
- Institutos
- Cuarteles
- Residencias de ancianos
- Guarderías
- Colegios y escuelas
- Incineradoras

c) Sector agropecuario

- Granjas avícolas
- Granjas porcinas
- Desmontadoras de algodón
- Secadores de pimiento
- Asadores
- Secadores de tabaco
- Invernaderos

d) Industria

- Cerámicas
- Queserías
- Industria química
- Embotelladoras de vino
- Fundiciones
- Fábricas de pastas
- Piscifactorías
- Panaderías
- Industria textil
- Pastelerías
- Fábricas de conserva
- Fábricas de embutidos
- Reciclado de pilas
- Gas para automoción
- Fábricas de caramelos

e) Automoción

- Carretillas
- Transporte público (taxis, autobuses).

Figura 2.13. Camión con bombonas y una bombona de butano.

Nota:

Normalmente las instalaciones están constituidas por dos depósitos. Uno de reserva y el otro en servicio, que se rellenará cuando se vacíe.

Tabla 2.18. Puntos de ebullición de diferentes gases.

Gas	Temperatura de ebullición
Butano	- 0,5 °C
Metano	- 161,7 °C
Etano	- 88,6 °C
Propano	- 42,1 °C

Tabla 2.20. Procedencia del petróleo que importó España el año 2010

País de procedencia	% Importado
Irán (OPEP)	14,6
Libia (OPEP)	13,0
Arabia Saudí (OPEP)	12,5
Nigeria (OPEP)	10,6
Irak (OPEP)	3,6
Angola (OPEP)	2,1
Otros OPEP	3,6
Rusia	12,7
México	11,3
Noruega	1,3
Otros	14,7

Nota:

El año 2010 se consumieron en España 67,1 millones de toneladas de productos petrolíferos, de los que sólo 122.000 toneladas (0,18%) se extrajeron en España. El coste total de la factura petrolera a la que se suman otras importaciones (gasóleos, querosenos, fuelóleos, etc.) se eleva a 51.000 millones de euros.

Tabla 2.17. Poder calorífico de diferentes gases.

Gases	Fórmula	Poder calorífico
Hidrógeno	H_2	2.160 kcal/kg
Metano	CH_4	1.335 kcal/kg
Etano	C_2H_6	1.960 kcal/kg
Óxido de carbono	CO	2.140 kcal/kg
Propano	C_3H_8	2.310 kcal/kg
Propileno	C_3H_6	2.350 kcal/kg
Butileno	C_4H_8	2.647 kcal/kg

2.5.9. Suministro de los gases licuados del petróleo (GLP)

Varias son las formas en que se realizan los suministros de gases licuados, según sea el lugar y sistema de consumo.

A continuación se presentan algunos de los procedimientos de suministro.

- A granel (camión cisterna) para el llenado de depósitos.
- Envasado (botellas y bombonas).
- Canalizado (red de tuberías).

Depósitos en la utilización (viviendas, comercio, industria, explotaciones agrícolas, etc.). Pueden ser: aéreos o enterrados.

Capacidad de los depósitos: desde 1.029 a 10.227 kg de carga útil.

2.5.10. Ventajas e inconvenientes de esta forma de energía

Tabla 2.19. Resumen de ventajas e inconvenientes sobre la utilización del petróleo.

PETRÓLEO	
Ventajas	**Inconvenientes**
• La más importante fuente de energía. • Producto básico actual para el desarrollo de la humanidad. • Muchísimas aplicaciones en la industria y los transportes. • Se obtienen muchos subproductos.	• Reservas limitadas. • Elevado efecto contaminante en su utilización. • Problemas de abastecimiento, transporte y almacenamiento. • Incertidumbre en su precio que es muy variable.

2.5.11. Procedencia del petróleo que importa España

El petróleo que llega a España procede de muchos países lo que facilita el suministro, en caso de crisis, dada la diversidad de los mercados.

2.5. Gas natural

El gas natural se ha convertido en una energía básica para el mundo moderno, que se utiliza como fuente de calor en cantidad de aplicaciones, empezando por nuestro propio hogar. Muy importante como fuente de energía para la generación de electricidad.

2.5.1. Características principales del gas natural y otros gases

Cada tipo de gas tiene sus características. La tabla 2.21 resume las características de los gases más consumidos.

Tabla 2.21. Características principales de los gases más consumidos.

Características	Gas natural	Butano	Propano
Fórmula química	CH_4	C_4H_{10}	C_3H_8
Poder calorífico	Entre: 13.200 y 11.900 kcal/kg	Entre 11.800 y 10.930 kcal/kg	Entre: 12.000 y 11.100 kcal/kg
Densidad del gas comparado con el aire	0,56	2.09	1,56
Peso de aire mínimo necesario para la combustión de 1 kg de combustible, en kg	17,2	15,5	15,6
Volumen del gas producido por 1 kg de líquido. Volumen específico del gas a 0 °C y 76 mm de columna de mercurio (cdHg)	1.220	372	530
Temperatura mínima de inflamación	650 °C	480 °C	490 °C
Presión normal de utilización para aparatos domésticos, en milibar (mbar)	20	28	37
Temperatura de ebullición a presión atmosférica normal a °C	-161	-1	-43
Temperatura crítica, en °C	-83	+151	+97
Presión crítica, en bar absolutos	46	38	45
Velocidad de propagación de la llama en el interior de un tubo de 25 mm de diámetro, en m/s	0,70	0,80	0,80

Tabla 2.22. Otros gases relacionados con el gas natural. Hidrocarburos saturados.

Hidrocarburos	Fórmula química
Metano	CH_4
Etano	CH_3
Propano	$CH_3 - CH_2 - CH_3$
Butano	$CH_3 - CH_2 - CH_2 - CH_3$
Pentano	$CH_3 - CH_2 - CH_2 - CH_2 - CH_3$
Hexano	$CH_3 - CH_2 - CH_2 - CH_2 - CH_2 - CH_3$

Figuras 2.14. a) Punto de control de un gasoducto. b) Acometida en una vivienda. c) Distribución de gas a los pisos.

Tabla 2.23. Países exportadores de gas.

País	Exportaciones 2007
Rusia	237.200
Canadá	107.300
Noruega	86.100
Argelia	59.400
Países Bajos	55.700
Turkmenistán	49.400
Qatar	39.300
Indonesia	32.600
Malasia	31.600
Estados Unidos	23.300
Las exportaciones de gas están expresadas en millones de metros cúbicos.	

Figura 2.15. Almacenado de gas en instalación en superficie.

Figura 2.16. Depósito y surtidor de propano.

2.5.2. Aplicaciones del gas natural

El gas natural supone uno de las principales formas de energía que utilizamos en la actualidad. Se utiliza en todos los sectores de la actividad que demandan energía térmica, como:

- Cocinar alimentos.
- Calefacción en la vivienda, industria y servicios.
- Generación de electricidad (ciclo normal o combinado).
- Tratamientos térmicos.
- Proceso de secado directo.
- Cocción de productos cerámicos.
- Hornos de fusión.
- Generación de vapor para procesos diversos.
- Cogeneración. Sistema por el que se produce energía térmica (vapor) y electricidad.
- Como materia prima en la producción de amoníaco para fertilizantes, por su alto contenido en hidrógeno.

El gas natural en el hogar tiene un amplio campo de aplicaciones, como son:

- Cocinar en cocinas con quemadores o vitrocerámicas para gas.
- Hornos para cocinar.
- Lavar y secar.
- Proporcionar agua caliente sanitaria (ACS). Una de sus ventajas es la de suministrar agua caliente de forma inmediata.
- Calefacción y climatización.
- Calentar superficies con calentadores especiales.

2.5.3. Ventajas e inconvenientes de esta forma de energía

La tabla 2.24 recoge las ventajas e inconvenientes respecto al consumo del gas natural, principal fuente de energía que se extrae de bolsas situadas en pozos subterráneos.

Tabla 2.24. Resumen de ventajas e inconvenientes sobre la utilización del gas natural.

GAS NATURAL	
Ventajas	**Inconvenientes**
• Facilidad de extracción. • Tecnología desarrollada. • Buen rendimiento térmico. • Muchas alternativas de aplicación. • Su poder contaminante es menor que el del carbón y el petróleo.	• Necesidad de transporte y almacenamiento relativamente caros. • Reservas limitadas. • Riesgo de accidente por fugas incontroladas.

2.5.4. Procedencia del gas natural que importa España

El gas natural que llega a España procede de un número muy reducido de países. El 60% llega a través de gasoducto del Magreb y el 40% restante en metaneros.

Tabla 2.25. Poder calorífico de combustibles gaseosos.

Combustible	Densidad kg/m^3	PCI kJ/kg	PCS kJ/kg	Combustible	Densidad kg/m3	PCI kJ/kg	PCS kJ/kg
Gas natural	(*)	39.900	44.000	Gas de agua	0,711	14.000	16.000
Gas de hulla	0,50		46.900	Gas ciudad	0,650	26.000	28.000
Gas de coquería	0,56	31.400	35.250	Gas de agua carburado	0,776	26.400	27.200
Gas de aire	-	10.000	12.000	Propano506 [(l)]	1,85 (g)	46.350	50.450
Hidrógeno	0'0899	12.0011	14.1853	Butano580 [(l)]	2,4 (g)	45.790	49.675

(*) Varía según el país de procedencia.
[(l)] (g) Densidad a 20 °C en estado líquido y gaseoso, respectivamente.
P.C. Medio del biogás = 5.554 kcal/m^3

PCI – Poder calorífico inferior.
PCS – Poder calorífico superior.

2.6. Energía nuclear

La energía nuclear en lo que se refiere a la generación de energía eléctrica es muy importante. Está por detrás de las energías de origen fósil y suponen en torno al 20% de la energía eléctrica generada.

2.6.1. Minerales radioactivos

Como se ha indicado, el mineral base es el uranio, que se encuentra disuelto en rocas sedimentarias. Es muy escaso y muy localizado en la naturaleza, y se encuentra en:

- Canadá, que es el primer productor mundial de uranio.
- Australia (22,5%).
- África (Níger, Namibia y otros).
- Estados Unidos (27,4%).
- Asia (Urbekistán, Kazanstán y otros).

En España había un yacimiento en Saelices El Chico (Salamanca), que se cerró el año 2000.

Del mineral uranio se obtienen tres isótopos: **U-238** (0,28%), **U-235** (0,71%) y **U-234** (0,01%).

El isótopo U-235 se enriquece (purifica o refina) para aumentar su concentración y hacerlo útil para reaccionar.

El uranio-235 se encuentra en la naturaleza, representando el 0,7% del uranio natural. El uranio-233 y el plutonio-239 se obtienen artificialmente. Estos tres isótopos son utilizados en las reacciones de fisión.

Para su empleo en reactor, se elabora en pequeñas pastillas de dióxido de uranio de unos milímetros, con una energía equivalente a una tonelada de carbón. Con las pastillas se forman barras de unos 4 metros de longitud, y con ellas se alimentan los reactores.

Comportamiento de la energía reactiva:

- La combustión de 1 g de carbón (destrucción), produce 8.083 cal (calorías).

Tabla 2.26. Procedencia del gas natural que importó España el año 2010

Zona de procedencia	% Importado
Argelia	32,6
Nigeria	20,0
Egipto	7,5
Libia	1,7
Trinidad y Tobago	9,0
Perú	1,7
Qatar	16,3
Yemen	1,0
Omán	0,5
Noruega	8,9
Otros	0,8

Nota:

El año 2010 se consumieron en España 404.042 GWh, de los que sólo 664 GWh (0,16%) se extrajeron en España. El coste de las importaciones se elevó a unos 7.700 millones de euros.

Tabla 2.27. Resumen de ventajas e inconvenientes sobre la utilización del uranio.

URANIO (energía nuclear)
Ventajas
• Hay grandes reservas de uranio. • Gran productividad en la transformación a energía eléctrica. • Tecnología desarrollada. • Usos medicinales. • Aplicaciones industriales. • Se obtiene electricidad a un coste muy competitivo.
Inconvenientes
• Energía no renovable • Elevado riesgo de contaminación en caso de accidente. • Residuos radiactivos peligrosos a corto y largo plazo. • Alto coste de las instalaciones y el mantenimiento de las instalaciones. • Problemas en el transporte y almacenamiento de residuos • Riesgo de uso no pacífico. • Armas nucleares

- 1 g de radio (Ra), material radioactivo produce sin combustión, 135 cal/hora.

Al cabo de una año, la energía desprendida equivaldría a 145 g de carbón, sin haber perdido de forma medible, merma de su energía.

Serán necesarios 1.590 años para que el peso quede reducido a la mitad.

1 gramo de uranio 235 genera la misma cantidad de energía calorífica que la combustón de 2.700 kg de carbón.

Las partículas alfa son emitidas a una velocidad del orden de 1,6 x 10^7 m/s (~16.000 km/s)

2.6.2. Fisión nuclear

La fisión nuclear fue descubierta por los científicos O. Hahn y F. Strassman en 1938 y consiste en un proceso por el cual los núcleos de ciertos elementos pesados (U-235) son bombardeados por neutrones que los rompen originando dos átomos de un tamaño aproximadamente la mitad del de uranio, liberando dos o tres neutrones que inciden sobre otros átomos de U-235 vecinos que se vuelven a romper, con lo que se origina una reacción en cadena que es controlada externamente.

2.6.3. Aplicaciones del uranio

El uranio tratado se utiliza con fines diversos:

- Fabricación de armamento nuclear para la guerra (bombas, cohetes, etc.).
- Medicina nuclear. Servicios de radiodiagnóstico, radiología y medicina nuclear.
- Aplicación de radioisótopos en los campos de medicina, agricultura e industria.
- Aparatos de detección y control.
- Aplicación de las radiaciones ionizantes en la industria agroalimentaria.
- Aplicación de energía nuclear a base de isótopos radioactivos, emisiones de radiaciones y radiaciones electromagnéticas.
- Centrales nucleares para generar electricidad.
- La fisión nuclear es la principal aplicación del uranio para usos civiles de la energía nuclear aplicada a centrales para la generación de electricidad, como una gran fuente de energía.

2.6.4. Centrales nucleares en el mundo

Hay países industrializados que generan electricidad a partir de energía nuclear, como es el caso de:

- Estados Unidos tiene 104 plantas.
- Francia, 58 plantas.
- Japón, 55 plantas.

2.6.5. Ventajas e inconvenientes de esta forma de energía

La energía nuclear tiene muchas más utilidades que la utilizada en la generación de electricidad, sin embargo, es en esta aplicación donde encuentran los mayores detractores.

2.6.6. Terremoto en Japón que afectó a la central nuclear de Fukushima.

El 11 de marzo de 2011 se produjo en Japón un gran terremoto (a las 14h46 hora de Japón) de magnitud 8,9 en la escala sismológica, seguido en tsunami (15h41), en la costa nordeste y que afectó a la central nuclear de Fukushima.

Central nuclear de Fukushima:
Año de construcción: 1966
Inicio de actividad: 1971
Tipo de central: reactor de agua en ebullición (BWR).
Número de reactores: 6
Capacidad total de generación: 4.696 MW
Generadores: 1 x 460 MW (igual potencia que Sta. María de Garoña)
4 x 784 MW
1 x 1.100 MW

Los reactores 1, 2, y 3 estaban en servicio y los reactores 4, 5 y 6 estaban en corte para una inspección periódica. Al detectarse el terremoto se apagaron automáticamente los reactores 1, 2 y 3 y paró la producción de electricidad. Al averiarse con el terremoto la red externa, se cortó el suministro eléctrico exterior y entraron a funcionar los grupos diésel de generación eléctrica, que pararon por inundación, al llegar el tsunami a las 15h41.

Al faltarles la refrigeración a los núcleos, hubo diversas explosiones, lo que dio lugar a escapes y a generarse a nivel internacional un pánico exagerado respecto a la energía nuclear. Hay que tener en cuenta, que la central soportó un gran terremoto y después un tsunami devastador en la zona, algo casi impensable, incluso para Japón.

Figura 2.17. Red de transporte de energía eléctrica obtenida por transformación de otras energías.

2.7. La energía eléctrica

La energía eléctrica no existe libre en la naturaleza. Se obtiene por transformación de otras energías en energía eléctrica.

2.7.1. Introducción

La electricidad es una energía que se obtiene por transformación de otras energías, mediante procesos más o menos eficientes. En toda transformación hay una pérdida de energía especialmente en los procesos térmicos. En la tabla 2.28 se citan las energías básicas y los procesos mediante los cuales se consigue generar energía eléctrica.

La energía eléctrica es básica para el mundo moderno actual, y es necesaria para el funcionamiento y desarrollo de todo tipo de actividad: doméstica, industrial, de servicios, agrícola, comunicaciones, entretenimiento, etc.

Tabla 2.28. Principales energías de las que se obtiene electricidad.

Energía básica	Procedimiento para generar electricidad
• Carbón	• Térmico
• Petróleo	• Térmico
• Gas natural	• Térmico normal o combinado
• Uranio	• Nuclear
• Agua de los ríos	• Hidráulico
• La luz del Sol	• Fotovoltaico
• El viento	• Eólico
• Biomasa	• Térmico

2.7.2. Impacto medioambiental

En la generación de la electricidad se emiten gases y residuos, algunos de ellos muy contaminantes, como sucede con el carbón, el petróleo, gas natural y el proceso nuclear.

a) Emisiones de dióxido de carbono
- *Contenido de carbono*: 0,39 kg de dióxido de carbono por kWh.

b) Residuos radioactivos Alta Actividad
- *Residuos radioactivos*: 0,42 mg (miligramos) por kWh.

Tabla 2.29. Procedencia de la energía eléctrica en el Sistema Eléctrico Español en 2008.

Tipo de energía	Porcentaje (%)
CC Gas natural	30,1
Renovable	20,7
Nuclear	19,3
Carbón	15,9
Cogeneración	8,1
Fuel/Gas	3,3
Cogeneración de alta eficiencia	1,7
Otras	0,9

Tabla 2.30. Impacto medioambiental de diferentes formas de generar electricidad, considerando todo el ciclo del combustible hasta convertirse en electricidad. (Toneladas por GWh)

Fuentes de energía	CO_2	NO_x	SO_x	Partículas	CO	Hidrocarburos	Residuos nucleares	TOTAL
Carbón	1.058,2	2,986	2,9721	1,626	0,267	0,102	-	1.0066,1
Gas natural ciclo combinado	824,0	0,251	0,336	1,176	TR	TR	-	825,8
Nuclear	8,6	0,034	0,029	1,176	0,267	0,001	3,641	12.3
Fotovoltaica	5,9	0,008	0,023	0,017	0,003	0,002	-	5,9
Biomasa	0	0,614	0,154	0,512	11,361	0,768	-	13,4
Geotérmica	56,8	TR	TR	TR	TR	TR	-	56,8
Eólica	7,4	TR	TR	TR	TR	TR	-	7,4
Solar térmica	3,6	TR	TR	TR	TR	TR	-	3,6
Hidráulica	6,6	TR	TR	TR	TR	TR	-	6,6

TR – Valores inapreciables
Fuente: US Department of Energy, Council for Renewable Energy Education.

Nota:

Una central eléctrica de ciclo combinado considerando la fase de funcionamiento puede emitir unos 345 g de CO2 por kWh. Pudiendo llegar hasta 402 g de CO2 por kWh si se tiene en cuenta todo el ciclo de vida (energía y materiales para construir la instalación, transporte, infraestructuras anexas, etc.

2.7.3. Generación eléctrica

La electricidad se obtiene a través de diferentes elementos y máquinas, como son:

- Pilas eléctricas. Generan corriente continua (CC)
- Células fotoeléctricas. Generan corriente continua (CC). Forman parte de los paneles fotovoltaicos.
- Dinamos. Generan corriente continua (CC).
- Alternadores. Generan corriente alterna (CA).

2.7.4. Corriente continua o corriente alterna

Edison quiso imponer la corriente continua (dinamo) como forma de energía eléctrica, pero se encontró con que se producían muchas pérdidas de energía en el transporte y la sección de los conductores tenía que ser muy grande. En esta batalla tecnológica le ganó la partida Nikola Tesla cuando consiguió generar corriente alterna (alternador) y elevar el valor de la tensión por medio de un transformador, con lo que al reducir la intensidad de la corriente (I), conseguía reducir el valor de la sección de los conductores y también las pérdidas en el transporte. Otro transformador en la llegada permite reducir la alta tensión de transporte a una baja tensión de utilización. Se estima que en el transporte de energía se pierde entre 6 y 8% de la energía generada.

2.8. Centrales eléctricas

Las centrales eléctricas son los lugares o centros en los que se transforman otras formas de energía, en energía eléctrica. Varios son los tipos de centrales para generar energía eléctrica y entre los más generalizadas están las que se relacionan a continuación.

En los primeros años del pasado siglo XX se construyeron muchas centrales hidráulicas de pequeño y mediana capacidad.

Entre los años 60 y 75 les tocó el turno a las centrales de fuel-oil.

En los años 70 y 80 se construyeron las centrales nucleares.

A partir del año 2000 se inició la construcción las centrales de gas natural (ciclo normal y ciclo combinado).

Las centrales generan electricidad a una tensión de 6 a 20 kV, que se eleva por transformador a 132, 220 o 440 kV, para realizar el transporte de la energía eléctrica con las menores pérdidas posibles.

2.8.1. Principales tipos de centrales eléctricas

Muchos son los procedimientos de transformación de otras energías en energía eléctrica, como son algunos de los tipos de centrales que se relacionan a continuación:

- Centrales hidráulicas (hidroeléctricas).
- Centrales térmicas.
- Centrales de ciclo combinado.
- Centrales eólicas.
- Centrales fotovoltaicas.
- Centrales nucleares.
- Centrales solares térmicas.
- Centrales incineradoras de residuos sólidos urbanos (RSU).
- Centrales de cogeneración empleando biomasa.
- Centrales de gasificación de ciclo combinado.
- Centrales con motores diesel.

2.8.2. Centrales hidráulicas

Las centrales hidráulicas utilizan la fuerza del agua de los ríos y de los pantanos, el agua del mar (mareas) para a través de una turbina, generar electricidad.

En la década los años 40 y 50 se dio un gran impulso a las grandes centrales hidráulicas. Desde principios de siglo habían proliferado las pequeñas y medianas centrales que aprovechaban pequeños saltos y desniveles.

a) Tres tipos de centrales

- *Centrales de alta presión*
  - Salto hidráulico: ≤ 200 m.
  - Caudal: 20 m^3/s (caudales relativamente pequeños).
  - Turbinas: Pelton y Francis.
  - Pequeños caudales que se canalizan con mucho desnivel.
- *Centrales de media presión*
  - Salto hidráulico: entre 20 y 200 m.
  - Caudal: 200 m^3/s por turbina.
  - Turbinas: Francis y Kaplan y Pelton para grandes saltos.
  - Saltos hidráulicos (pantanos).
- *Centrales de baja presión*
  - Salto hidráulico: ≤ 20 m.
  - Caudal: hasta más de 300 m^3/s
  - Turbinas: Francis y Kaplan.
  - Grandes caudales.

Nota:

Las turbinas modernas tienen un rendimiento o eficiencia que va desde 0,85 a 0,95.

Figura 2.18. Presa sobre el río Ebro donde nace el canal de Lodosa (Navarra), con una minicentral que aprovecha el agua sobrante para generar electricidad. Potencia instalada: 4.200 kW.

Nota:

La generación hidráulica se desarrolla en el capítulo 8, dado que se trata de una energía renovable.

Figura 2.19. Central térmica de Guardo (Palencia) que quema carbón (proceso térmico).

b) Minicentrales

Centrales de pequeña y media potencia que aprovechan pequeños desniveles en el cauce de los ríos y canales.

Turbinas

Son los dispositivos que transforman la energía potencial del agua en energía mecánica rotativa con la que se mueve el alternador que genera corriente eléctrica.

c) Centrales eléctricas de bombeo

Estas centrales permiten elevar el agua de una presa a la superior por bombeo, cuando la demanda de energía en la red general es baja. Cuando la demanda de energía eléctrica es mayor, con el agua almacenada se genera electricidad y así se refuerza la producción eléctrica.

2.8.3. Centrales térmicas de carbón

Las centrales térmicas consumen carbón como energía base

Queman carbón para generar vapor de agua y por medio de turbinas de vapor producir energía mecánica rotativa con la que se accionan generadores eléctricos (alternadores).

Entre los años 70 y 85 del pasado siglo XX se construyeron centrales térmicas que queman carbón.

El problema principal de este procedimiento de generación está en su elevada emisión de gases contaminantes (CO_2) y azufre (S) que dan lugar al llamado efecto invernadero y a la lluvia ácida.

Tabla 2.31. Ventajas e inconvenientes de las centrales térmicas de carbón.

Ventajas	Inconvenientes
• El carbón es el principal producto energético que se transforma en electricidad. • La tecnología de las centrales de carbón han mejorado su rendimiento y reducido los efectos de los humos y residuos. • Son una salida para que ciertas minas sigan en actividad y haya trabajo para la zona.	• Quema un producto no renovable. • Los humos de combustión son muy contaminantes, especialmente CO_2 y S. • Efectos medioambientales en la zona de extracción y almacenado. • Transporte y almacenado. • Efectos nocivos para la zona en que se ubica la central

2.8. 4. Centrales de gas natural

Como es obvio, estas centrales consumen gas natural.

Pueden ser de dos tipos:

a) De ciclo normal: turbina de gas.

La turbina de gas (motor) acciona un alternador que genera corriente eléctrica.

b) De ciclo combinado: turbina de gas y turbina de vapor producido por los humos y gases calientes de la turbina de gas.

Con la misma energía se consigue accionar dos alternadores.

Tabla 2.32. Ventajas e inconvenientes de las centrales térmicas de gas natural.

Ventajas	Inconvenientes
• Fácil extracción. • Buen rendimiento energético en el proceso de transformación a energía eléctrica. • Llega a las centrales a través de la red de gasoductos.	• Quema un gas que no es renovable. • Problemas de transporte (gasoductos y barcos). • Menos contaminante que el carbón o el petróleo. • Problemas e incertidumbres en su suministro y precio.

Figura 2.20. Central de ciclo combinado que quema gas natural.

2.8.5. Centrales nucleares

Mediante la fisión nuclear controlada se consigue generar calor que produce vapor de agua, con el que por medio de una turbina de vapor, generar energía mecánica que se transforma en electricidad por medio del alternador eléctrico.

Cuando se inició la crisis del petróleo en los años 70 y 80, se desarrollaron muchas de las centrales nucleares que utilizan como producto energético el uranio y el plutonio y que todavía están en servicio.

La energía eléctrica se consigue a partir de un ciclo o cadena de transformaciones, como se relaciona a continuación.

El calor que se desarrolla en el proceso de fisión, se emplea para producir vapor de agua con el que se accionen turbinas y estas a alternadores, generadores de corriente eléctrica.

Las centrales nucleares tienen reactores que en su interior alojan el material combustible, generalmente uranio o plutonio, con otros elementos reguladores de la fisión. Esta fisión se produce al bombardear los átomos de uranio con neutrones en el interior del reactor nuclear contenido en un recipiente o vasija de acero inoxidable de 12,5 mm de espesor y aislado con gruesas paredes de hormigón. La fisión se controla por barras de carburo de boro o de grafito.

Figura 2.21. Vista general de una central nuclear con dos reactores.

El agua que circula en el núcleo a una presión de 150 bar se calienta a 325 °C, pasando a vapor de agua.

En un intercambiador, el vapor de agua proveniente del reactor, calienta agua de un segundo circuito hasta hacerlo vapor que es conducido a las turbinas para moverlas y con éllas, los alternadores eléctricos.

Un tercer circuito, pasa el vapor de agua que sale de las turbina a líquido, para reconducirlo de nuevo al intercambiador y volverlo a vapor de agua, y así, repetir el ciclo.

El agua del tercer circuito se refrigera (enfría) en las llamativas torres de refrigeración, que todos conocemos por su espectacularidad, y que tiran a la atmósfera vapor de agua.

El proceso que sigue el uranio, es el siguiente:

- Extracción del mineral.
- Tratamiento del mineral para su conversión en hexafluoruro de uranio.
- Enriquecimiento en el isótopo U-235.
- Fabricación del elemento combustible para su utilización.

[1] Consecuencias del accidente en la central nuclear de Chernobil (Ucrania) ocurrido el 26 de abril de 1986.

- Se contaminaron 150.000 km^2, en Rusia, Ucrania y Bielorrusia.
- 7 millones de personas afectadas.
- Han muerto en torno a 200.000 personas.
- Incremento de cáncer de tiroides, leucemias y otras enfermedades.

Causas: tecnología obsoleta, mala construcción del reactor, falta de control, deficiente estado de conservación y mantenimiento, poca fiabilidad de los equipos, dejadez por parte de los técnicos. Una suma de fallos en cadena.

La primera central nuclear en España empezó su actividad el año 1968, y estaba ubicada en Almonacid de Zorita (Guadalajara).

En la actualidad están en servicio en España 8 reactores con potencia total instalada de 7.742,32 MWe

Las centrales de España son de dos tipos:

- PWR Agua ligera a presión.
- BWR Agua ligera en ebullición.

Este tipo de centrales es poco contaminante hacia la atmósfera, mucho menos que las centrales térmicas que queman carbón, gas o petróleo, pero si lo es, en lo que respecta a su proceso y a los residuos que genera.

Sus problemas principales son:

- Posibles escapes de radioactividad.
- Fallo en el control del proceso radioactivo que lo lleve a temperaturas muy elevadas y a la fusión del reactor (caso Chernobil) [1].
- Residuos radiactivos que hay que evacuar y almacenar durante muchos años.

Cada vez son menores los riesgos y los residuos generados. Existe una preocupación por este problema que hace que las centrales nucleares sean tan temidas y mal vistas.

Se están haciendo estudios y aplicaciones para que los residuos radiactivos no tengan una vida superior a 100 años.

También está en estudio la destrucción de los residuos en el mismo lugar en que se generan, aunque sea a costa de un menor rendimiento de la central.

Respecto al proceso, cada vez es más seguro y las nuevas centrales van incorporando todos los adelantos que garanticen la seguridad de las personas y el medio ambiente.

El centro de almacenado de residuos en España, se encuentra en El Cabril (Córdoba).

Las entidades que controlan y regulan la aplicación de la energía atómica, son:

En España: Consejo de Seguridad Nuclear (CSN).

Ley 15/1980 de 22 de abril.

En Europa: Organización Europea de Energía Atómica (EURATOM).

Internacional: Organismo Internacional de la Energía Atómica (OIEA), con sede en Viena.

Agencia de la Energía Nuclear de la OCDE (NEA).

Las ventajas de la aplicación de la energía nuclear para generar electricidad están en:

- No dependencia del petróleo.
- Disponemos de uranio.
- El precio de la energía eléctrica obtenida por este procedimiento, es más barata.

La energía que se obtiene en España por este procedimiento viene a ser una quinta parte del total generado.

a) Centrales nucleares de España

La tabla 2.33 recoge todas las centrales que hay en España y su lugar de implantación y capacidad de producción de las mismas.

Tabla 2.33. Centrales nucleares activas en España.

CENTRAL NUCLEAR	JOSÉ CABRERA
Localidad	Almonacid de Zorita (Guadalajara)
Año de puesta en marcha	1968
Potencia instalada	160 MW
En la actualidad	Fuera de servicio
CENTRAL NUCLEAR	SANTA Mª DE GAROÑA
Localidad	Santa Mª de Garoña (Burgos)
Año de puesta en marcha	1971
Potencia instalada	466 MW
CENTRAL NUCLEAR	ALMARAZ 1 y 2
Localidad	Navalmoral de la Mata (Cáceres)
Año de puesta en marcha	1971 y 1983
Potencia instalada	973,5 MW y 982,6 MW
CENTRAL NUCLEAR	ASCÓ I y II
Localidad	Ascó (Tarragona)
Año de puesta en marcha	1983 y 1985
Potencia instalada	979, 05 MW y 976,24 MW
CENTRAL NUCLEAR	COFRENTES
Localidad	Cofrentes (Valencia)
Año de puesta en marcha	1984
Potencia instalada	1025,4 MW
CENTRAL NUCLEAR	VANDELLÓS 2
Localidad	Hospitalet de L'Infant (Tarragona)
Año de puesta en marcha	1988
Potencia instalada	1057 MW
CENTRAL NUCLEAR	TRILLO 1
Localidad	Trillo (Guadalajara)
Año de puesta en marcha	1988
Potencia instalada	1066 MW

Tabla 2.34. Ventajas e inconvenientes sobre la utilización del uranio para generar electricidad.

URANIO (energía nuclear)	
Ventajas	Inconvenientes
• Gran productividad en la transformación a energía eléctrica. • Tecnología muy desarrollada en los reactores actuales. • Se obtiene electricidad a un coste muy competitivo. • Reduce dependencia energética exterior.	• Elevado riesgo de contaminación en caso de escapes o accidentes radioactivos o por efectos de terremotos y maremotos. • Residuos radiactivos peligrosos a corto y largo plazo. • Alto coste de las instalaciones y el mantenimiento de las mismas. • Problemas en el transporte y almacenamiento de residuos

Figura 2.22. Generador eólico o aerogenerador.

Nota:

La generación eólica se desarrolla en el capítulo 8, dado que se trata de una energía renovable.

Figura 2.23. Grandes paneles fotovoltaicos.

Nota:

La generación fotovoltaica se desarrolla en el capítulo 7, dado que se trata de una energía renovable.

Figura 2.24. Instalación experimental en la Plataforma Solar de Almería (PSA).

2.8.6. Centrales eólicas

Dentro de la generación eléctrica a partir de una energía alternativa renovable están las centrales eólicas que transforman la fuerza del viento en energía eléctrica.

2.8.7. Centrales fotovoltaicas

La fuente de energía está en la luz solar que mediante muchos pequeños generadores fotovoltaicos (células fotovoltaicas) se consigue transformar en energía eléctrica.

Este sistema de aprovechamiento de la luz solar tiene muchas aplicaciones puntuales, especialmente donde falta el suministro de la red eléctrica, como ya se verá en el capítulo dedicado a esta tecnología.

2.8.8. Centrales solares térmicas

Se trata de centrales experimentales y de pequeña y mediana potencia, en las que se aprovecha el calor del Sol, para por medio de helióstatos (espejos) concentrarlo sobre una pequeña superficie y calentar un fluido que pueda utilizarse para generar electricidad, como son los sistemas que se presentan a continuación.

a) Concentradores de calor de foco lineal

b) Disco parabólico concentrador de calor sobre motor "Stirling"

2.8.9. Centrales de cogeneración

El término y procedimiento de cogeneración empezó a utilizarse en EE.UU. en la década de los años 70 para la producción conjunta en una o más etapas, de energía mecánica que se transforma en eléctrica accionando un alternador y energía térmica aprovechando el calor de los humos de la combustión.

Ejemplo de instalación: turbina que quema gas natural con la que se acciona un generador eléctrico, y con humos muy calientes, se calienta agua a determinada presión y temperatura que se aprovecha para determinados procesos industriales (cocción de neumáticos) y otros usos, como calefacción.

Las centrales de cogeneración pueden quemar gas natural u otros gases, así como residuos y otros combustibles.

2.8.10. Centrales con motores diesel

Se trata de generadores accionados por motores de combustión que consumen carburante diesel. Estas centrales son de pequeña potencia, para suministrar energía eléctrica en aquellos lugares alejados de redes eléctricas de suministro.

Las potencias de estos grupos generadores pueden ser superiores a los 2.000 kW.

2.8.11. Pilas eléctricas

La generación eléctrica por pilas eléctricas se debe a una transformación de la energía química en energía eléctrica.

En 1800, Alessandro Volta dio a conocer su experimento al que se llamó "pila", por estar constituido el generador eléctrico por una serie de discos de zinc y cobre apilados de forma alternativa intercalando entre ellos un fieltro impregnado en agua o salmuera.

Se trata de pequeños generadores en tensión e intensidad que pueden utilizarse solos o acoplados en serie (aumentar la tensión), en paralelo (para aumentar la intensidad de corriente), o en acoplamiento mixto, serie-derivación (para aumentar el valor de la tensión y la intensidad).

Figura 2.25. Instalación experimental en la Plataforma Solar de Almería (PSA).

Son muchísimos los receptores que utilizan pilas eléctricas, como todos conocemos: móviles, relojes, mando a distancia, linternas, aparatos de radio, etc.

Hay diversos tipos y modelos de pilas, tales como:

- Salinas.
- Alcalinas.
- Litio.
- Níquel-cadmio.
- Níquel-metal de hidruro.
- Pilas botón de mercurio.
- Pila botón de litio.

Figura 2.26. Grupo electrógeno accionado por motor diesel.

Pilas ecológicas son aquellas que apenas contienen mercurio y otros productos que pueden contaminar el medio ambiente.

Hay pilas que pueden ser recargables.

Cuidados que debemos prestar en el uso de las pilas:

- No dejar las pilas en los receptores cuando no se usan.
- Sustituir las pilas por otras de las mismas características.
- No tirar las pilas agotadas a la basura.
- Evacuar las pilas a los puntos de recogida.
- No manipular el interior de las pilas.
- Hay pilas que contienen mercurio y son altamente contaminantes.
- Utilizar en lo posible pilas verdes.

2.9. Análisis de las diferentes energías

Como se viene diciendo, España tiene un problema muy grave con la falta de energías, lo que da lugar a que el 80% de la misma provenga de importación.

En este capítulo hemos analizado las principales energías utilizadas, para que tengamos una visión general antes de iniciar el estudio de las energías alternativas, para que nos permita analizar con claridad, la conveniencia o no de unas energías sobre las otras, sean renovables o no renovables, contaminantes o no contaminantes.

Para analizar el valor real de un tipo de energía, se deben considerar diferentes parámetros, como:

a) Respecto a la energías básicas

A tener en cuenta:

- Procedencia (interior o exterior).
- Coste de la materia.
- Estabilidad de los precios (sin grandes fluctuaciones)
- Seguridad de un suministro continuo.
- Emisión de contaminantes en su proceso de transformación.

Figura 2.27. Pilas eléctricas (generación química)

Tabla 2.35. Principales energías básicas.

Energías	
Petróleo	• Imprescindible. • Necesario para la automoción, la industria y para la obtención de muchos subproductos. • Energía no renovable. • Energía contaminante (efecto invernadero). • Se importa.
Carbón	• Necesario para la industria y para la generación de electricidad. • Una parte del carbón consumido es de importación.
Gas natural	• Imprescindible. • Necesario para la industria, los servicios, la generación de electricidad y los hogares. • Energía contaminante (efecto invernadero). • Se importa.
Uranio	• Diversas aplicaciones. • Generación de electricidad. • Riesgo de accidentes varios, muy peligrosos. • Se importa.

b) Respecto a la electricidad:

Tengamos en cuenta, que la electricidad es una energía transformada desde otra fuente de energía.

- Tipo de fuente de energía básica.
- Procedencia de la energía básica (interior o exterior).
- Tipo de central.
- Generación continua o discontinua.
- Niveles de contaminación o riesgo.
- Coste de un MW.
- Vida útil de la central.

Tabla 2.35. Principales procedimientos de generación eléctrica.

Procedimiento	Características del proceso
Térmico	• Consumen carbón y gas natural. • Suministro exterior. • Buena eficiencia, con buen precio para el MW producido. • Suministro del combustible, incierto. • Oscilación en el precio. • Generación continua, en función de la demanda. • Contaminan (efecto invernadero).

Procedimiento	Características del proceso
Nuclear	• Consumen uranio. • Suministro exterior. • Muy buena eficiencia, con buen precio para el MW producido. • Suministro más seguro. • Precio más regular. • Generación continua, en función de la demanda. • Riesgo de fugas y problemas con los desechos.
Hidráulico	• Consumen agua. • Energía renovable. • Suministro interior. • Las posibilidades dependen la pluviometría. • Buena eficiencia, con muy buen precio para el MW producido. • Limitación de instalación de centrales. • No produce emisiones contaminantes.
Eólico	• Consume viento. • Energía renovable. • No tiene dependencia exterior. • No es una generación continua ni en función de la demanda. • No produce emisiones contaminantes.
Fotovoltaico	• Funciona con la luz solar. • Energía renovable. • No tiene dependencia exterior. • No es una generación continua ni en función de la demanda. • No produce emisiones contaminantes.
Biomasa	• Aprovecha recursos naturales. • Energía renovable. • Las emisiones contaminantes son menores que la de los combustibles fósiles.
Geotermia	• Aprovecha el calor interno de la Tierra.
Mareomotriz	• Aprovecha las mareas (desplazamientos de grandes masas de agua) y la fuerza de las olas.
R.S.U.	• Aprovecha los residuos sólidos urbanos.

2.10. Mix Energético de España para el año 2020

Las renovables representarán más del 20% del ***Mix Energético en España en 2020***. España prevé que en 2020 la participación de las renovables en nuestro país será de 22,7% sobre la energía final y un 42,3% de la generación eléctrica. Este superávit podrá ser utilizado, a través de los mecanismos de flexibilidad previstos en la Directiva de renovables, para su transferencia a otros países europeos que resulten deficitarios en el cumplimiento de sus objetivos.

La aportación de las energías renovables al consumo final bruto de energía en España se estima para el año 2020 en un 22,7%, casi tres puntos superior al objetivo obligatorio fijado por la Unión Europea para sus estados miembros, mientras que la aportación de las renovables a la producción de energía eléctrica alcanzará el 42,3%, con lo que España también superará el objetivo fijado por la UE en este ámbito (40%).

Los datos están contenidos en el anticipo del Plan de Renovables 2011-2020, enviado por el ***Ministerio de Industria, Turismo y Comercio*** a la ***Comisión Europea*** en cumplimiento de la propia Directiva comunitaria sobre la materia (2009/28/CE), que contempla objetivos obligatorios de energías renovables para la UE y para cada uno de los Estados miembros en el año 2020, y la elaboración por parte de éstos de planes de acción nacionales para alcanzar dichos objetivos.

Cada país miembro de la UE ha notificado a la Comisión, antes del 1 de enero de 2010, una previsión en la que se indica:

- Su estimación del exceso de producción de energía procedente de fuentes renovables con respecto a su trayectoria indicativa que podría transferirse a otros Estados miembros, así como su potencial estimado para proyectos conjuntos hasta 2020.
- Su estimación de la demanda de energía procedente de fuentes renovables que deberá satisfacer por medios distintos de la producción nacional hasta 2020.

Plan español de Energías Renovables 2011-2020

El Plan de Acción Nacional de Energías Renovables 2011-2020 se encontraba en el momento de la preparación de esta obra en proceso de elaboración, por lo que tanto el escenario como los objetivos para cada una de las tecnologías renovables durante este periodo pueden ser objeto de revisión. Para la formación del escenario del mapa energético en 2020, se ha tenido en cuenta la evolución del consumo de energía en España, el alza de los precios del petróleo en relación a los mismos en la década de los noventa y la intensificación sustancial de los planes de ahorro y eficiencia energética.

Las conclusiones principales del informe notificado a la Comisión Europea son las siguientes:

- En una primera estimación, la aportación de las energías renovables al consumo final bruto de energía sería del 22,7% en 2020—frente a un objetivo para España del 20% en 2020—, equivalente a unos excedentes de energía renovable de aproximadamente de 2,7 millones de toneladas equivalentes de petróleo (tep).
- Como estimación intermedia, se prevé que en el año 2012 la participación de las energías renovables sea del 15,5% (frente al valor orientativo previsto en la trayectoria indicativa del 11,0%) y en 2016 del 18,8% (frente a al 13,8% previsto en la trayectoria).
- El mayor desarrollo de las fuentes renovables en España corresponde a las áreas de generación eléctrica, con una previsión de la contribución de las energías renovables a la generación bruta de electricidad del 42,3% en 2020.

3 Introducción a las energías renovables

Energías renovables son aquellas que se producen de forma continua y son inagotables a escala humana, tales como: solar (luminosa y térmica), eólica, hidráulica, biomasa, geotérmica, mareomotriz y otras.
Respetan el medio ambiente y son menos contaminantes que las energías de origen fósil.
En la generación de electricidad las energías de origen fósil contaminan hasta 31 veces más que las energías renovables.
Lo ideal sería que las energías renovables fueran limpias y una alternativa real a las energías de origen fósil (contaminantes).

El girasol representa a las energías limpias y no contaminantes

3.1. Introducción

Las energías renovables fueron aprovechadas desde la antigüedad, por lo que no se trata de una aplicación reciente y que se haya puesto de moda, especialmente por grupos ecologistas.

Las energías renovables, como se ha dicho, fueron aprovechadas desde la antigüedad en variantes muy concretas, como son algunos de los ejemplos que se citan a continuación:

- Las hogueras para calentarse, cocinar y alumbrarse. Primeramente con ramas y leña (biomasa) y después también con carbón.
- Aplicando la arquitectura solar pasiva, desde 500 años d C, aprovechando el calor del Sol para calentar ciertos locales y dependencias de la casa en la época fría del invierno.
- Molinos accionados por la fuerza del viento, desde hace 3.000 años.
- Norias hidráulicas desde hace 2.000 años.
- Navegación por mares y ríos de barcos empujados por la fuerza del viento desde la antigüedad.
- Aprovechamiento de las mareas, el calor del Sol, la fuerza del agua de los ríos, etc.

Cuando llegó la máquina de vapor y posteriormente el motor eléctrico se abandonó el uso de ciertas energías alternativas. Ha sido a partir de los años 70 del pasado siglo XX, cuando por causa de problemas energéticos en el mercado mundial (precios y suministros), se ha visto un renacer de las energías renovables y alternativas, pero ahora con nuevas tecnologías que hacen que sus rendimientos sean mejores y más competitivos.

Están los que piensan que las energías renovables, por sí solas, son la solución a la demanda de energía que tiene el mundo moderno.

Las energías alternativas son buenas y conviene aprovecharlas al máximo de sus posibilidades y siempre que sean competitivas, con el fin de reducir el consumo de otras energías más contaminantes y que pueden ser escasas.

Tabla 3.1. Energías renovables limpias.

Fuente de energía	Procedencia
Solar (térmica)	Calor del Sol que llega a la Tierra.
Solar (luminosa)	Luz del Sol que llega a la Tierra.
Eólica	Fuerza del viento.
Hidráulica	Agua de los ríos y corrientes de agua dulce.
Mareomotriz	Mares y océanos.
Geotérmica	Calor del interior de la Tierra.
Undimotriz	Olas del mar y de los océanos.

3.2. Energías renovables

Se consideran energías renovables aquellas que se pueden reutilizar de nuevo y son inagotables.

Las energías renovables se clasifican atendiendo a sus características principales, como son su grado o nivel de contaminación a que den lugar en su lugar de procedencia, obtención y utilización.

Las energías renovables se presentan como una alternativa frente a las energías convencionales algunas de ellas muy contaminantes.

Los países con escasos recursos energéticos, buscan por todos los medios el aprovechamiento de las energías alternativas, para reducir la dependencia exterior.

Otro de los objetivos que se persigue con las energías renovables es reducir los altos niveles de contaminación atmosférica.

Tabla 3.4. Contribución de las energías renovables en el año 2010 al mercado energético español.

Energías	ktep	%
Biomasa	9.640	58,1
Hidráulica >50 MW	2.121	12,8
Hidráulica entre 10 y 50 MW	542	3,3
Minihidráulica <10 MW	594	3,6
Geotérmica	3	0,0
Solar termoeléctrica	180	1,1
Solar térmica	335	2,0
Solar fotovoltaica	19	0.0
Eólica	1.852	11,2
R.S.U.	681	4,1
Biocombustibles	500	3,0
Biogás	150	0,9
TOTAL	**16.639 ktep**	**100%**

Total de energía: 16.639 ktep (kilotoneladas equivalentes de petróleo)

En el año 2010, la demanda de energía eléctrica fue de 259.940 GWh, un 2,9% más que el año 2009. El año 2010 ha sido más lluvioso (el mejor desde 1997) y ventoso que otros años, lo que ha significado que el 14% de la energía eléctrica generada sea de procedencia hidráulica y el 16% de procedencia eólica (30% del total).

Tabla 3.2. Fuentes de energía renovables.

Tipo de energía	Procedencia
Energía eólica.	El viento.
Energía solar térmica.	El Sol.
Energía fotovoltaica.	El Sol.
Energía hidráulica.	El agua.
Energía proveniente de la biomasa.	Materias agrícolas diversas.
Energía mareomotriz.	El mar.
Energía geotérmica.	El calor del interior de la Tierra.
Energía de gradiente térmico oceánico.	El mar.
Otras energías.	Origen diverso.

Tabla 3.3. Energías renovables contaminantes.

Fuente de energía	Procedencia
Biomasa	A partir de materia orgánica (madera, vegetales, semillas, residuos y desechos varios).

Tabla 3.5. Evolución que han tenido las energías renovables en España en los últimos años.

	1990	1999	2000	2001	2002	2003	2004	2010
Minihidráulica (< 10 MW)	184	360	376	406	361	460	417	575
Hidráulica (> 10 MW)	2.019	1.886	2.159	3.122	1.627	3.073	2.297	2.536
Eólica	1	232	403	596	826	1.037	1.338	3.914
Biomasa	3.753	3.602	3.630	3.704	3.922	4.062	4.107	9.208
Biogás	-	114	125	134	170	257	275	455
Biocarburantes	-	0	51	51	121	184	228	2.200
R.S.U.	-	261	261	344	352	352	395	395
Solar térmica	22	28	31	36	41	47	54	376
Solar fotovoltaica	0	1	2	2	3	3	5	52
Solar termoeléctrica	0	0	0	0	0	0	0	509
Geotermia	3	5	8	8	8	8	8	8
TOTAL	**5.983**	**6.489**	**7.047**	**8.402**	**7.430**	**9.483**	**9.124**	**20.228**

Figuras 3.1. Pantanos de Aguilar de Campoo (Palencia) y Riaño (León).

Figuras 3.2. "Molinos" para el aprovechamiento de la fuerza del viento.

Plan de Acción Nacional de Energías Renovables
Objetivos año 2020
Energía total: 400.420 GWh

Energías alternativas:
Se consideran energías alternativas aquellas que se obtienen a partir de materias primas no agotables o que se reponen por vía natural.

A este grupo pertenecen las energías que tienen su origen en: la fuerza del viento (eólica), la fuerza del agua (hidráulica), las mareas y las olas en el mar y el calor del interior de la Tierra (geotérmica).

3.3. Principales energías renovables

Las principales energías renovables por su importancia en la cantidad de energía generada y por sus muchas aplicaciones son:

3.3.1. Energía hidráulica
Dentro de las energías renovables, la energía hidráulica es la segunda con mayor peso por la cantidad de energía producida, le precede la energía eólica.

La energía hidráulica se basa en el aprovechamiento de la fuerza del agua aplicada a turbinas para generar electricidad.

3.3.2. Energía eólica
Se trata de la primera forma de energía renovable más productiva, está por delante de la energía hidráulica. Tiene fuerte impacto ambiental y se encuentra sometida a las variaciones de la fuerza del viento.

3.3.3. *Energía fotovoltaica*
Otra de las tecnologías más extendidas es la fotovoltaica que aprovecha la energía de la luz del Sol (fotones). Se trata de una forma de conseguir energía que se puede suministrar a muchas aplicaciones que necesitan pequeñas fuentes de energía eléctrica.

3.3.4. Biomasa
Es un combustible formado por materia orgánica renovable de origen vegetal resultante de procesos de transformación natural o artificial en residuos biodegradables o cultivos energéticos.

Además de desprendernos de desechos urbanos, agrícolas e industriales se puede conseguir energía y combustible.

3.3.5. *Energía térmica*
La energía térmica que recibimos del Sol se recupera principalmente con el fin de producir agua caliente sanitaria (ACS), para calefacción y para máquinas de refrigeración.

No es suficiente para calentar toda la masa de agua, pero ayuda a reducir el gasto energético de las energías convencionales.

3.3.6. Energía del mar

Del mar se obtiene la energía de las mareas y de las olas, transformando la fuerza del agua y de las olas para generar energía eléctrica.

Figuras 3.3. Paneles fotovoltaicos para generar electricidad.

3.4. Ventajas de las energías renovables

Las energías renovables que tanta aceptación tienen en amplios sectores de la sociedad, tienen ventajas importantes, que se relacionan a continuación.

3.4.1 Relación de ventajas:

- Son energías gratuitas.
- Están en la naturaleza.
- Algunas de ellas son muy abundantes.
- Se pueden obtener en lugares remotos donde se necesita energía.

3.5. Inconvenientes de las energías renovables

Las energías renovables también tienen inconvenientes que es conveniente que se conozcan, para que se pueda opinar basándose en datos concretos.

Figuras 3.4. Masas arbóreas (biomasa).

3.5.1. Relación de inconvenientes:

- En algunos casos, las tecnologías no están suficientemente desarrolladas.
- Producen impactos visuales y ambientales importantes.
- Su poder de generación de energía es relativamente pequeño, para la superficie que ocupan.
- No tienen continuidad en la generación de la energía que se ve interrumpida por causas diversas.
- Tienen dificultad en almacenar la energía que producen.
- Suelen estar a expensas de ayudas oficiales, tanto en la inversión necesaria para su instalación, como sobre la energía que generan.
- Pueden causar encarecimiento de algunas materias primas (aceites, pan, azúcar y otros productos de primera necesidad.
- La energía obtenida suele ser más cara que la proveniente de las energías no renovables tradicionales.

3.6. Comparación entre energías renovables y no renovables

La tabla 3.6 recoge las ventajas e inconvenientes de las energías renovables y no renovables.

Figura 3.5. Paneles térmicos solares (captadores) sobre los tejados de viviendas.

Figuras 3.6. El mar.

Figura 3.7. Salinas en el Cabo de Gata (Almería). Aprovechamiento del calor del Sol.

Tabla 3.6. Ventajas e inconvenientes de estas dos formas de energías.

a) Energías renovables	
Ventajas	Inconvenientes
• No son contaminantes, o tienen bajo nivel de contaminación. • La energía base es gratuita. • Son muy polivalentes, al poder conseguir energía en lugares recónditos, y ser aplicada a casos puntuales. • A estas energías se las denomina *blancas* o *limpias y también verdes.* • Son bastante respetuosas con el medio ambiente. • Aprovechan energía que está ahí. • Reducen la dependencia respecto a los combustibles tradicionales de origen fósil. • Generan actividad y puestos de trabajo.	• Pueden tener impacto ambiental elevado. • Las instalaciones para recuperar la energía son relativamente costosas. • Algunas de ellas no tienen continuidad, se interrumpe el suministro (luz y calor del Sol, el viento, las mareas). Son variables y no previsibles, las mareas, sí. • Las instalaciones tienen bajos rendimientos. • Ocupan mucha superficie. • En algunos casos, su tecnología está en desarrollo. • El suministro de energía debe ser complementado con energía procedente de fuentes contaminantes.
b) Energías no renovables	
• Las principales energías no renovables son: petróleo, carbón, gas natural y nuclear. • Tienen buen rendimiento y continuidad en su aplicación. • Suponen más del 80% de la energía consumida por el hombre. • Tienen un amplio campo de aplicación (generación de electricidad, transporte, industria y servicios). • En la actualidad estas energías son básicas. • Se obtienen subproductos muy importantes para la industria.	• Son bastante, a muy contaminantes. • Estas formas de energía tienen fecha de caducidad. • En su proceso de transformación generan gran cantidad de CO_2. • Muchos países no son productores de estas materias, y las tienen que importar. • Periódicamente se producen crisis energéticas (dificultad de suministro y precios elevados), que dan lugar a incertidumbres en los mercados.

3.7. Plataforma solar de Almería

La Plataforma Solar de Almería (PSA), situada en el denominado desierto de Tabernas, en la provincia de Almería, lleva más de 25 años de actividad, realizando estudios y desarrollos sobre el aprovechamiento del calor proporcionado por el Sol.

En la actualidad, el centro está dirigido por CIEMAT (Centro de Investigaciones Energéticas, Medioambientales y Tecnológicos) perteneciente al Ministerio de Educación y Ciencia.

Es el mayor centro europeo de investigación en energía solar de concentración, que desarrolla estudios y aplicaciones.

Con esta Plataforma solar, España es país pionero en la investigación de la energía termoeléctrica y otras aplicaciones de la energía solar.

Los estudios son muy costosos debido a los materiales, al tiempo necesario en lograr objetivos, y a la incertidumbre de los resultados.

El calor se concentra por medio de espejos (helióstatos) que se mueven siguiendo el camino del Sol, para conseguir la máxima radiación térmica. También se utilizan otros sistemas de captación y concentración del calor.

Figura 3.8. Campo de ensayos de diferentes tipos de espejos (helióstatos) en PSA (Almería).

3.8. Instituciones que fomentan la generación de energías renovables

Varías son las instituciones que fomentan el empleo de energías renovables, bien sea a nivel nacional o autonómico, de acuerdo con la normativa en vigor.

AEE. Asociación Española Eólica.
APPA. Asociación de Productores de Energía Renovable.
AEH2. Asociación Española de Hidrógeno.
ÁPICE. Asociación Española de Pilas Combustibles.
ASIF. Asociación de la Industria Fotovoltaica.
IDAE. Instituto de la Diversificación y Ahorro de la Energía.
PSA. Plataforma Solar de Almería.
CIEMAT. Centro de Investigaciones Energéticas, Medioambientales y Tecnológicos.

3.9. Lo que representan las energías renovables

La utilización de energías renovables tiene un volumen muy pequeño de aportación al global de la energía total consumida, si exceptuamos las energías hidráulica y eólica (en España), tal como podemos apreciar en la tabla 3.7.

Consumo de carburantes a nivel mundial

Tabla 3.7. Estimación de la energía consumida a escala mundial.

Producto energético	% sobre el total
Petróleo	37,5
Carbón	25,5
Gas	23,5
Hidroelectricidad	6,5
Electronuclear	6 [1]
Eólica, solar, geotérmica y biomasa	1 [1]

[1] Como puede apreciarse, las energías renovables sólo suponen el 7,5% de la energía mundial consumida. La suma de las energías eólica, solar (térmica y fotovoltaica), geotérmica y biomasa, escasamente llegan al 1% del total.

3.10. Importancia de las fuentes de energía renovable

El 20% de la electricidad generada en el mundo [1] tiene su origen en fuentes de energía renovables, y de esta, el 90% es de origen hidráulico.

- 5,5% procede de la biomasa.
- 1,5% de la geotérmica.
- 0,5 de la eólica.
- 0,05% de la solar.

[1] Similar porcentaje para España, pero con un reparto diferente.

Tabla 3.8. Reparto en porcentaje de las energías renovables en el mundo.

Energías renovables	Porcentaje
Biomasa	2,9%
Eólica	1,9%
Hidráulica	1,4%
RSU	0,5%
Biocarburantes	0,4%
Biogás	0,2%
Fotovoltaica	0,2%
Solar térmica	0,1%
Geotérmica	0,01%
Solar termoeléctrica	0,004%

Nota:

Las energías renovables suponen en torno al 40% del total y corresponden a las cuadrículas sombreadas en la tabla.

Tabla 3.9. Relación de energías y su participación en la generación de energía eléctrica.

Energías	GWh	%
Gas natural	141.746,68	35,4
Eólica	79.283,16	19,8
Nuclear	55.658,38	13,9
Carbón	33.635,28	8,4
Hidráulica	33.234,86	8,3
Solar termoeléctrica	15.215,96	3,8
Solar fotovoltaica	14.415,12	3,6
Resto de renovables	10.410,92	2,6
Productos petrolíferos	8.809,24	2,2
Centrales hidráulicas de bombeo	8.008,40	2

4 El Sol fuente de energía y de vida

El Sol además de proporcionar a la Tierra luz y calor, da lugar a que se produzcan agentes atmosféricos, como el viento (energía eólica) por las diferencias de presiones y temperaturas en diferentes zonas y que junto con la Luna se produzcan las mareas, las olas y las corrientes marinas.
Colabora al ciclo del agua con la evaporación y la formación de nubes que el viento desplaza, dando lugar a las lluvias, nieve, granizo, rocío.
Participa en el desarrollo de las plantas (fotosíntesis) con la luz y el calor (biomasa).
La Tierra tiene vida gracias a la energía que nos proporciona el Sol.

Puesta de Sol en Punta Humbría (Huelva)

4.1. Introducción

El Sol es fuente de energía y de vida para nosotros y nuestro planeta Tierra.

4.1.1. Importancia del Sol

En las culturas antiguas el Sol era considerado un gran dios, ya que de él venía la vida, cosa que es verdad. Sin el Sol no sería posible la vida sobre nuestro planeta.

En las mitologías, el Sol siempre estaba considerado como un elemento importante en la vida de los hombres. Los antiguos griegos tenían al Sol como el dios Helios. Para los romanos era el dios Apolo.

Los egipcios lo consideraban el dios principal (res o Amun-amun-Ra). El Sol simboliza el oro y la Luna la plata.

Para muchas culturas, el Sol era el dios más importante de su Olimpo.

En las culturas precolombinas, el Sol era una divinidad muy importante, igual que para ciertas culturas orientales.

4.1.2. El Sistema Solar

El Sistema Solar se formó hace 4.600 millones de años y se estima que todavía tenga energía para otros 5.000 millones de años más, sin reducir su emisión de energía de forma apreciable.

El Sistema Solar esta constituido por:

- El Sol (centro del sistema) que tiene luz propia.
- Ocho planetas: Mercurio, Venus, La Tierra, Marte, Júpiter, Saturno, Urano y Neptuno.

Algunos de los planetas son rocosos y pequeños, y con densidad alta: Mercurio, Venus, la Tierra, Marte, y su giro es lento. Júpiter, Saturno, Urano y Neptuno, son gigantes gase*osos*, muy grandes y ligeros, formados de gas y hielo, y su giro es rápido.

El sistema Solar forma parte de una galaxia, y gira alrededor del núcleo galáctico.

4.2. El Sol

El Sol es una estrella que está en el centro del Sistema Solar.

Ejerce una fuerza muy grande de atracción gravitatoria sobre todos los planetas del sistema, que los hace desplazarse en su entorno.

El Sol emite luz y calor, elementos imprescindibles para la vida sobre la Tierra. Es la principal fuente de energía y almacena el 99% del total del Sistema Solar.

4.2.1. Características principales del Sol

En este apartado se recogen las características del Sol y que afectan directamente a la materia objeto de estudio (luz, calor, biomasa, etc.).

Tabla 4.1. Resumen de las principales características del Sol.

Características	Valores
Edad	4.600 millones de años.
Período de rotación alrededor del núcleo de la galaxia	225.000.000 años.
Diámetro	1.391.980 km
Volumen	$1{,}412 \times 10^{27}$ m^3 (1.300.000 veces el volumen de la Tierra).
Masa	$1{,}99 \times 10^{30}$ kg (332.946 veces la masa de la Tierra).
Densidad	150 veces la del agua.
Giro del Sol	El Sol gira una vez cada 27 días cerca del ecuador, pero una vez cada 31 días más cerca de los polos.
Temperaturas	En el centro: 16.000.000 K En la corona: 1.000.000 K En la superficie: 5.000 K
Distancia desde la Tierra	Mínima: 147.100.000 km Media: 150.000.000 km Máxima: 152.100.000 km
Energía	$3{,}83 \times 10^{26}$ J/s
Energía recibida sobre la atmósfera exterior de la Tierra	1.367 W/m^2
Velocidad de la luz	300.000 km/s

4.2.2. Composición química del Sol

- 71% de hidrógeno.
- 27% de helio.
- 2% elementos pesados (al menos 70).

Estas materias se generan en las profundidades del Sol.

Tabla 4.2. Radiaciones del Sol.

Radiación	Longitud de onda (micras)	Componente energético
Ultravioleta	$< 0{,}38$	7%
Visible	0,38 – 0,76	47%
Infrarrojo	$> 0{,}76$	46%

Nota:

La longitud de onda que transporta mayor energía es de 47 micras, y está dentro del espectro de radiación visible.

4.2.3. Luz

Es una manifestación de la energía. La encontramos natural, que corresponde a la que irradia el Sol, o artificial como la que emiten los diferentes tipos de lámparas.

La luz se transmite a frecuencias y longitud de onda diferentes, resultando luz visible y luz invisible. Las radiaciones visibles se caracterizan por ser capaces de estimular el sentido de la vista y estar comprendidas dentro de una franja de longitud de onda muy estrecha, aproximadamente entre 380 y 780 nm (namómetros).

Tabla 4.3. Zonas de espectros de la luz.

Zona no visible (espectro)	Zona visible para el hombre ⇐ (espectro visible) ⇒	Zona no visible (espectro)
INFRARROJOS	**380 y 760 nm**	**ULTRAVIOLETAS**

(1 mμ (milimicra) = 1 nm (nano-metro) = 10^{-9} m).

Las diferentes longitudes de onda nos permiten apreciar los objetos y los colores.

4.2.4. Constante solar

Cantidad de energía (perpendicular a la superficie) que llega al exterior de la atmósfera terrestre, directamente desde el Sol, por unidad de superficie y unidad de tiempo.

La radiación media según World Radiation Center es de 1.367 W/m^2.

La radiación media solar según la NASA es de 1.353 W/m^2.

Figura 4.4. Zonas del espectro visible correspondiente a los distintos colores.

Color	Longitud de onda
Violeta	380 a 450 nm
Azul	450 a 490 nm
Verde	490 a 550 nm
Amarillo	550 a 590 nm
Naranja	590 a 630 nm
Rojo	630 a 760 nm

4.3. El Sol fuente de energía y de vida

Gracias al Sol y la atmósfera que nos rodea, hay vida en la Tierra. El Sol nos suministra luz y calor imprescindibles para el desarrollo de las plantas, los animales, y el hombre; además del agua, y la formación de minerales y productos energéticos.

El Sol nos proporciona además, energía constante, y que es posible aprovecharla con instalaciones adecuadas, para su conversión en energía eléctrica y calorífica.

Tabla 4.5. Los dispositivos de aprovechamiento de la energía del Sol.

Energía del Sol	Dispositivo de transformación
Calor (calentar agua)	Panel solar.
Luz (generar electricidad)	Panel o módulo fotovoltaico.
Luz y calor	La propia naturaleza.
Luz y calor	Biomasa

El Sol nos proporciona luz y calor, dos energías fundamentales para que haya vida sobre el planeta Tierra, dado que proporcionan unos niveles de temperatura que permiten la vida sobre la tierra para las plantas, los animales y el hombre.

Que se produzca el fenómeno de la evaporización del agua (nubes, lluvia, nieve, ríos, lagos y mares.).

Que haya atmósfera en la que se den movimientos del aire (viento), al darse diferentes temperaturas del mismo al calentarse por el calor del Sol.

Proceso de fotosíntesis (plantas).

La luz y el calor además de ser imprescindibles para la vida, también proporcionan energía por medio de paneles fotovoltaicos (generar electricidad) a partir de la luz, y paneles solares (calentar agua) a partir del calor.

Los ríos nos proporcionan energía hidráulica.

El viento (energía eólica) tiene su aprovechamiento a partir de la generación eléctrica eólica y otras aplicaciones.

La naturaleza nos proporciona energía a partir de la biomasa.

Las plantas y su fosilización posterior durante millones de años, han dado lugar a los combustibles fósiles (carbón, petróleo y gas natural).

Figura 4.1. Bonitas flores y paisaje que nos suministra la naturaleza por efectos de la luz (fotosíntesis) y del calor.

4.3.1. Muestras de la influencia del Sol en la Tierra

En las figuras 4.1 y 4.2 se representa la influencia positiva del Sol en el desarrollo de la vida y actividad sobre el planeta Tierra, aprovechando las energías que nos suministra.

4.3.2. Energías que tienen su origen en efectos del Sol

El Sol no se limita al suministro de las energías luminosa y térmica, sino que por la influencia de estas dos energías nos suministran otras energías adicionales, como son:

- La lluvia y la nieve y con ellas los ríos que nos proporcionan ***energía hidráulica***, para diferentes aplicaciones, especialmente electricidad.
- El viento (***energía eólica***), que con su fuerza mueve los aerogeneradores, molinos y barcos.
- La ***fuerza de las olas*** (energía undimotriz), con la que se mueven turbinas y estas a alternadores.
- Las ***mareas*** (energía mareomotriz), aprovechando las diferencias (entre pleamar y bajamar y viceversa) de niveles, para a través de una turbina, generar electricidad.
- La ***biomasa***, a partir de la cual se genera calor con fines diversos.

Estas energías se estudian a lo largo de esta obra.

Figura 4.2. Animales que comen hierba, vegetación y agua.

4.4. Luz emitida por el Sol hacia la Tierra

La luz es una radiación electromagnética y es una forma de energía.

La velocidad de la luz es de 300.000 km/s.

4.4.1. Camino que sigue la luz que sale del Sol

La luz que nos envía el Sol sufre pérdidas en el camino:

- 50% es reflejada por las nubes y la atmósfera.
- 40% se pierde por reflexión especialmente por las superficies del agua de los mares y océanos.
- El 10% útil, lo emplean las plantas y otros elementos de la naturaleza.

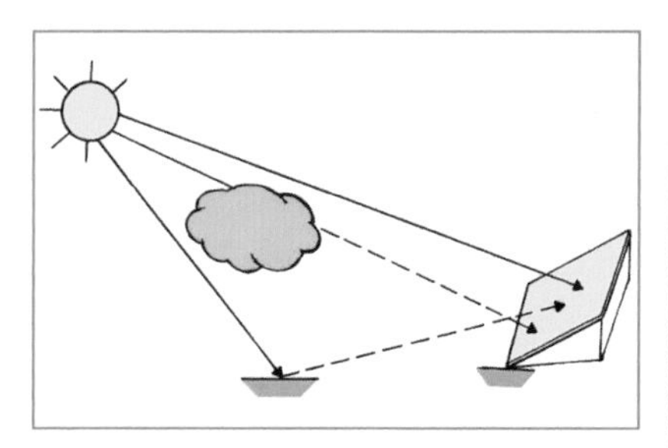

Figura 4.3. Radiación que llega a la superficie de la Tierra.

Por ejemplo, la radiación que llega a un panel fotovoltaico puede ser:

- Radiación directa (sin interferencia).
- Radiación difusa (con interferencia de nubes o niebla).
- Radiación albedo (reflejada).

4.4.2. Radiación solar global

Los rayos solares llegan a la superficie de la Tierra después de pasar por la atmósfera. Una parte de la radiación se devuelve al espacio exterior otra parte llega directamente, otra difusa tras pasar por las nubes y otro parte llega reflejada tal como se aprecia en la figura 4.3.

4.4.3. Potencia luminosa que llega a la superficie de la Tierra

En día soleado y a pleno sol, la energía irradiada sobre la superficie de la tierra es de aproximadamente 1 kW/m^2. Esta energía puede ser recuperada en parte, y transformada en energía eléctrica por medio de paneles o módulos fotovoltaicos.

4.5. Calor emitido por el Sol hacia la Tierra

Empezaremos por el calor que llega a la superficie terrestre, que lo podemos aprovechar como fuente de energía que es, para calentar agua, e incluso para generar vapor de agua cuando se concentra sobre sistemas especiales.

El Sol es el principal emisor de energía hacia el planeta Tierra. La energía solar llega a través de la atmósfera terrestre que hace de regulador.

Al exterior de la atmósfera llegan 1.353 W/m^2 (constante solar). La energía solar que llega a la superficie terrestre no es uniforme debido a que a lo largo de las 24 horas del día, se dan día y noche, días claros y días cubiertos, estaciones y otras circunstancias, la potencia media que llega es de 342 W/m^2.

4.6. Aprovechamiento de las energías del Sol

Las dos formas de energía que transmite el Sol (luz y calor), son aprovechadas por nuestro planeta para:

- Que tengamos vida sobre la tierra.
- Proporcionarnos calor.
- Participar en la fotosíntesis de las plantas.
- Que haya agua sobre la tierra (evaporización).
- Ayudar al crecimiento de las plantas y la formación de las materias.
- Participar en el clima (frío, calor, lluvia, viento, etc.).

4.6.1. Principales aprovechamientos de la energía.

Los principales aprovechamientos respecto a la consecución de energía, son:

a) El calor

El calor se recupera a través de paneles o captadores solares térmicos, y por medio de construcciones especiales.

b) Tipos de paneles de recuperación del calor solar en función de la temperatura que alcanza el agua.

· Paneles solares de baja temperatura

Se trata de instalaciones convencionales utilizadas para calentar agua o para ser parte de una instalación de calefacción. Su temperatura no supera los 90 °C.

Este es el tipo normal de paneles e instalaciones que se instalan en viviendas, piscinas, hoteles, edificios, etc.

· Paneles solares de media temperatura

Paneles preparados para concentrar el calor y en los que la temperatura máxima que alcanza el fluido está comprendida entre 80 y 250 °C.

· Paneles solares de alta temperatura

Requieren dispositivos especiales para concentrar el calor por medio de espejos (heliόstatos) y su temperatura es superior a los 250 °C.

Dependiendo de la temperatura y presión el agua líquida pasa a ser vapor de agua.

· Instalación para aprovechamiento térmico concentrado

Las centrales solares térmicas aprovechan el calor de la radiación solar para calentar un líquido concentrando el calor a elevada temperatura, por medio de una serie de espejos (heliόstatos), convenientemente dispuestos y orientados, en una superficie pequeña, generando vapor que se aplica a una turbina de vapor y por su medio se acciona un generador eléctrico. Es un equivalente a una central termoeléctrica, pero que no tiene ningún tipo de contaminación y la energía base es gratuita.

Se trata de instalaciones puntuales y en ensayo, que no están generalizadas, por lo que no suponen todavía una aportación importante de energía.

En España hay un centro experimental en la provincia de Almería.

La torre de ensayo de prototipos sobre los que incide el calor reflejado por los espejos (heliόstatos), y concentración de calor sobre un conducto de agua para producir vapor de agua.

c) La luz

La energía de la luz se recupera a través de paneles fotovoltaicos.

Se utilizan paneles o módulos fotovoltaicos que contienen células fotoeléctricas en las que se genera electricidad (corriente continua).

Figura 4.4. Paneles solares térmicos planos de baja temperatura, para calentar agua sanitaria.

Figura 4.5. Plataforma Solar de Almería perteneciente al CIMAT.

Figura 4.6. Ejemplos de aprovechamiento de las energías procedentes del Sol.

4.7. Calorimetría

La calorimetría es la parte de la física dedicada a la medida de las cantidades de calor que intervienen en distintos fenómenos. Medición del calor que se desprende o absorbe en los procesos biológicos, físicos o químicos.

El calor es la energía en tránsito (movimiento) entre dos cuerpos o sistemas como efecto de una diferencia de temperatura entre ellos.

También definimos como calor a la sensación que se experimenta al recibir directa o indirectamente la radiación solar o aproximarnos a un foco de calor.

4.7.1. Calor específico

Es la cantidad de calor necesario para elevar 1 °C la temperatura de una unidad de un cuerpo o materia determinada.

El calor específico se da en calorías (cal) por gramo (g) y por grado centígrado (1 °C).

4.7.2. Cantidad de calor

Es el calor necesario para elevar la temperatura de un cuerpo de un valor determinado y viene dado por la fórmula siguiente:

$$Q = m \cdot c\,(t - t_o)$$

$$Q = m \cdot c \cdot \Delta t$$

$$\Delta t = (t - t_o)$$

Q – Cantidad de calor, en calorías (cal).
M – Masa, en gramos (g).
c – Calor específico del cuerpo.
t_o – Temperatura inicial, en °C.
t – Temperatura final, en °C.

Tabla 4.6. Calores específicos medios entre 0 y 100 °C.

Sustancia	Calor específico Cal/g · °C	Sustancia	Calor específico Cal/g · °C
Agua	1,0000	Hierro	0,1115
Aluminio	0,2170	Mercurio	0,0333
Calcio	0,1705	Plata	0,0555
Cinc	0,0925	Plomo	0,0314
Cobre	0,0952	Latón	0,0951
Estaño	0,0552	Hielo	0,4759
Glicerina	0,565	Níquel	0,106
Acetona	0,528	Latón	0,093

4.7.3. Equivalencia mecánica del calor

La energía de un julio (J), genera 0,24 calorías (cal).

1 J = 0,24 cal.

1 kcal = 1.000 cal = 4,180 J

1 cal = 4,18 J

1 kWh = 1.000 W x 3.600 s = 3.600.000 J

1 kWh = 0,24 x 3.600.000 = 864.000 cal = 864 kcal

Ejemplo de aplicación:

Determinar el calor necesario en kWh, para calentar 100 litros de agua en un calentador (termo) para pasar de 20 a 60 °C.

- Masa (m): *m = 100 l = 100 kg = 100.000 g; m' = 100 kg*
- Diferencia de temperatura $(t - t_o)$: $(t - t_o) = 60 - 20 = 40$ °C

$Q = m \cdot c\,(t - t_o) = 100.000 \times 1\,(60 - 20) = 4.000.000$ cal

$Q = m' \cdot c\,(t - t_o) = 100 \times 1\,(60 - 40) = 4.000$ kcal

4.8. Escalas de temperatura

Se emplean tres escalas preferentemente, que son:

- Escala absoluta.
- Escala centígrada.
- Escala Fahrenheit.

Tabla 4.7. Diferentes escalas de temperatura.

Para pasar de:		a:	Fórmula
Grado Centígrado	°C	Grado Kelvin	$K = °C + 273$
Grado Kelvin	K	Grado Centígrado	$°C = K - 273$
Grado Centígrado	°C	Grado Centígrado	$°C = \frac{°F - 32}{180} \cdot 100$
Grado Fahrenheit	°F	Grado Fahrenheit	$°F = \frac{°C \cdot 180}{100} + 32$

Tabla 4.8. Equivalencias entre las tres escalas de temperatura.

	Escala absoluta	Escala centígrada	Escala Fahrenheit
Ebullición del agua	373 K	100 °C	212 °F
Congelación del agua	273 K	0 °C	32 °F
Cero absoluto	0 K	-273 °C	-460 °F

4.9. Código Técnico de Edificación y la energía solar térmica

Hay que buscar por todos los medios reducir el consumo de energías tradicionales finitas y contaminantes. El aprovechamiento del calor que nos proporciona gratuitamente el Sol está recogido en la normativa actual, que es muy clara en este tema de las energías renovables, para reducir consumo de energías tradicionales, tal como se recoge en Real Decreto 314/2006.

4.9.1. Contenido del Real Decreto 314/2006

Real Decreto 314/2006, de 17 de marzo, por el que se aprueba el **Código Técnico de la Edificación (CTE).**

Artículo 15. ***Exigencias básicas de ahorro de energía***

Exigencia básica HE 1: Limitación de demanda de energía.

Exigencia básica HE 2: Rendimiento de las instalaciones térmicas.

Exigencia básica HE 3: Eficiencia energética de las instalaciones de iluminación.

Exigencia básica HE 4: Contribución solar mínima de agua caliente sanitaria.

Exigencia básica HE 5: Contribución fotovoltaica mínima de energía eléctrica.

(Se estudia el capítulo 5).

Tabla 4.9. Relación entre la demanda de agua caliente sanitaria ACS) dentro de la edificación y la contribución energética obligatoria se recoge en la tabla .

Demanda de agua (litros/dia)	% de contribución mediante energías renovables [1]				
	I	II	III	IV	V
50-5.000	30	30	50	60	70
5.000-6.000	30	30	50	65	70
6.000-7.000	30	35	61	70	70
7.000-8.000	30	45	63	70	70
8.000-9.000	30	52	63	70	70
10.000-12.500	30	65	70	70	70
12.500-15.000	30	70	70	70	70
15.000-17.500	35	70	70	70	70
17.500-20.000	45	70	70	70	70
> 20.000	52	70	70	70	70

[1] Se considera que el sistema de apoyo necesario será mediante gasóleo, propano, gas natural u otros.

Tabla 4.10. Relación entre la demanda de agua caliente sanitaria (ACS) dentro de la edificación y la contribución energética obligatoria en el supuesto de que el apoyo se realice con electricidad.

Demanda de agua (litros/dia)	% de contribución mediante energías renovables [2]				
	I	II	III	IV	V
50-1.000	50	60	70	70	70
1.000-2.000	50	63	70	70	70
2.000-3.000	50	66	70	70	70
3.000-4.000	51	69	70	70	70
4.000-5.000	58	70	70	70	70
5.000-6.500	62	70	70	70	70
> 6.000	70	70	70	70	70

[2] Se considera que el sistema de apoyo es con electricidad.

Tabla 4.11. En el caso de piscinas cubiertas.

Demanda de agua (litros/dia)	% de contribución mediante energías renovables				
	I	II	III	IV	V
	30	30	50	60	70

Así por ejemplo, en la zona III, la contribución de energía renovable será del 50%.

4.10. Energía y horas de sol en España

En la figura 4.6 se muestran para cada provincia españolа la energía recibida en un año y las horas de sol en un año.

Para cada una de las provincias:

- Cifra superior: energía en kWh que incide por m² de superficie horizontal en un año.
- Cifra inferior: horas de sol en un año.

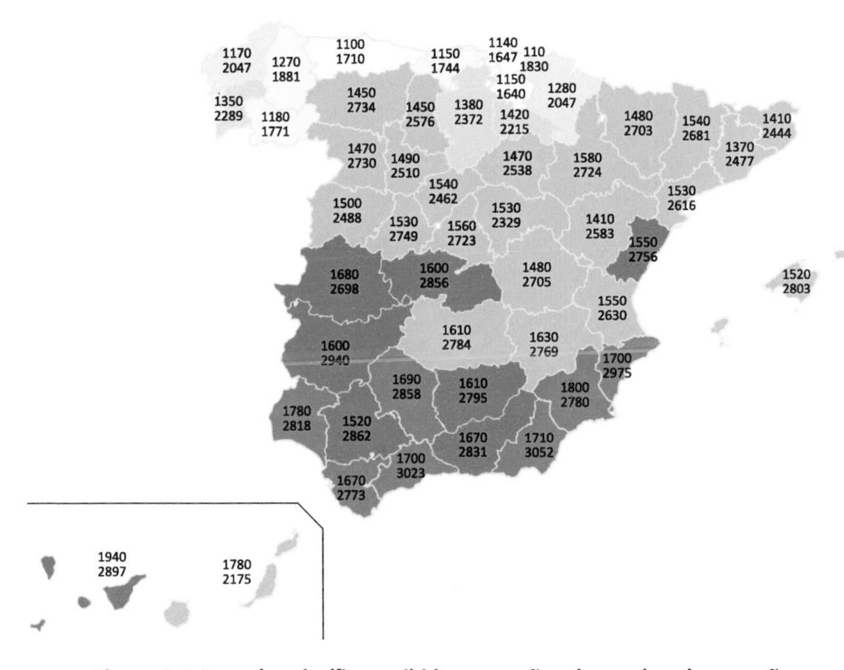

Figura 4.7. Energía calorífica recibida en un año y horas de sol en un año.

Nota:

El consumo por persona de agua caliente sanitaria a 45 °C, se estima en 40 litros/día

5 Código Técnico de la Edificación (CTE)

El Código Técnico contiene un Documento Básico de Ahorro de Energía donde se establecen las exigencias básicas en eficiencia energética y energías renovables que deben cumplir los nuevos edificios y los que se reformen o rehabiliten.
El Documento Básico consta de las siguientes secciones que tocan muy directamente a la materia que se estudia en esta obra:

- *HE.1: Limitación de demanda energética (calefacción y refrigeración).*
- *HE.2: Rendimiento de las instalaciones térmicas.*
- *HE.3: Eficiencia energética de las instalaciones de iluminación.*
- *HE.4: Contribución solar mínima de agua caliente sanitaria.*
- *HE.5: Contribución fotovoltaica mínima de energía eléctrica.*

Edificios en construcción

5.1. Introducción

El nuevo ***Código Técnico de la Edificación (CTE)*** señala la obligación de recuperar una parte de la energía que nos llega desde el Sol, en determinados casos y construcciones. El contenido de la CTE resulta muy importante en lo que respecta a esta obra, por la reglamentación del aprovechamiento energético solar como es el calor y la luz solar en diferentes elementos de construcción.

Ámbito de aplicación:

· Edificios de nueva construcción.

· Modificaciones, reformas o rehabilitaciones de edificios existentes con una superficie útil superior a 1.000 m^2 donde se renueve más del 25% del total de sus cerramientos.

Se incluyen en el campo de aplicación:

a) Aquellas edificaciones que por sus características de utilización deben permanecer abiertas.
b) Edificios y monumentos protegidos oficialmente por ser parte de un entorno declarado o en razón de su particular valor arquitectónico o histórico, cuando el cumplimiento de tales exigencias pudiese alterar de manera inaceptable su carácter o aspecto.
c) Edificios utilizados como lugares de culto y para actividades religiosas.
d) Construcciones provisionales con un plazo previsto de utilización igual o inferior a dos años.
e) Instalaciones industriales, talleres y edificios agrícolas no residenciales.
f) Edificios aislados con una superficie útil total inferior a 50 m^2.

5.2. Contenido del Real Decreto 314/2006

Real Decreto 314/2006, de 17 de marzo, por el que se aprueba el **Código Técnico de la Edificación (CTE).**

Artículo 15. **Exigencias básicas de ahorro de energía**

Exigencia básica HE 1: Limitación de demanda de energía.

Exigencia básica HE 2: Rendimiento de las instalaciones térmicas.

Exigencia básica HE 3: Eficiencia energética de las instalaciones de iluminación.

Exigencia básica HE 4: Contribución solar mínima de agua caliente sanitaria.

Exigencia básica HE 5: Contribución fotovoltaica mínima de energía eléctrica.

5.3. Instalación de energía solar obligatoria en nuevos edificios

Desde el pasado mes de septiembre del 2006, en que entró en vigor el nuevo Código Técnico de la Edificación (CTE), todas las nuevas construcciones en las que se emplee agua caliente (viviendas, hospitales, hoteles, polideportivos, etc.) deberán instalar sistemas solares térmicos. En el resto de edificios será obligatorio el uso de la energía solar fotovoltaica para producir electricidad, y que podrá venderse a una compañía eléctrica.

La Unión Europea ha marcado como objetivo para el año 2010, lograr que el 12% del consumo energético proceda de fuentes de energía renovable.

Las instalaciones aunque son costosas en su origen, se amortizan a los pocos años y pueden tener subvenciones de las Administraciones Públicas para su instalación.

La vida útil de un panel solar térmico está en torno a 25-35 años.

La energía solar térmica es un apoyo a las energías convencionales utilizadas y que suponen un ahorro de hasta el 60% del coste total de las mismas.

5.4. Mapa climático de España. Radiación solar global

Se definen 5 zonas climáticas según la radiación solar media en España.

Zonas climáticas solares:

El número de paneles a instalar para conseguir la misma potencia, dependerá del lugar de España donde se instalen. Para facilitar su cálculo exacto para las diferentes partes del país, el territorio está dividido en zonas climáticas solares.

Tabla 5.1. Zonas climáticas de España en función de la radiación solar.

Zona	Horas de sol h/año	Insolación global kWh/m^2
II	1.500 – 1.700	Aprox. 1.030
III	1.700 – 1.900	Aprox. 1.150
IV	1.900 – 2.100	Aprox. 1.230
V	2.100 – 2.300	Aprox. 1.370
VI	2.300 – 2.500	Aprox. 1.490
VII	Más de 2.500	Aprox. 1.610

Tabla 5.2. Radiación solar en función de la zona climática.

Zonas climáticas	Radiación solar global	
	MJ/m^2	KWh/m^2
I	H < 13,7	H < 3,8
II	13,7 ≤ H < 15,1	3,8 ≤ H < 4,2
III	15,1, ≤ H < 16,6	4,2 ≤ H < 4,6
IV	16,6 ≤ H < 18,0	4,6 ≤ H < 5,0
V	H > 18,0	H > 5,0

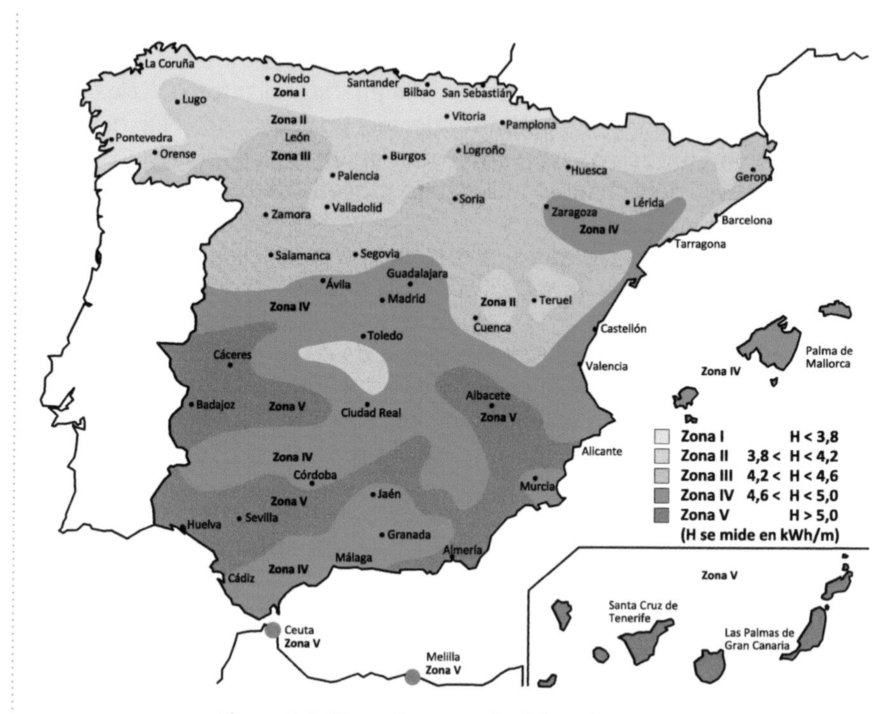

Figura 5.1. Mapa de zonas climáticas de España.

5.5. Exigencias básicas de Ahorro de Energía (HE)

1. El objeto del requisito básico “Ahorro de energía” consiste en conseguir un uso racional de la energía necesaria para la utilización de los edificios, reduciendo a límites sostenibles su consumo y conseguir asimismo que una parte de este consumo proceda de fuentes de energía renovable, como consecuencia de las características de su proyecto, construcción, uso y mantenimiento.
2. Para satisfacer este objetivo, los edificios se proyectarán, construirán, utilizarán y mantendrán de forma que se cumplan las exigencias básicas que se establecen en los apartados siguientes.
3. El documento básico “DB-HE Ahorro de energía” especifica parámetros objetivos y procedimiento cuyo cumplimiento asegura la satisfacción de las exigencias básicas y la superación de los niveles mínimos de calidad propios del requerimiento de ahorro de energía.

5.6. Exigencias básicas de ahorro energético

5.6.1. Exigencia básica HE 1: Limitación de demanda energética

Los edificios dispondrán de una envolvente de características tales que limite adecuadamente la demanda energética necesaria para alcanzar el bienestar térmico en función del clima de la localidad, del uso del edificio y del régimen de verano y de invierno, así como por sus características de aislamiento e inercia, permeabilidad al aire y exposición a la radiación solar, reduciendo el riesgo de aparición de humedades de condensación superficial e intersticiales que puedan perjudicar sus características y tratando adecuadamente los puentes térmicos para limitar las pérdidas o ganancias de calor y evitar problemas higrotérmicos en los mismos.

5.6.2. Exigencia básica HE 2: Rendimiento de las instalaciones térmicas

Los edificios dispondrán de instalaciones térmicas apropiadas destinadas a proporcionar el bienestar térmico de sus ocupantes, regulando el rendimiento de las mismas y de sus equipos. Esta exigencia se desarrolla actualmente en el vigente Reglamento de Instalaciones Térmicas en los Edificios, RITE, y su aplicación quedará definida en el proyecto del edificio.

5.6.3. Exigencia básica HE 3: Eficiencia energética de las instalaciones de iluminación

Los edificios dispondrán de instalaciones de iluminación adecuadas a las necesidades de los usuarios y a la vez eficaces energéticamente, disponiendo de un sistema de control que permita ajustar el encendido a la ocupación real de la zona, así como de un sistema de regulación que optimice el aprovechamiento de la luz natural, en las zonas que reúnan unas determinadas condiciones.

5.6.4. Exigencia básica HE 4: Contribución solar mínima de agua caliente sanitaria

En los edificios con previsión de demanda de agua caliente sanitaria o de climatización de piscina cubierta, en los que así se establezca en esta CTE, una parte de las necesidades energéticas térmicas derivadas de esa demanda se cubrirá mediante la incorporación en los mismos de sistemas de captación, almacenamiento y utilización de energía solar de baja temperatura adecuada a la radiación solar global de su emplazamiento y a la demanda de agua caliente del edificio. Los valores derivados de esta exigencia básica tendrán la consideración de mínimos, sin perjuicio de valores que puedan ser establecidos por las administraciones competentes y que contribuyan a la sostenibilidad, atendiendo a las características propias de su localización y ámbito territorial.

5.6.5. Exigencia básica HE 5: Contribución fotovoltaica mínima de energía eléctrica

En los edificios que así se establezca en esta CTE se incorporarán sistemas de captación y transformación de energía solar, en energía eléctrica, por procedimientos fotovoltaicos para uso propio o suministro a la red. Los valores derivados de esta exigencia básica tendrán la consideración de mínimos, sin perjuicio de valores más estrictos que puedan ser establecidos por las administraciones competentes y que contribuyan a la sostenibilidad, atendiendo a las características propias de su localización y ámbito territorial.

DB-HE: ***Documento Básico de Ahorro de Energía***

La exigencia básica HE 5 determina que los edificios de determinada superficie instalarán módulos de generación fotovoltaica para contribuir a generar electricidad y así reducir el consumo de electricidad generada por energías contaminantes, especialmente las que tienen su origen en combustibles fósiles.

Tabla 5.3. Superficie a partir de la cual se deben instalar elementos fotovoltaicos.

Tipos de uso	Límite de aplicación
Hipermercado	5.000 m^2 construidos
Multitienda y centros de ocio	3.000 m^2 construidos
Nave de almacenamiento	10.000 m^2 construidos
Edificios administrativos	4.000 m^2 construidos
Pabellones de recintos feriales	10.000 m^2 construidos
Hoteles y hostales	100 plazas
Hospitales y clínicas	100 camas

Tabla 5.5. Coeficiente climático (C).

Zonas climáticas	C
I	1
II	1,1
III	1,2
IV	1,3
V	1,4

5.7. Superficies afectadas por la exigencia básica HE 5

Contribución fotovoltaica mínima de energía eléctrica.

Usos o límites a partir de los cuales resulta de aplicación la exigencia.

a) Potencia pico a instalar en estos casos

$$P = C \cdot (A \cdot S + B)$$

P – Potencia pico a instalar, en kWp.
C – Coeficiente de uso, según tabla.
B – Coeficiente de uso, según tabla.
S – Superficie total construida, en m^2.

Tabla 5.4. Valores del coeficiente de uso (A y B).

Tipos de uso	A	B
Hipermercado	0,001875	-3,13
Multitienda y centros de ocio	0,004688	-7,81
Nave de almacenamiento	0,001406	-7,81
Edificios administrativos	0,001223	1,36
Pabellones de recintos feriales	0,001406	-7,81
Hoteles y hostales	0,003516	-7,81
Hospitales y clínicas	0,000740	3,29

5.8. Observaciones con carácter general

- La potencia mínima pico a instalar será de 6,25 kWp
- El inversor tendrá una potencia mínima de 5 kW.
- Respecto a la superficie (S) a considerar en el caso de edificios construidos en un mismo recinto será:
  - Mismo uso: suma de la superficie de todos los edificios del recinto.
  - Distintos usos: se aplicará la fórmula de la potencia aunque las superficies sean inferiores a las indicadas en la tabla correspondiente. La potencia pico mínima a instalar será la suma de las potencias pico de cada uso, siempre que sea superior a 6,25 kWp.

La potencia eléctrica mínima podrá disminuirse o suprimirse por causas justificadas, como son las siguientes:

- Cuando se cubra la producción eléctrica estimada que correspondería a la potencia mínima mediante el aprovechamiento de otras fuentes de energía renovable.
- Cuando el emplazamiento no cuente con suficiente acceso al sol por barreras externas al mismo y no puedan aplicar soluciones alternativas.

- En rehabilitación y construcción nueva de edificios, cuando existan limitaciones impuestas por la normativa urbanística aplicable.
- Cuando así lo determine el órgano competente en materia de protección histórico- artístico.

Figura 5.2. Vivienda equipada con paneles solares térmicos.

5.8.1. Consideraciones a tener en cuenta:

Según el Ministerio de la Vivienda, la implantación de sistema de ER [1] y otras medidas de reducción del consumo de energía que incluye el CTE supondrán un ahorro energético por edificio de entre el 30% y el 40%, y una reducción de emisiones de CO_2 de entre 40% y el 55% por el equivalente a la energía aprovechada.

Según los expertos, el aumento del costo de las viviendas por la colocación de sistemas ER supondrá entre un 3% y un 1%, lo que resulta perfectamente asumible.

[1] Sistemas ER – Sistemas de energías renovables.

6 Energía Solar Térmica

La energía solar térmica es fundamental para la vida de nuestro planeta, además, es muy abundante y llega a todas las partes, aunque de manera irregular. Es muy importante que la aprovechemos, y con más razón, en los lugares en que es más abundante.
A la superficie de la Tierra llaga en torno a 1.000 W/m2.
Los captadores de calor permiten la recuperación de esta energía, que es muy abundante en nuestro país, aplicándolo a las viviendas (agua caliente sanitaria y calefacción) y para otros fines como se estudian en este capítulo.

Puesta de sol

6.1. Evolución de esta tecnología

El aprovechamiento del calor que nos proporciona el Sol siempre ha sido una constante para los hombres de cada época, especialmente en la construcción de edificios para viviendas, buscando que el Sol irradie las habitaciones y se aproveche el calor y la luz natural el máximo número de horas diurnas posible.

También desde la antigüedad, se han realizado ensayos de concentración del calor como lo atestiguan los hechos conocidos de encender fuegos a distancia mediante espejos, que concentraban calor en un punto.

Entre los siglos XVII a XX se realizaron muchos experimentos, pero sin más objetivo que el espectáculo.

En 1953 se inician estudios para el posible aprovechamiento del calor del Sol, en algunas universidades de EE.UU. En 1954 se descubre la fotopila de silicio.

Figura 6.1. Helióstatos (espejos) que recogen y concentran el calor del Sol.

En 1973, con el encarecimiento del petróleo y las posteriores y casi continuas crisis de suministros energéticos se relanza la utilización de esta energía que está a nuestro alcance. Se trata de una tecnología relativamente reciente y que está en desarrollo.

La Administración fomenta el empleo de está tecnología en el Real Decreto 314/2006, de 17 de marzo, por el que se aprueba el **Código Técnico de la edificación (CTE)**.

Se trata de una energía cuyo suministro gratuito está asegurado para los próximos 5.000 millones de años. Es una fuente inagotable y limpia, cuya recuperación se encuentra con ciertos problemas, como son el de su suministro irregular (estaciones del año), dificultad de recuperación, e instalaciones caras y aspecto ambiental.

6.2. Clasificación de las instalaciones por su temperatura

6.2.1. La energía solar

De la radiación solar que incide sobre la superficie de la Tierra se consigue un aprovechamiento en forma de energía renovable y limpia. La potencia de irradiación solar se estima en aproximadamente 1.000 W/m^2.

Clasificación de las energías solares:

- **Energía pasiva**

Se aprovecha la energía pasiva de forma directa, sin ningún tipo de mecanismo, como lo hacen los animales, las plantas, nuestras viviendas, la naturaleza a través de la temperatura ambiente.

- **Energía solar térmica**

Se utiliza para producir agua caliente a baja temperatura, para agua caliente sanitaria (ACS) y fluido para calefacción, por medio de captadores térmicos.

- **Energía solar fotovoltaica**

Se utiliza para generar electricidad por medio de células fotoeléctricas que se excitan por los fotones contenidos en la radiación solar a través de paneles fotovoltaicos.

- **Energía solar termoeléctrica**

A partir de un fluido a alta temperatura (aceite térmico), producir vapor, y generar electricidad por medio de un alternador accionado por una turbina (ciclo termodinámico).

- **Energía solar híbrida**

Por medio de paneles híbridos se consigue generar electricidad y producir calor. También, combina la energía solar con la combustión de biomasa, combustibles fósiles, energía eólica o cualquier otra energía alternativa.

- **Energía eólica solar**

Estas instalaciones funcionan con el aire calentado por el Sol, que asciende por una chimenea, y que con su fuerza acciona unos ventiladores que están situados en su parte alta y que a su vez mueven los generadores para producir electricidad.

Las instalaciones que aprovechan el calor solar se pueden agrupar en tres grupos, atendiendo a la temperatura que alcanzan.

1. Baja temperatura (hasta 150 °C)

A este nivel de temperatura corresponden a instalaciones clásicas que utilizan paneles solares térmicos para calentar agua, distinguiendo dos colectores de placa plana, que son:

a) ***Colector selectivo***

Aprovecha muy bien el calor que recibe y puede llegar a alcanzar los 100 °C.

b) ***Colector no selectivo***

El aprovechamiento del calor es inferior al selectivo y en estos paneles el calor no supera los 80 °C.

2. Media temperatura (desde 150 a 600 °C)

Estas temperaturas se obtienen por semi-cilindros parabólicos de concentración que concentran el calor que le llega sobre un tubo con agua que se calienta a temperaturas muy elevadas que generan vapor de agua. Para aprovechar el calor del Sol, el conjunto concentrador de calor gira en torno al camino que describe el Sol.

Se emplea para proporcionar agua caliente para calefacción y para generar electricidad.

3. Alta temperatura (desde 600 a más de 2.000 °C)

Estas elevadas temperaturas se obtienen a través de helióstatos que concentran el calor recibido sobre un punto concreto o central de calor con el que se consiguen elevadas temperaturas que permiten generar vapor de agua con el que por medio de turbinas de vapor se genera electricidad.

Figura 6.2. Paneles térmicos, para calentar agua sanitaria a baja temperatura.

Figura 6.3. Instalación de captación de media temperatura.

Figura 6.4. Instalaciones CIEMAT en Tabernas (Almería). Torre de pruebas sobre la que se concentra calor que envían los espejos helióstatos, alta temperatura.

6.3. Captadores solares térmicos

Los paneles solares térmicos son los elementos mediante los cuales se aprovecha el calor que recibimos del Sol. También están las instalaciones que utilizan las llamadas bombas de calor para aprovechar el calor del ambiente, del agua o de la tierra.

6.3.1. Captadores o paneles solares térmicos

A los captadores también se les conoce como placas solares o paneles solares, que son elementos o cajas rectangulares que están constituidos por una tapa o cubierta transparente (cristal o plástico) sobre la que inciden los rayos solares y que recogen en el interior el calor por efecto invernadero, y que se aprovecha para calentar el agua circulante que pasa por una red de tubos. El calor recogido por el fluido térmico se utiliza para calentar el agua sanitaria contenida en un depósito acumulador por medio de un intercambiador que puede estar situado dentro del acumulador o fuera del mismo, como se presenta en los esquemas que se estudian a continuación.

El agua o fluido que recoge el calor está preparado con anticongelante para soportar las bajas temperaturas a que pueda estar sometido en períodos fríos.

En la instalación de placas solares se puede colocar uno o varios paneles.

Por lo general, las placas solares no son suficientes para aportar y suministrar todo el calor que necesita la utilización a lo largo del año, especialmente en la época fría del año. Para compensar esta deficiencia, se incorpora a la instalación una aportación de calor que puede provenir de resistencias eléctricas, o calentadores de gas o fuel.

6.3.2. Forma constructiva de los captadores solares térmicos

Los paneles o captadores solares de baja temperatura están constituidos por:

- Caja recubierta de aislante lateral y posterior.
- Parrilla de tubos.
- Placa absorbente.
- Cubierta transparente.

El calor que incide sobre la placa absorbente se concentra en el interior del recipiente intercambiador calentando el agua que circula por la parrilla de tubos.

El agua o fluido calefactor que circula por los paneles, se mueve en circuito cerrado, bajo dos formas de funcionamiento:

a) Circulación natural.
b) Circulación forzada.

El agua es impulsada por una bomba accionada por un motor eléctrico.

La figura 6.5 muestra la forma constructiva de un captador solar térmico, constituido por las partes siguientes:

- Caja envolvente que por efecto invernadero concentra el calor solar.
- Aislante que evita la pérdida de calor que se concentra en el interior de la caja.
- Placa absorbente que acumula calor.
- Cubierta transparente, sobre la que incide el calor solar y que deja pasar al interior de la caja.
- Parrilla de tubos con fluido que se calienta con el calor acumulado en el interior de la caja.

Ejemplos de captador solares térmicos:

La figura 6.6 muestra dos ejemplos de empleo de captadores de calor térmico planos, situados en sendos parking de hoteles, que cumplen con la misión de proteger a los vehículos del calor solar, y proporcionar agua caliente sanitaria (ACS).

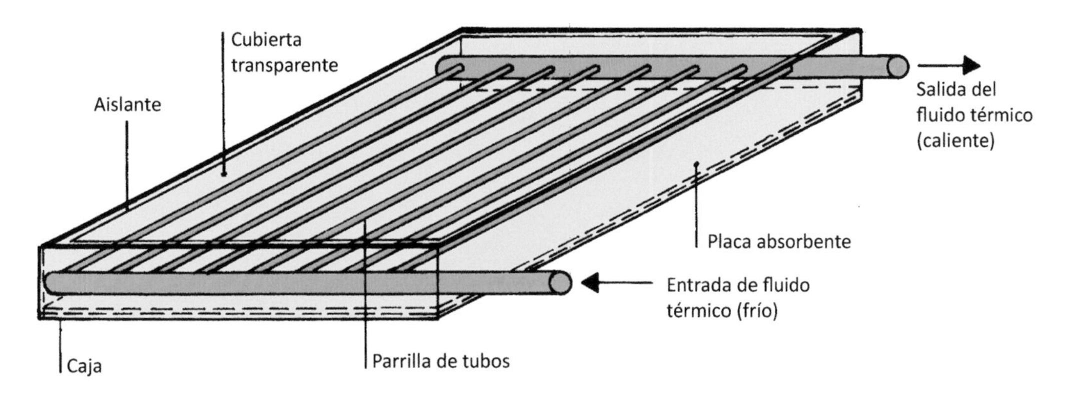

Figura 6.5. Partes que constituyen un captador solar térmico.

6.3.3. Formas de conectar los paneles o captadores

Tres formas de conectar los paneles cuando son varios los que integran una instalación:

a) Acoplamiento serie de captadores

Características de este acoplamiento:

- Elevar la temperatura final del agua a costa de bajar el rendimiento de la instalación.
- No es aconsejable poner más paneles en serie, que los que señale el fabricante.
- Los últimos paneles aportan menos calor que los primeros.

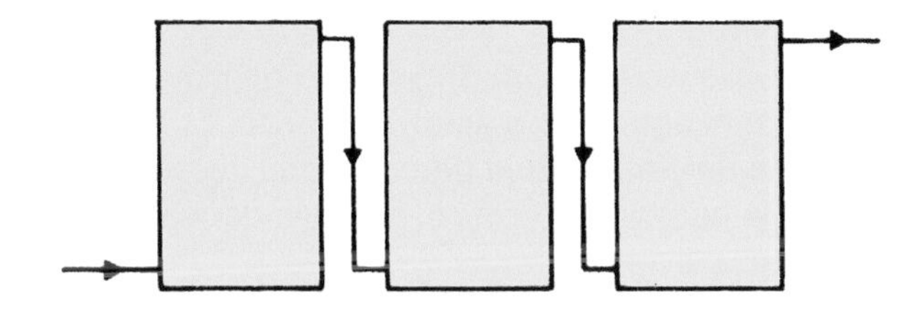

Esquema 6.1. Acoplamiento serie de captadores solares térmicos.

b) Acoplamiento en paralelo de captadores

Características de este acoplamiento:

- El número de paneles o captadores en serie será el que determine el fabricante y las características de los mismos.
- La finalidad de este acoplamiento es la aumentar el caudal de agua caliente a suministrar al intercambiador de calor.
- Hay que asegurar igual o similar funcionamiento para el agua en todos los paneles.

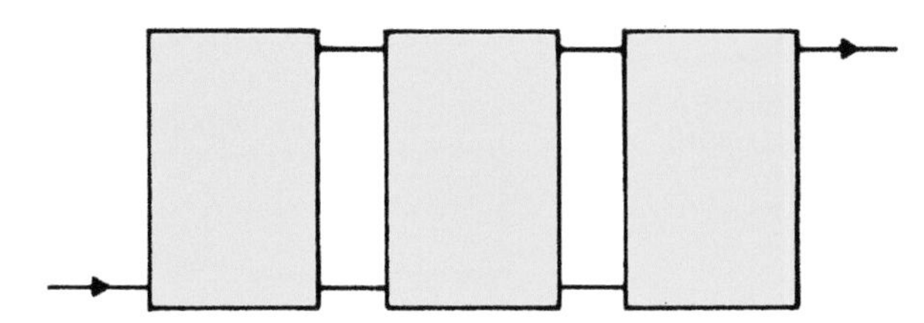

Esquema 6.2. Acoplamiento paralelo de captadores solares térmicos.

Figura 6.6. Dos ejemplos de empleo de captadores solares para calentar agua.

Figura 6.7. Instalaciones térmicas solares en hotel y vivienda unifamiliar.

Figura 6.8. Captadores solares térmicos para calentar agua caliente sanitaria en una instalación deportiva y en un hotel.

c) Acoplamiento mixto de captadores

Características de este acoplamiento:

- Las que corresponden a los paneles agrupados en serie y paralelo.
- Su finalidad es la de aumentar el caudal y la cantidad de calor aportada.

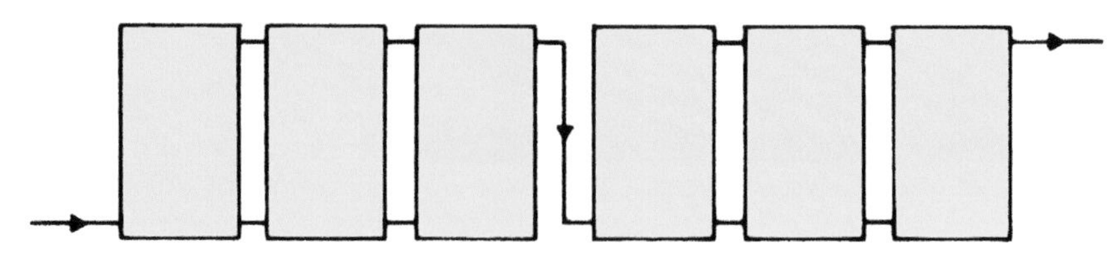

Esquema 6.3. Acoplamiento mixto de paneles (serie-derivación).

6.4. Aplicaciones de los captadores solares térmicos de baja temperatura

Tres son las principales aplicaciones de los paneles solares térmicos:

6.4.1. Calentar agua

Calentar agua en edificios de viviendas, hospitales, hoteles, servicios, etc. Se trata de agua caliente sanitaria (ACS) para uso y servicio de las personas, para su aseo personal, lavado y fregado.

6.4.2. Calefacción

Como medio de ayuda a reducir el consumo de otros tipos de energía (carbón, petróleo, gas y electricidad). Se trata de instalaciones de calefacción que utilizan el agua como agente térmico. Es posible que el calor que se recupera del Sol por una determinada instalación no sea suficiente para cubrir la necesidad térmica, pero ayudará a reducir la factura energética de nuestra vivienda.

6.4.3. Calentar el agua de una piscina

Las instalaciones de paneles térmicos también se utilizan para calentar el agua de piscinas. El fluido de los captadores de la piscina es el agua de la piscina, por lo que se trata de un circuito cerrado en el que el intercambio de calor se hace en la piscina.

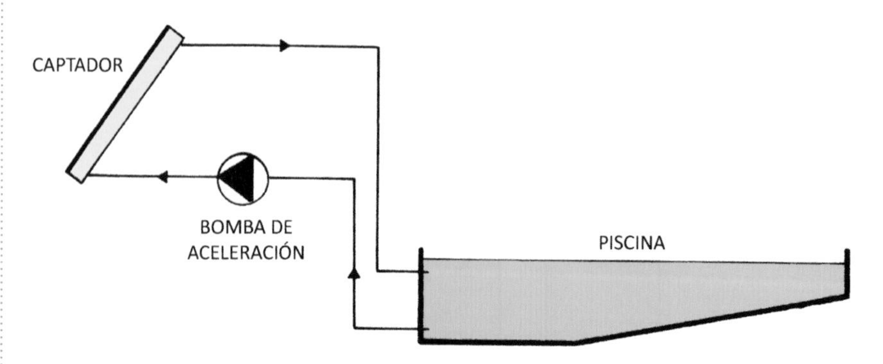

Esquema 6.4. Instalación para calentar el agua de una piscina.

6.4.4. Calentar el agua para duchas en instalaciones deportivas

Aunque parezca un aprovechamiento menor, no lo es, ya que abarata la factura energética en este apartado.

6.5. Otras aplicaciones industriales del calor

6.5.1. Cocinas solares

El calor que nos envía el Sol se concentra mediante un casquete esférico, cuya construcción se asemeja a una forma parabólica, con interior brillante (pulido como un espejo), centrando el calor en un punto concreto, que permite cocinar sin hacer fuego. Este dispositivo permite cocinar en el campo (cuando hay Sol) y de forma especial, en los lugares secos del planeta, en los que hay escasez de leña.

Figura 6.9. Paellera solar parabólica.

6.5.2. Salinas

Obtener sal del agua salina procedente del mar o de fuentes salinas por efecto de la evaporización aportando calor solar.

En zonas planas y parceladas en piscinas, se rellenan estas de agua salada, para que por efecto del calor y la evaporización consiguiente, se vaya el agua y quede la sal. En este caso, la energía aportada al proceso es totalmente gratuita.

6.5.3. Evaporación del agua y fenómenos atmosféricos

La evaporización del agua de los mares, lagos y ríos por efecto del calor del Sol al incidir sobre el agua que cubre dos tercios de la superficie de la Tierra. El agua en forma de vapor pasa a la atmósfera (nubes), para crear otros fenómenos de la naturaleza, como la lluvia, humedad, niebla, nieve, granizo, rocío.

Figura 6.10. Salinas marinas.

6.6. Generación de electricidad aprovechando el calor del Sol

El calor del Sol podemos aprovecharlo para generar electricidad. Los procedimientos no son simples y por tanto costosos, sin embargo, las posibilidades están ahí, y conviene conocerlas. Cada vez tiene más importancia el aprovechamiento de la energía solar térmica para generar electricidad por diferentes procedimientos, como son los que se presentan a continuación.

6.6.1. Motores de calor

El motor de calor fue ideado en 1816 por el reverendo Robert Stirling, con el objetivo de buscar una alternativa a los motores o máquinas de vapor que con sus explosiones tantas pérdidas de vidas ocasionaban.

Se puede denominar como un motor de combustión externa, el calor se aplica externamente. Su mecanismo consiste en dos pistones (cilindros), uno para disipar calor y desplazar aire caliente hacia la sección fría o viceversa. En la práctica funciona como un intercambiador de calor y se le denomina regenerador. El otro pistón entrega la fuerza para aplicar par al cigüeñal.

Figura 6.11. Mar y nubes.

Figura 6.12. Generador que transforma calor solar en electricidad.

Figura 6.13. Instalación de ensayo del sistema que calienta un fluido térmico que se calienta con el calor que llega desde el Sol. Planta Solar de Almería (PSA), perteneciente a CIEMAT.

Figura 6.14. Torre experimental en las instalaciones PSA de Tabernas (Almería).

Generación de electricidad:
Mediante un espejo parabólico se concentra el calor procedente del Sol, que se reenvía al motor Stirling, que con su movimiento acciona un generador eléctrico. Si se quiere generar electricidad cuando se oculta el Sol, puede darse calor al motor por medio de un quemador que consume gas natural.

6.6.2. Generación termoeléctrica

En la actualidad ya funcionan varias centrales eléctricas que transforman el calor que llega desde el Sol, en vapor de agua con el que se mueven turbinas que accionan generadores eléctricos (alternadores).

Estas instalaciones, que utilizan colectores cilindro-parabólicos recogen el calor que es dirigido hacia un tubo que contiene un fluido térmico. Son varios los fluidos ensayados (agua/vapor, aceite y sales fundidas). Se busca un fluido que mejore el intercambio térmico actual, para que la eficiencia del proceso sea mejor.

6.6.3. Generación por torre de concentración de calor

Mediante este procedimiento se concentra el calor recogido por helióstatos sobre una torre para calentar un fluido térmico con el que producir vapor de agua por intercambio calorífico, y por medio de una turbina de vapor, accionar un alternador que genera corriente eléctrica.

6.7. Aprovechamiento del calor solar en instalaciones frigoríficas

Aunque parezca curioso, el calor solar se aprovecha en grandes instalaciones de refrigeración que necesitan calor en su ciclo de funcionamiento, para producir "frío", lo que se hace en máquinas de absorción y adsorción que tienen ciclos bastante parecidos. Estas máquinas se emplean para refrigerar grandes superficies.

6.8. Principales elementos de una instalación solar térmica

Las instalaciones de captadores solares térmicos se utilizan preferentemente para calentar agua caliente sanitaria (ACS). En este apartado se estudian las principales aplicaciones de esta forma de recuperación de energía.

6.8.1. Elementos principales de una instalación solar térmica

Los elementos principales de este tipo de instalación son:

- Paneles solares térmicos.
- Soportes sobre los que se fijan los paneles.
- Tuberías con dispositivos de aislamiento, protección y toma de información (presión, temperatura y caudal).
- Intercambiador de calor.
- Bomba de circulación.
- Vaso de expansión.
- Válvulas antirretorno.

- Válvula de seguridad.
- Válvulas manuales de aislamiento.
- Acumulador de agua caliente.
- Circuito de agua caliente

El intercambiador de calor es después de los paneles solares térmicos el principal elemento de la instalación.

Tabla 6.1. Símbolos que representan los elementos principales de una instalación de ACS.

Placa solar térmica	Intercambiador de calor	Depósito acumulador de calor	Depósito acumulador de calor con resistencia eléctrica de apoyo

a) Instalación con intercambiador directo de calor sobre el depósito acumulador

El sistema termosifón es una de las principales aplicaciones para viviendas unifamiliares. Normalmente no incorporan bomba de circulación de fluido. Es el propio fluido el que se eleva en el captador al alcanzar temperatura y empuja al fluido que está más frío y que entra por la parte inferior del captador.

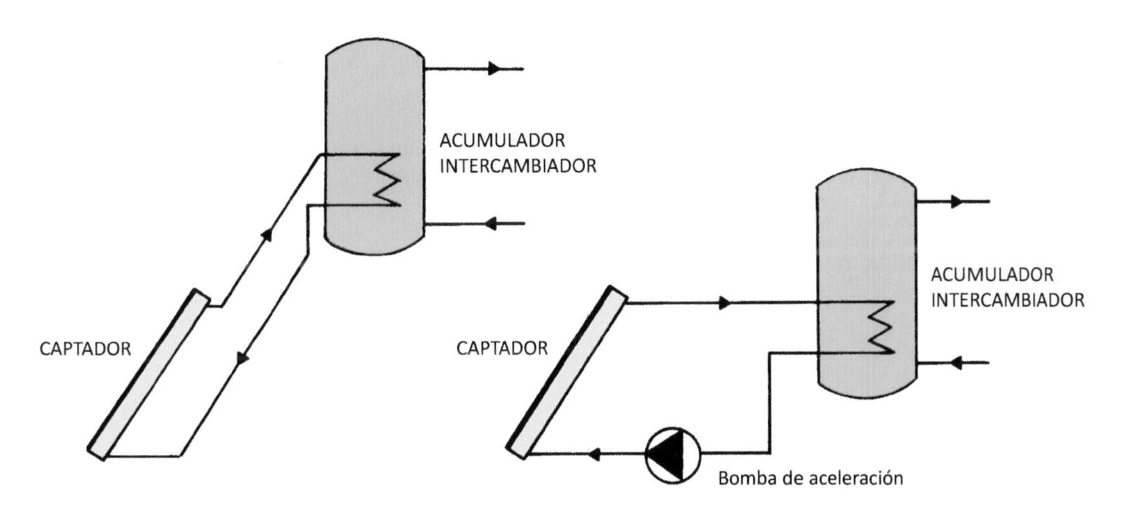

a) Sin bomba de aceleración. b) Con bomba de aceleración.

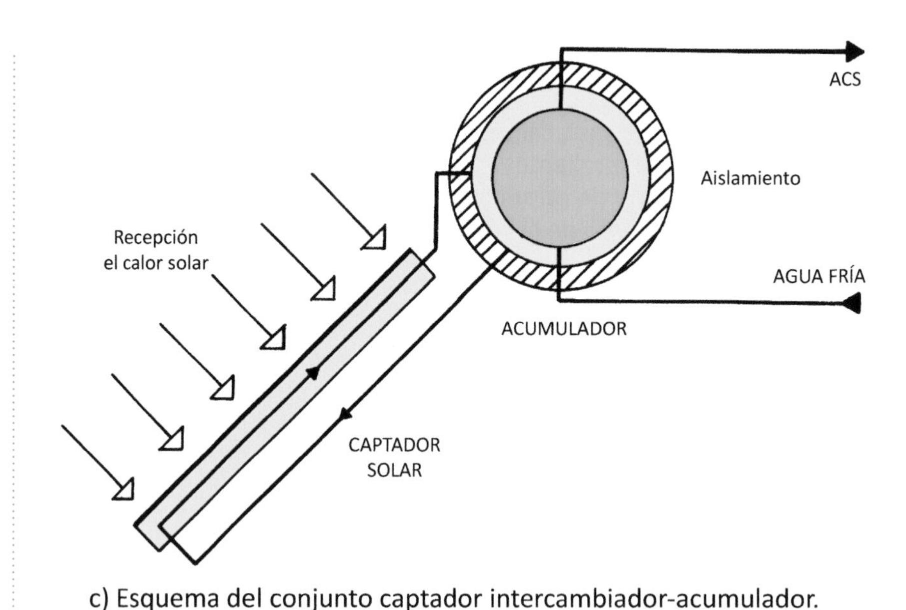

c) Esquema del conjunto captador intercambiador-acumulador.

Esquema 6.5. Instalación de un panel con intercambiador y depósito incorporado.

b) Instalación con intercambiador en el propio depósito acumulador

Se trata de un tipo de instalación muy empleado para medianos caudales.

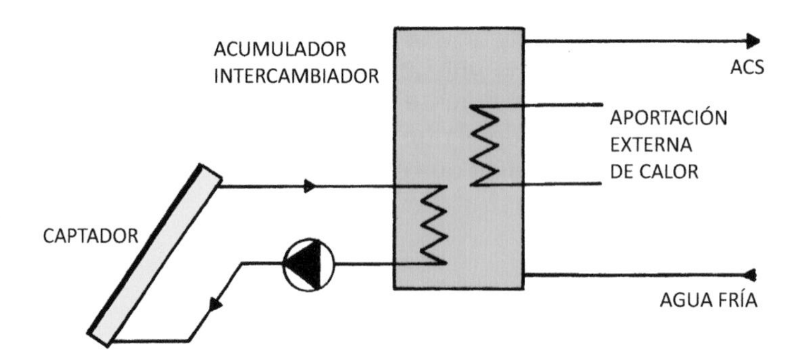

Esquema 6.6. Instalación de un captador con intercambiador en el depósito acumulador.

6.9. Instalaciones de captadores solares térmicos

Circuitos básicos de instalaciones de captadores solares:

6.9.1. Instalación solar térmica con varios captadores

Se trata de una instalación que incluye varios captadores, intercambiador incorporado en el depósito acumulador y depósito acumulador auxiliar de apoyo, además, valvulería y elementos de medida y control. El apoyo térmico se hace con energía eléctrica.

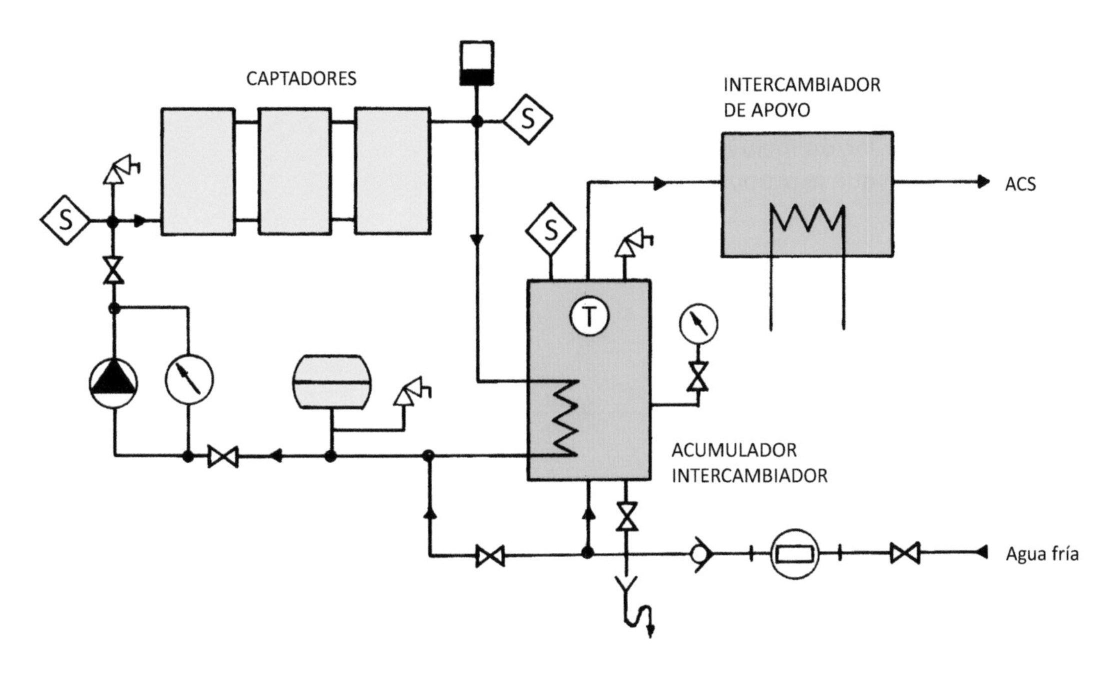

Esquema 6.7. Esquema para un equipo de ACS con intercambiador de calor interno.

6.9.2. Aplicación indirecta a través de un intercambiador

El intercambiador está fuera del depósito acumulador. Entre el depósito y el intercambiador se instala una bomba de aceleración.

> **Nota:**
> *El calor aportado por las placas solares no suele ser suficiente, por lo que se requiere aportar calor adicional, como puede ser por resistencia eléctrica.*

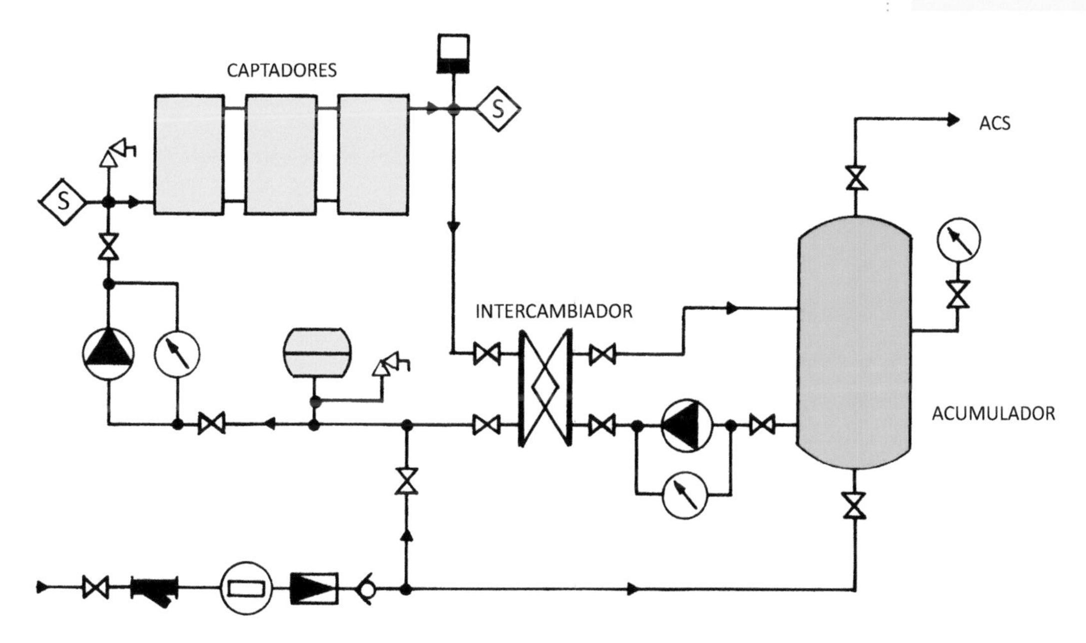

Esquema 6.8. Instalación indirecta de un intercambiador.

6.9.3. Sencilla instalación con depósito de recuperación del calor que proporciona el captador solar

Instalación que incluye los elementos principales. El depósito de acumulación incluye el intercambio térmico del captador y el aporte de calor que le proporciona una resistencia eléctrica.

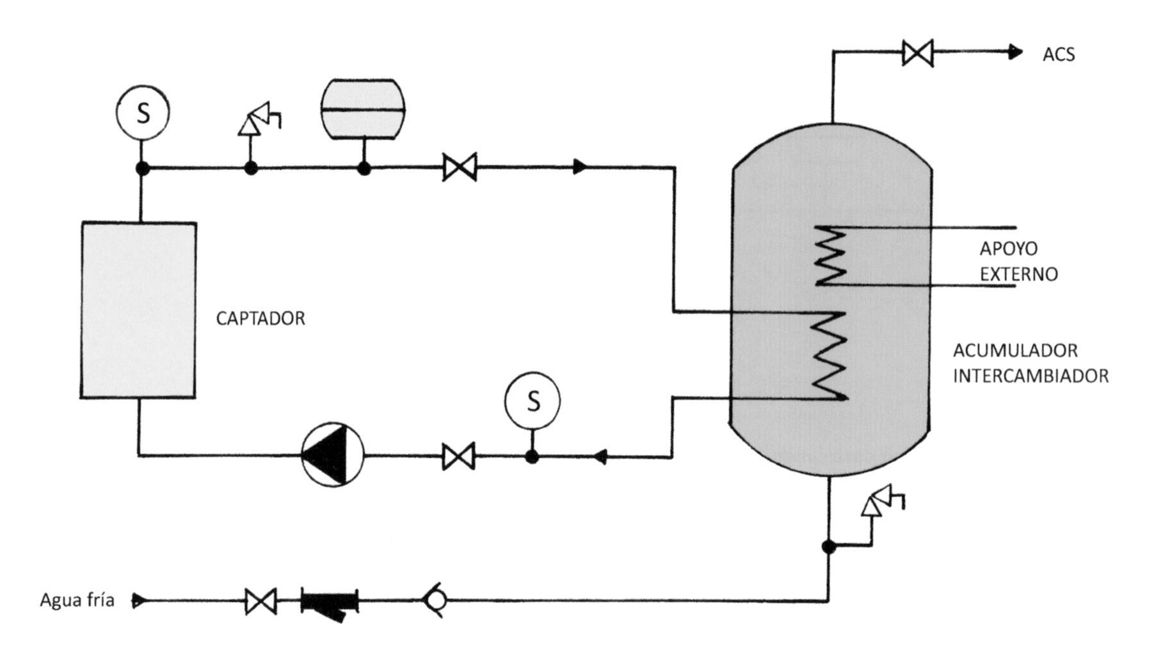

Esquema 6.9. Instalación con depósito de recuperación de calor solar.

6.9.4. Principales elementos de la tubería hidráulica

Tabla 6.2. Resumen de símbolos utilizados en el esquema.

Símbolo	Denominación
	Válvula de cierre o aislamiento.
	Filtro para tuberías.
	Válvula antirretorno.
	Purgador de aire.

Símbolo	Denominación
	Manómetro. Indicador de presión.
T	Indicador de temperatura. Sonda de temperatura.
	Desagüe conducido.
	Válvula de seguridad contra sobrepresión.
	Regulador de presión.
	Contador de caudal de agua.
	Bomba de circulación de agua.
	Vaso de expansión.

6.9.5. Todas las aplicaciones de una captación solar térmica en una vivienda

Existen varios tipos de sistemas para generar agua caliente sanitaria (ACS), como son: apoyo a calefacción y climatización de piscinas. También puede emplearse para alimentar una máquina de refrigeración por absorción, que emplea calor en lugar de electricidad para producir frío con el que se puede acondicionar el aire de los locales.

6.9.6. Nivel de automatismo de este tipo de instalaciones

Estas instalaciones pueden ser simples en su funcionamiento o integradas en un automatismo, para lo que se requiere instalar sensores de temperatura.

El automatismo de estas instalaciones permite una mejor adaptación a las situaciones que se dan en el tiempo de emisión de calor solar y durante la demanda de agua caliente.

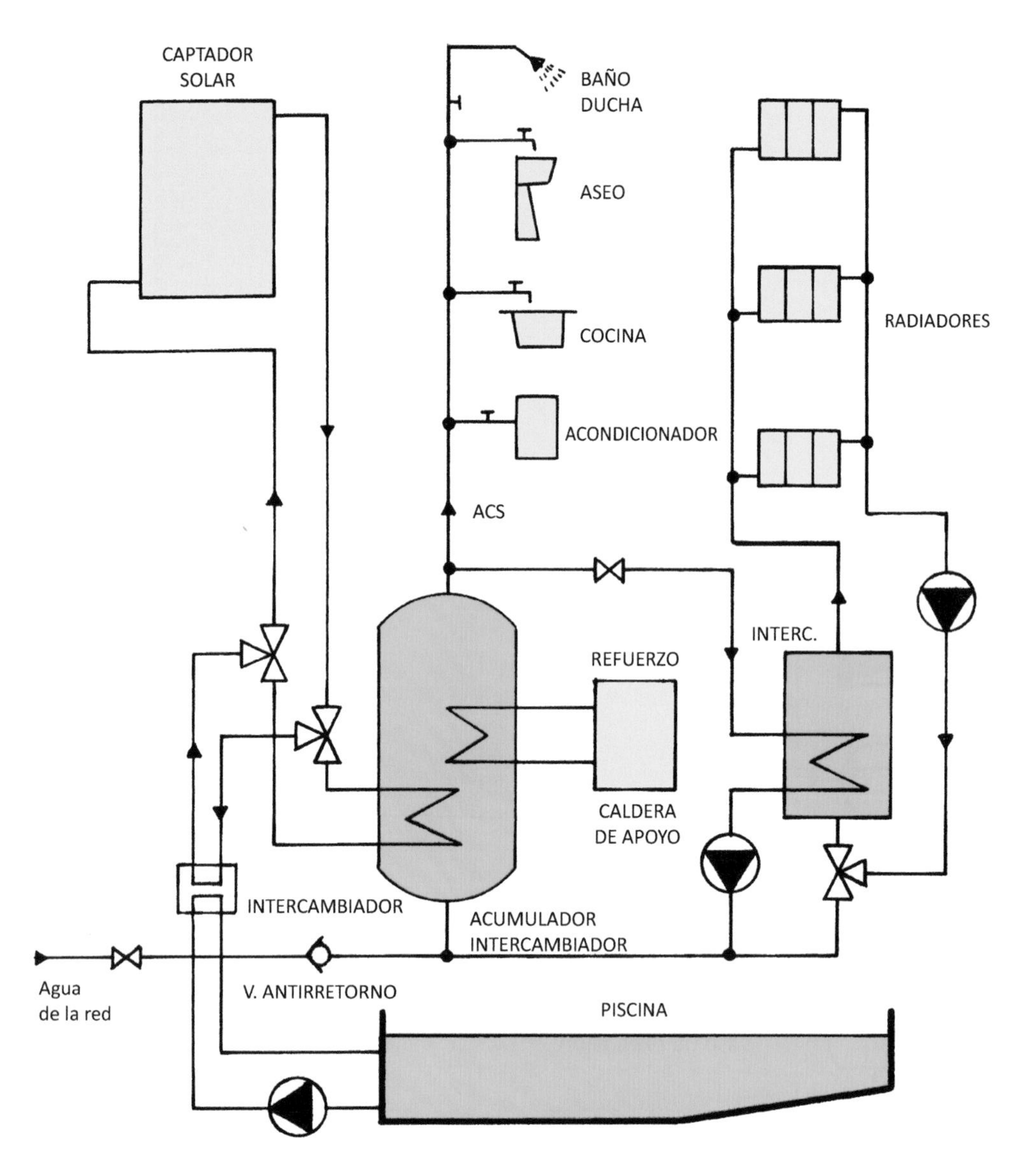

Esquema 6.10. Todas las aplicaciones reunidas en un mismo esquema.

6.9.7. Presentación comercial de captadores solares

(Ver Figura 6.14)

6.10. Mantenimiento de instalaciones térmicas solares

Después de su puesta en servicio, estas instalaciones se revisaran periódicamente para asegurar su correcto funcionamiento y aprovechamiento de la energía solar.

Entre las acciones principales están las siguientes:

- Mantener limpio el panel o paneles solares.
- Reparar las fugas que puedan darse en la instalación.
- Evitar la oxidación de los soportes.

- Revisar el fluido del circuito del captador solar térmico.
- Asegurar el correcto estado de los aislamientos térmicos de las tuberías, para evitar pérdidas de calor.
- Si la instalación tiene un determinado automatismo, asegurarse que los sensores de temperatura estén en buen estado.
- Proteger adecuadamente la instalación contra los efectos atmosféricos, especialmente contra el frío (fuertes heladas).

Figura 6.14. Captadores termosifón y placa.

6.11. Ventaja del aprovechamiento del calor del Sol

- Es una energía gratuita.
- Reduce el gasto energético.
- Abarata el gasto de calentar agua.
- Sistema limpio y no contaminante.
- Sistema silencioso.
- Sin apenas mantenimiento.
- Instalaciones sencillas que pueden estar enteramente automatizadas.
- Instalación obligatoria en ciertas circunstancias.

Las instalaciones se pueden dedicar a:
- Instalaciones de agua caliente sanitaria (ACS).
- Instalación para calefacción.
- Instalación para la calefacción de la piscina.
- Instalaciones de ACS + Calefacción.
- Instalación es de ACS + calefacción + calefacción de la piscina.

6.12. Código Técnico de la Edificación (CTE) aplicado

Se recomienda la lectura del Real Decreto 314/2006, de 17 de marzo, sobre el **Código Técnico de la Edificación (CTE).**

Artículo 15. **Exigencias básicas de ahorro de energía**

Exigencia básica HE 4: Contribución solar mínima de agua caliente sanitaria.

Exigencia básica HE 5: Contribución fotovoltaica mínima de energía eléctrica.

Este Código trata sobre las exigencias de equipamiento que deben tener las nuevas edificaciones, para aprovechar la energía solar térmica y luminosa.

6.13. Datos para cálculo de instalaciones solares térmicas

A continuación se recogen datos interesantes respecto a la radiación solar térmica procedente del Sol, y su aprovechamiento para calentar agua.
- El Sol emite energía en forma de radiación electromagnética.
- La distancia media entre el Sol y la Tierra es de 149.597.871 km.
- La luz recorre la distancia entre el Sol y la Tierra en 8 min y 19 s.
- La constante solar es de 1.353 W/m^2.

1.410 W/m^2 en el perihelio (principio de julio).
1.320 W/m^2 en el afelio (principios de enero).

- Para los cálculos se estima una radiación en torno a 1.000 W/m^2.

La trayectoria de la Tierra respecto al Sol es elíptica, de ahí la variación de calor en las diferentes épocas del año.

6.13.1. Necesidades de agua caliente sanitaria (ACS)

La tabla 6.2 muestra la estimación de ACS en función del tipo edificio, local, número de personas, camas o usuarios de los mismos.

Tabla 6.2. Estimación de consumo de agua caliente sanitaria, a 45 °C.

Actividad	litros/día	Utilizadores
Viviendas unifamiliares	40	Por persona
Viviendas multifamiliares	30	Por persona
Hospitales y clínicas	80	Por cama
Hoteles (4*)	100	Por cama
Hoteles (3*)	80	Por cama
Hoteles/Hostales (2*)	60	Por cama
Hostales/Pensiones (1*)	50	Por cama
Camping	60	Por persona
Residencias	80	Por cama
Vestuarios/Duchas colectivas	20	Por servicio
Escuelas	5	Por alumno
Cuarteles	30	Por persona
Fábricas y talleres	20	Por persona
Oficinas	5	Por persona
Gimnasios	30 a 40	Por usuario
Lavanderías	5 a 8	Por kilo de ropa
Restaurantes	8 a 15	Por comida
Cafeterías	2	Por almuerzo

Fuente: IDAE

6.13.2. Superficie de los paneles solares térmicos

Para captadores planos adecuadamente adaptados, la superficie aconsejada es de:

- Entre 0,5 y 1,5 m^2/persona, para pequeñas instalaciones individuales.
- Entre 0,3 y 0,5 m^2/persona, para instalaciones que abastezcan a más de 100 personas.

Consumo de agua/día: 50 litros por persona a una temperatura de 55 °C.

6.13.3. Datos necesarios para el dimensionado de una instalación

- Destino de la instalación (ACS, calefacción o piscina). Cuantos más datos se den, mejor.

- Datos geográficos: latitud del lugar y ubicación de los paneles de captación.
- Datos climatológicos: radiación sobre los captores térmicos, temperatura exterior, horas útiles de Sol.
- Masa de agua a tratar.
- Temperatura de entrada del agua de la red.
- Demanda energética.
- La aportación del sistema solar.
- Rendimiento de la instalación.
- Demanda de energía térmica total anual.
- Energía solar térmica aportada total anual.
- Aportación media anual (%).

6.13.4. Método a seguir para realizar el cálculo

Datos	Cálculos
· Demanda de energética · Temperatura del agua de la red (entrada) · Temperatura del agua caliente (ej.: 45 °C)	· Carga calorífica necesaria por mes.
· Radiación horas útiles de sol	· Intensidad radiante por mes.
· Intensidad radiante · Temperatura ambiente media. · Temperatura de uso necesaria · Rendimiento del captador	· Calor útil que se obtiene con el equipo captador por mes.
· Con los cálculos anteriores, se procederá a determinar la superficie necesaria de captación y el volumen del acumulador.	**· Cálculo de la superficie necesaria de captación.** **· Volumen de acumulación solar necesaria**

6.13.5. Elementos principales de la instalación

- Paneles solares térmicos.
- Intercambiadores de calor.
- Acumuladores de calor.
- Circuitos de tuberías.
- Válvulas de regulación.
- Aportación externa de calor.
- Dispositivos de regulación y control.

6.13.6. Potencia térmica de la instalación (Pt)

$$Pt = \eta \cdot S \cdot G_{ref} \text{ (W)}$$

η – Rendimiento del sistema de captación.
S – Superficie del sistema de captación, en m^2.
G_{ref} – Irradiación solar de referencia, igual a 1.000 W/m^2.

6.13.7. Demanda térmica energética que la instalación tiene que aportar

Carga de energía calorífica necesaria para calentar ACS.

$$Q = m \cdot Ce \cdot (Ts - Te) \cdot n$$

Q – Carga o energía calorífica necesaria, en kJ/mes.

m – Masa de ACS diaria consumida, en kg/día ≈ litros/día.
Ce – Calor específico del agua (4,18 kJ/°C kg).
Ts – Temperatura de salida del agua caliente, en °C.
Te – Temperatura de entrada del agua de la red de agua, en °C.
n – Número de días del mes.

6.13.8. Los parámetros de diseño de la instalación cumplirán los siguientes criterios

- La superficie total de captación (S), en m², cumplirá:

$50 \leq M/S \leq 80$

- El volumen de acumulación solar (V), en litros, cumplirá:

$0{,}8 \leq V/M \leq 1$

- Cuando por razones justificadas no de captación se instale la superficie de captación inicialmente diseñada, el volumen de acumulación solar cumplirá:

$50 \leq V/S \leq 80$

S – Superficie total de captación, en m².
M – Consumo diario de agua caliente sanitaria, en litros/día.
V – Volumen de acumulación solar a instalar, en litros.

Nota:

El volumen de acumulación (V) será aproximadamente igual a la carga de consumo diario (M).

6.13.9. Efectividad del intercambiador (ε)

Las instalaciones calientan el agua de la red de forma indirecta, utilizando intercambiadores de calor. El panel térmico calienta un fluido y este el agua de la red

$$\varepsilon = \frac{Q_{real}}{Q_{máx}} = \frac{Ts - Te}{Ts1 - Te1}$$

6.13.10. Volumen del acumulador

El volumen del acumulador se determina en función de la superficie de los captadores, y también:

- 100 litros/persona, para pequeña instalación.
- 60 litros/persona, para gran instalación.

6.13.11. Principales magnitudes y unidades empleadas en esta tecnología

a) Energía

1 julio = 0,24 cal
1 kcal = 1.000 cal
1 kWh = 0,24 x 1.000 W x 3.600 s = 840.000 cal = 864 kcal
1 termia (te o th) = 1.000 kcal

b) Presión

1 bar = 0,987 atm
1 atm = 1,01325 bar
1 atm = 10,33 mca (**m**etros **c**olumna de **a**gua)
1 mca = 0,0968 atm
1 kg/m² = 9,679 · 10^{-5} atm

c) Temperatura
- Grados centígrados o Celsius (°C).
- Grados Kelvin (K).
- Grados Fahrenheit (°F).

d) Potencia

1 W = 1 J · 1 s

1 kW = 1.000 W

7 Energía Fotovoltaica

La energía fotovoltaica tiene su origen en la luz del Sol (fotones), y puede transformarse en energía eléctrica por medio de células fotoeléctricas. No es muy grande la energía que se recupera, pero sí es importante en muchas aplicaciones puntuales, especialmente para lugares aislados, en los que es difícil o muy caro, acceder a las redes eléctricas de distribución de la energía eléctrica. La energía primaria (la luz) es una energía renovable y gratuita y nada contaminante. El problema principal de esta energía está en que sólo hay luz natural durante el día y por tanto, posibilidad de generación eléctrica. Para suministrar energía eléctrica cuando no hay luz solar (noche) hay que acumular la energía generada durante el día en baterías eléctricas.

Panel fotovoltaico que alimenta un parquímetro.

7.1. Introducción a la generación fotovoltaica

La historia de la generación fotovoltaica de electricidad a partir de protones (luz solar), es larga y ha sido en los últimos cincuenta años cuando se ha desarrollado.

En el efecto fotovoltaico tiene lugar una conversión de la energía luminosa, en energía eléctrica.

En 1832 el francés Alexandre Edmond Becquerel descubre el efecto fotovoltaico causado por la luz natural o solar.

En 1873 se descubre el efecto fotovoltaico en sólidos por el inglés Willoughby Smith.

En 1877, Willian Grylls Adam y Richard Evans Day crearon la primera célula fotovoltaica de selenio.

El fenómeno fotovoltaico fue observado por el físico Hertz en 1887, pero fue interpretado por Hallwachs en 1889.

En 1905, Albert Einstein descubre y presenta la teoría de la naturaleza de la luz y el efecto fotovoltaico basado en los Cuantos de Planck.

En 1954, los laboratorios Bell de EE.UU. construyeron el primer módulo fotovoltaico.

En 1955, se inicia el estudio y construcción de paneles espaciales en EE.UU.

En 1958, primer satélite equipado con panel fotovoltaico.

A partir de 1980 se inicia la fabricación de paneles fotovoltaicos que se destinan a aplicaciones diversas, especialmente para alimentar con energía eléctrica, viviendas aisladas.

Años 90, construcción de parques de generación fotovoltaica, que se impulsan a partir del año 2000.

El Sol es la fuente de energía que suministra ***fotones*** (luz) a la Tierra y que a través de ***células fotovoltaicas*** que están en los paneles o módulos fotovoltaicos se transforma en ***electricidad***.

En este capítulo se hace un repaso sobre los módulos o paneles fotovoltaicos, para pasar a continuación a estudiar los diferentes componentes que forman parte de este tipo de instalaciones, así como las instalaciones básicas de generación de electricidad.

Los esquemas se presentan de una forma didáctica para que el lector pueda entender esta tecnología y su aplicación práctica. Son muchas las aplicaciones de la generación fotovoltaica, desde pequeñas generaciones, hasta centrales eléctricas fotovoltaicas.

7.2. Paneles fotovoltaicos

Los paneles fotovoltaicos son los elementos que transforman la luz del Sol (protones), en energía eléctrica a partir de células fotoeléctricas.

7.2.1. Efecto fotovoltaico

El efecto fotovoltaico conocido por sus siglas FV, se basa en un proceso mediante el cual una célula fotovoltaica transforma la luz solar (fotones) energía eléctrica, conocida comúnmente como electricidad.

Los fotones pueden tener diferentes longitudes de onda, y por tanto, diferente energía. Solamente los fotones que son absorbidos por la célula FV generan electricidad al transferir su energía a un electrón de un átomo de la célula. Con la energía aportada, el electrón es capaz de abandonar su posición en el átomo e iniciar un movimiento formando parte de una corriente eléctrica.

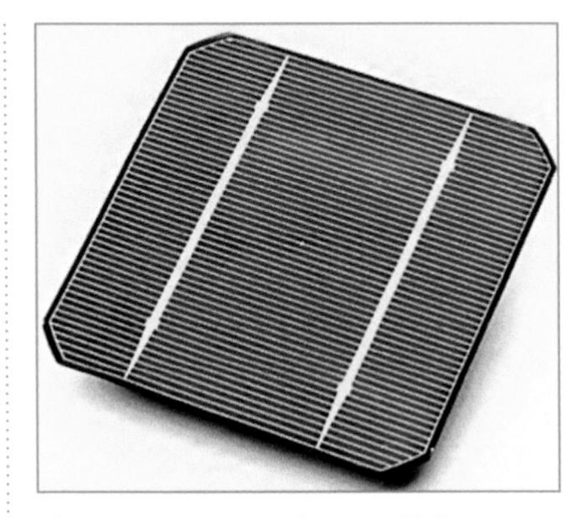

Figura 7.1. Aspecto de una célula fotovoltaica. Un panel FV contiene muchas células.

7.2.2. Células fotovoltaicas

La célula fotovoltaica es un semiconductor formado por una fina capa de un material semiconductor, generalmente silicio, con un tratamiento especial. La célula fotoeléctrica es un verdadero generador eléctrico, muy pequeño, que proporciona un tensión en torno a 0,5 V y una pequeña corriente eléctrica.

El efecto fotovoltaico es una propiedad que tienen ciertos semiconductores construidos a base de silicio, para transformar la luz solar (fotones), en electricidad.

El elemento fotovoltaico (célula solar fotovoltaica) está constituido por dos materiales (P y N) de diferente conductividad, que son:

Zona P. Las impurezas de boro en el silicio generan la zona P, que es positiva por tener un electrón menos que el silicio.

Zona N. Las impurezas de fósforo en el silicio generan la zona N, que es negativa por tener un electrón más que el silicio.

Al cerrar el circuito estando sometido el elemento a la radiación solar, se crea una corriente desde N hacia P, tal como se indica en la figura 2.10.

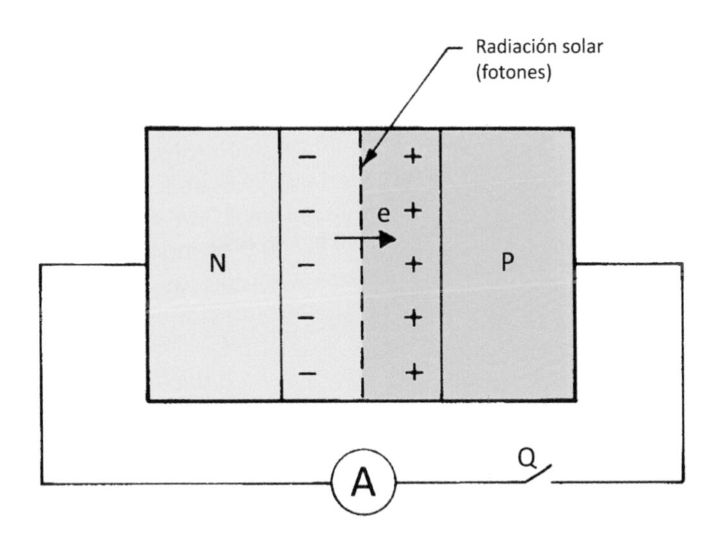

Figura 7.2. Efecto fotovoltaico.

7.2.3. Paneles fotovoltaicos

Los paneles fotovoltaicos están constituidos por células fotovoltaicas construidas con cristales de silicio, distinguiendo tres clases, de más a menos eficiencia o rendimiento y precio, como son:

Panel de silicio monocristalino. Eficiencia de 12 - 16%.
Panel de silicio policristalino. Eficiencia de 10 - 12%.
Panel de silicio amorfo. Eficiencia entre 6 - 8%.

Tabla 7.1. Principales características de módulos fotovoltaicos.

Módulos fotovoltaicos	
Monocristalinos	• Su comportamiento uniforme le proporciona un buen rendimiento. • Tiene una difícil fabricación. • Se reconoce por su coloración azulada oscura y metálica. • Su rendimiento es de 14 a 18%.
Policristalinos	• Precio más barato que los monocristalinos. • Las técnicas de preparación del silicio policristalino para el uso fotovoltaico son menos exigentes que los que necesita los de tipo monocristal. • Son muy empleados. • Difícil fabricación que se obtiene de silicio fundido y que está dopado con boro. • Comportamiento uniforme con resultado de buen conductor. • Eficiencia 10 a 15%. • Se reconoce por su color característico: azulado oscuro tirando a metálico.
Amorfos	• Su composición presenta un alto grado de desorden. • Tiene un proceso más simple de fabricación y resulta más barato. • Se deposita en forma de lámina sobre vidrio o plástico. • Estos módulos son eficientes ante una baja iluminación o radiación solar. • Se reconoce por su color marrón homogéneo • Calculadoras y relojes. • Tiene un rendimiento más bajo que los dos anteriores, en torno al 10%.

Otros módulos fotovoltaicos	
Módulos de película delgada	• Necesita poco material activo. • Tienen un sencillo método de fabricación. • Tiene un rendimiento de aproximadamente 5%.
Módulos de arseniuro de galio (Ga-As)	• Material muy eficiente. • Elevado costo de fabricación. • Tiene un rendimiento del 28%.

7.2.4. Características generales de los paneles fotovoltaicos

En la actualidad se fabrican paneles con eficiencia de hasta 18%.

En módulos fotovoltaicos de concentración se llega hasta 27% de eficiencia.

La materia base de los paneles fotovoltaicos es la sílice.

Los paneles tienen una eficiencia media de 12%, lo que supone, 120 W/m^2 en condiciones óptimas de irradiación.

Puede variar entre 100 W/m^2 en invierno y 250 W/m^2 en verano.

La vida útil de un panel viene a ser de 30 años. La radiación solar se transmite a través de fotones que al incidir sobre la superficie del cristal de silicio (célula fotoeléctrica) se genera una corriente eléctrica (intensidad de corriente) bajo una diferencia de potencial (tensión).

Los paneles están constituidos por células individuales en las que se produce una diferencia de potencial de aproximadamente 0,4 V por célula.

Las células fotoeléctricas se unen entre sí en conexiones serie, paralelo o mixta para obtener una tensión de 12 V o 24 V.

El panel o módulo fotovoltaico está constituido por un número determinado de células fotovoltaicas que pueden estar conectadas en grupos serie-paralelo para conseguir una determinada tensión (V), y una intensidad (A), para suministrar una determinada potencia (W).

Los hay de diferentes potencias y medidas atendiendo a sus características y que se elegirán en función de las necesidades de la aplicación.

Los paneles fotovoltaicos pueden conectarse entre ellos de diferentes formas según interese más tensión, más intensidad o los dos casos al mismo tiempo.

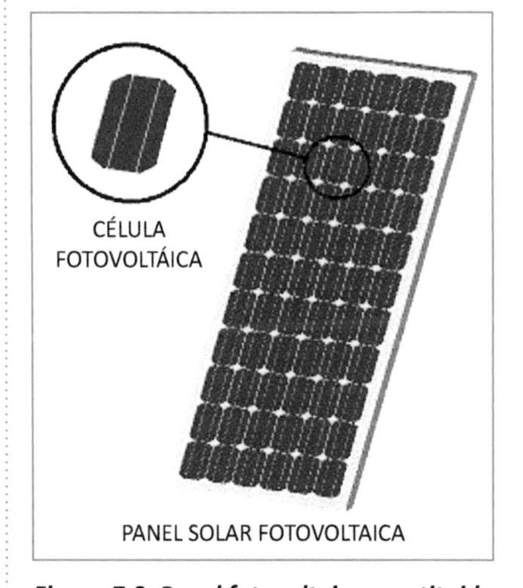

Figura 7.3. Panel fotovoltaico constituido por 6 x 11 = 66 células fotoeléctricas.

7.2.5. Constitución de un panel o módulo fotovoltaico

① Caja o envoltura protectora cerrada.
② Tapa protectora que permite y facilita la entrada de luz solar (fotones).
③ Células fotoeléctricas conectadas interiormente en grupos serie-paralelo.
④ Caja de conexiones del panel fotoeléctrico.

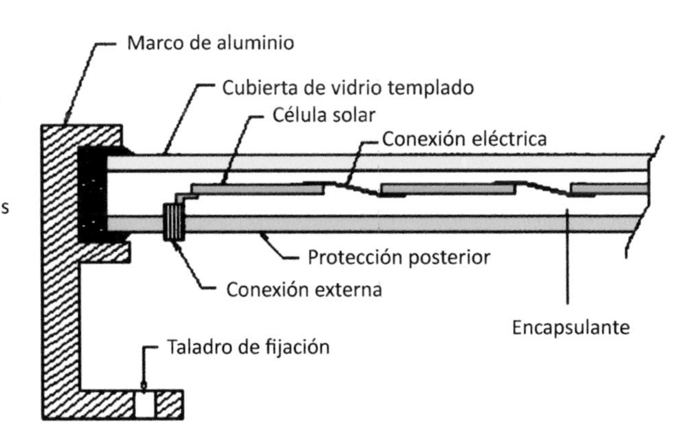

Figura 7.4. Esquema de un panel fotovoltaico.

El módulo fotovoltaico está constituido por:

- Marco o caja metálica abierta por una cara o superficie mayor.
- Elementos o células de silicio.
- Material encapsulante.
- Recubrimientos anterior y posterior.
- Conexiones eléctricas

7.2.6. Generación y rendimiento de los módulos fotovoltaicos

En España, la estimación de generación de potencia para un módulo fotovoltaico es de 115 Wp a 135 Wp por metro cuadrado (m^2) de módulo.

El rendimiento está comprendido entre 11,5% y 13,5% de su potencia teórica, con tendencia a subir progresivamente al mejorar la eficiencia constructiva de los paneles fotovoltaicos.

7.2.7. Placa de características de los módulos fotovoltaicos

Un módulo fotovoltaico se identifica por su placa de características, como por ejemplo: con los datos que se relacionan en la tabla 6.2, están tomados de un catálogo de módulos fotovoltaicos.

Tabla 7.2. Principales características que definen un módulo fotovoltaico.

Magnitudes	Denominación	Valores (ejemplo)
	Fabricante	XXX
	Denominación industrial del módulo.	XL 180-12
	Tipo de módulo	Silicio monocristalino
	Comportamiento a 800 W/m^2 a 25 °C	
$P_{máx}$	Potencia en el punto de máxima potencia	125 W
V_{mpp}	Tensión a potencia máxima	32,7 V
I_{mpp}	Corriente a potencia máxima	3,8 A
V_{oc}	Tensión en circuito abierto	40,9 V
I_{sc}	Corriente de cortocircuito	4,2 A
	Comportamiento bajo condiciones estándar de prueba	
$P_{máx}$	Potencia en el punto de máxima potencia	165 W
$V_{máxpp}$	Tensión a potencia máxima.	35,3 V
I_{mpp}	Corriente a potencia máxima	4,7 A
V_{oc}	Tensión en circuito abierto	44,1 V
I_{sc}	Corriente de cortocircuito	5,2 A
	Medidas de la célula	1.610 x 810 x 34
	Peso	25 kg
	Mínimo valor del fusible en serie	10 A
	Garantía del producto	3 años
	Superficie del módulo	1,2 m^2
	Garantía de prestaciones	10 años sobre el 90%. 25 años sobre el 80%.
	Otros datos que así lo considere el fabricante	

7.2.8. Conexión de los paneles fotovoltaicos

Varias son las formas de conexión entre paneles atendiendo a la demanda del circuito, como puede ser aumentar los valores de la tensión o de la corriente.

a) Funcionamiento individual

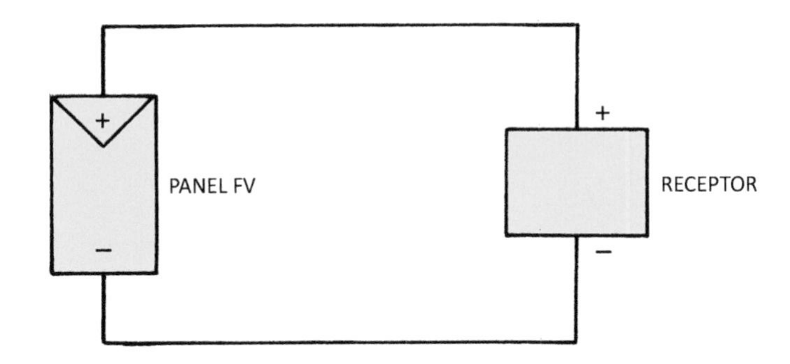

Esquema 7.1. Panel fotovoltaico en solitario.

Características: la tensión y corriente corresponderán a las nominales del panel.

b) Conexión serie

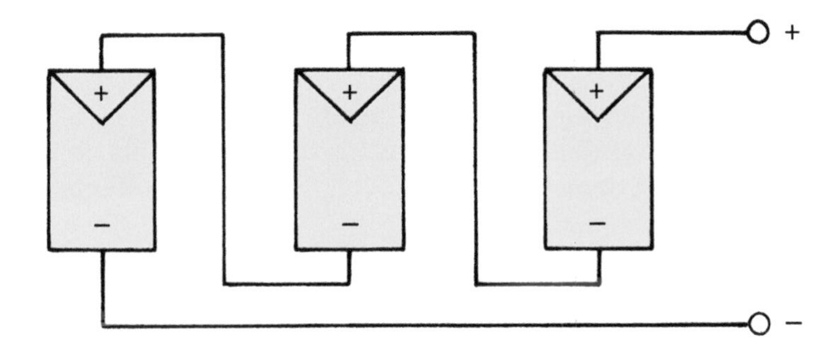

Esquema 7.2. Conexionado serie de paneles fotovoltaicos.

Características de este acoplamiento: aumentar de la tensión (U) de suministro.

c) Conexión paralelo

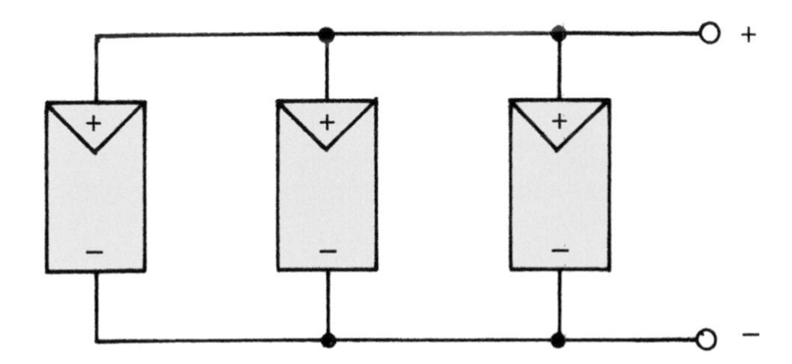

Esquema 7.3. Conexionado paralelo de paneles fotovoltaicos.

Figura 7.5. Presentación de diferentes tipos de paneles fotovoltaicos colocados en bastidor.

Nota:

Se ha polemizado mucho sobre la generación solar fotovoltaica como una alternativa para generar electricidad. Hay que reconocer que en el momento actual y con los rendimientos obtenidos, la aportación es casi insignificante. Se necesitan grandes superficies y una inversión importante, para conseguir potencias poco significativas respecto a la demanda.
Es una solución muy interesante en casos puntuales, que han ayudado a resolver muchos problemas de abastecimiento eléctrico por encontrase la red de energía eléctrica muy lejana o por que llegar hasta la utilización era difícil o costoso, o ambos.

Características de este acoplamiento: aumentar la intensidad (I) de suministro.

d) Conexión mixta (serie paralelo)

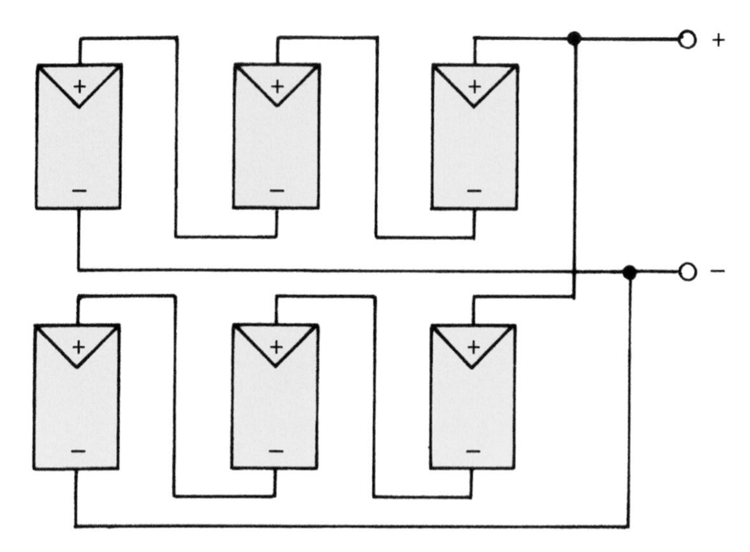

Esquema 7.4. Conexionado mixto (serie-paralelo) de paneles fotovoltaicos.

Características de este acoplamiento: aumentar la tensión (U) y la intensidad (I) de suministro.

7.2.9. Presentación de paneles fotovoltaicos

Aclaramos, que a los módulos fotovoltaicos se les denomina de formas diferentes, también, como paneles fotovoltaicos o simplemente placas fotovoltaicas.

7.2.10. Generación industrial de electricidad

La generación eléctrica de origen fotovoltaico se obtiene en centrales solares fotovoltaicas, también denominadas "parques fotovoltaicos" o "huertas solares". Los elementos de captación están constituidos por grandes bastidores que contienen muchos módulos fotovoltaicos, conectados en alguna de las formas que se indican en el apartado 6.

En la figura 7.6 se presentan dos grandes bastidores de los muchos que constituyen la central solar de generación fotovoltaica.

7.3. Elementos que constituyen las instalaciones fotovoltaicas

Las instalaciones fotovoltaicas incluyen diferentes elementos complementarios para controlar los circuitos, almacenar energía, y suministrarla a la utilización.

1. Paneles o módulos fotovoltaicos

Elementos generadores de energía eléctrica. Generan corriente continua (CC).

Este elemento lo tienen todas las instalaciones fotovoltaicas.

2. Regulador de corriente

Regula la instalación en lo que se refiere a la recepción de la corriente que recibe de los módulos fotovoltaicos, su envío y devolución del acumulador de baterías (carga y descarga) y el suministro al circuito utilizador.

Hay instalaciones que aplican directamente la corriente generada, por lo que no hacen uso de regulador de corriente.

3. Acumulador eléctrico de la energía eléctrica generada

Acumula y guarda la energía generada a lo largo de un tiempo, para suministrarla cuando no haya generación o esta sea inferior a la demanda. Está constituido por una o más baterías, su regulación la hace el regulador de corriente.

Conexión de baterías:

a) Acoplamiento serie

Finalidad: aumentar la tensión.

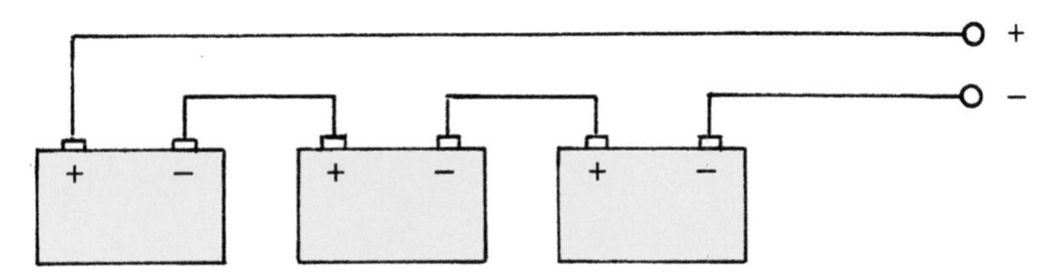

Esquema 7.5. Acoplamiento serie de baterías.

b) Acoplamiento derivación o paralelo

Finalidad: aumentar la tensión y la carga de corriente almacenada.

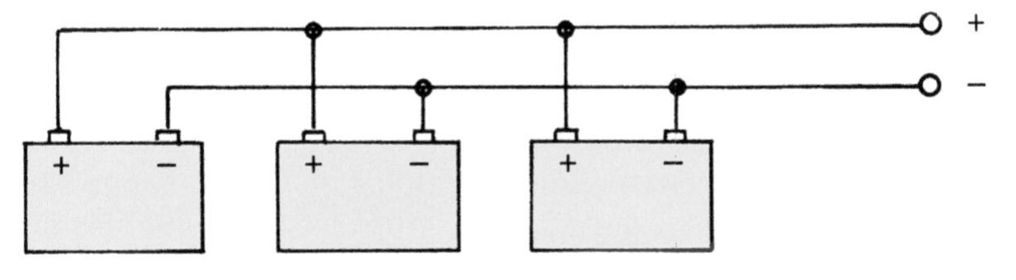

Esquema 7.6. Acoplamiento derivación de baterías.

c) Acoplamiento mixto (serie-derivación

Finalidad: aumentar la tensión y la carga de corriente almacenada.

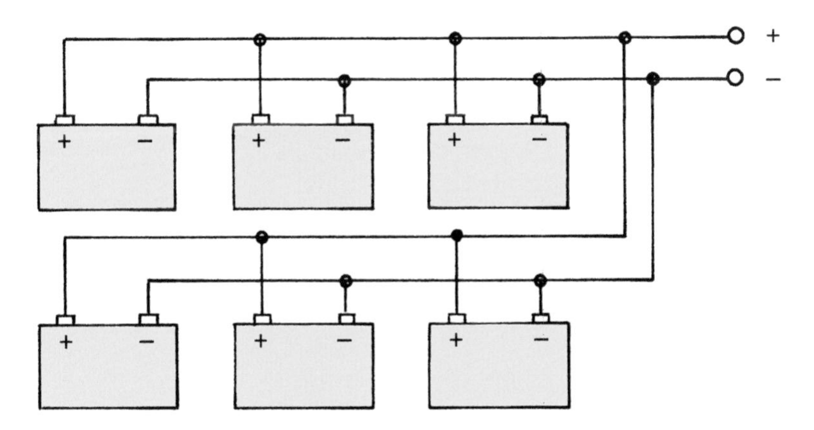

Esquema 7.7. Acoplamiento mixto de baterías.

Figura 7.6. Grandes bastidores con muchos paneles fotovoltaicos.

Figura 7.7. Cartel que señala que el "parque fotovoltaico" tiene una capacidad de generación de 9 MW.

Nota:

El parque fotovoltaico construido en Almaraz (Cáceres) el año 2008, tiene una potencia instalada de 22,1 MW. Se trata de un gran parque fotovoltaico (central fotovoltaica).

La figura 7.7 muestra un cartel con la potencia del parque fotovoltaico de Zaratán (Valladolid), con una potencia instalada de 9 MW.

Un solo reactor de una central nuclear proporciona una potencia del orden de 1.000 MW. Imaginemos la cantidad de placas que hay que instalar para conseguir la generación eléctrica equivalente, con el inconveniente de que durante la noche no hay generación eléctrica y es mucho menor en los días nublados y del invierno con días más cortos.

Un solo aerogenerador (de gran potencia, en curso de ensayos) puede suministrar hasta 3,5 MW, la potencia de muchos parques fotovoltaicos de media potencia.

4. Inversor

Equipo convertidor o rectificador al que se denomina inversor, y que adapta la forma de la corriente (CC) a la necesidad de la utilización en corriente alterna (CA) y tensión.

5. Otros complementos de estas instalaciones

- Aparamenta general.
- Instalación y aparatos de protección y maniobra.
- Contadores de energía.
- Transformadores.

Tabla 7.3. Representación de los principales elementos de una instalación de generación fotoeléctrica:

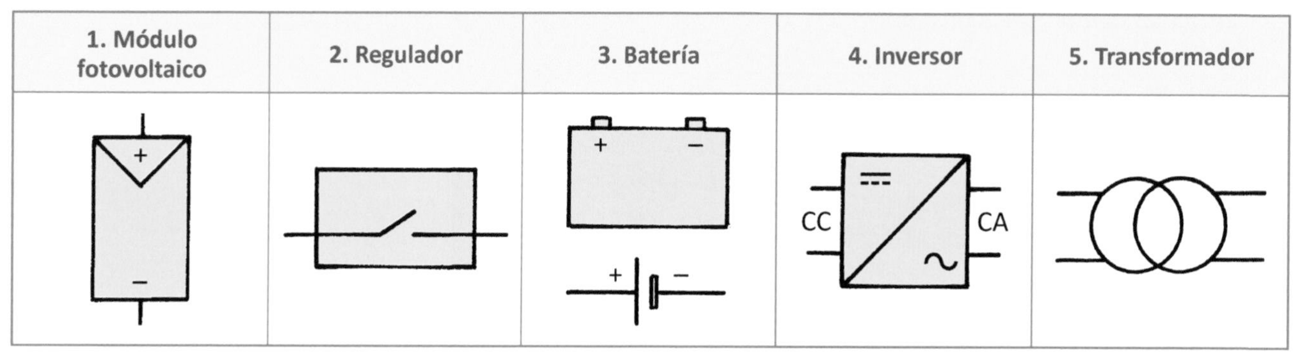

1. Módulo fotovoltaico	2. Regulador	3. Batería	4. Inversor	5. Transformador

Nota: En los circuitos que se estudian a continuación se representan estos símbolos.

7.4. Circuitos básicos de generación fotovoltaica

A continuación se presentan varios ejemplos de instalaciones básicas con módulos fotovoltaicos que son suficientes para entender los principios de esta tecnología.

La particularidad principal de estas instalaciones cuando se aplica directamente la corriente continua generada, está en que al ser de baja tensión (12, 24 o 48 V), las corrientes (intensidades) son elevadas, por lo que los conductores tendrán una elevada sección y las caídas de tensión pueden ser importantes si los aparatos están situados a cierta distancia. Por lo que se tendrá en cuanta lo siguiente:

- Implantar los paneles orientados al Sur y con inclinación adecuada, para aprovechar al máximo la irradiación solar, tanto en el caso de que sean fijos, como orientables.
- Acortar la instalación todo lo que sea posible.
- Reunir todos los aparatos y no alejarlos.
- Evitar las caídas de tensión.
- Que los aparatos instalados en el lado de CC, que no sean para CA.
- Cuando la tensión sea superior a 48 V, la instalación tendrá conductor de protección.

Circuito 1. Generación fotovoltaica básica.

Circuito básico con acumulación de la energía generada y no consumida.

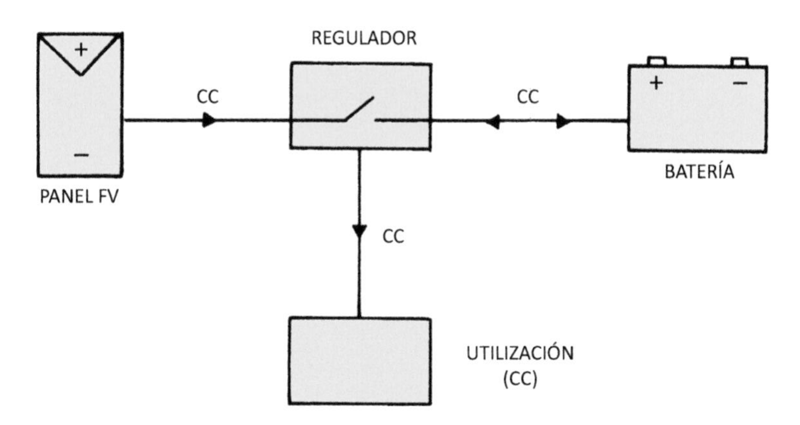

Esquema 7.8. Instalación aislada básica con receptores de corriente continua (CC).

① Placa fotovoltaica.
② Regulador. Envía la corriente eléctrica generada hacia:
- Acumulador (batería).
- De acumulador a consumo.
- De placa a consumo.

③ Acumulador (CC/CC).
④ Consumo (receptor/es) en corriente continua (CC).

Circuito 2. Generación fotovoltaica.
Utilización en corriente alterna (CA).

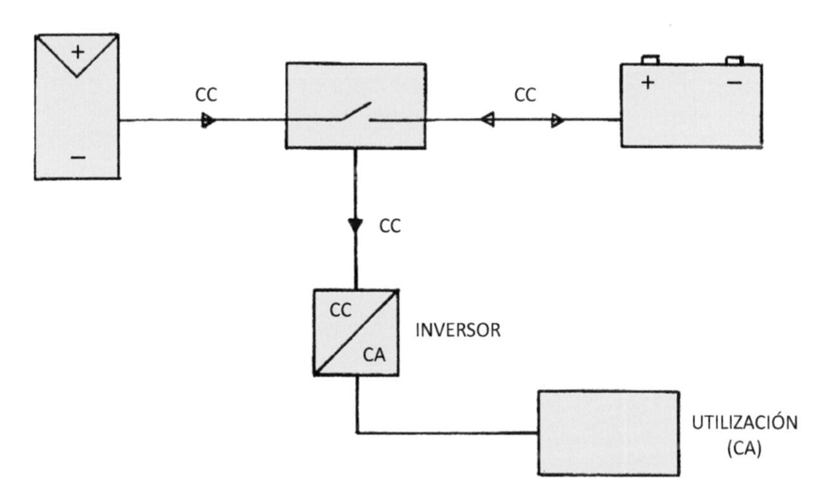

Esquema 7.9. Instalación aislada básica con receptores de corriente alterna (CA).

① Placa fotovoltaica.
② Regulador.
③ Acumulador (batería).
④ Inversor [1] de CC en CA.
⑤ Consumo (receptor/es) en corriente alterna (CA).

[1] El inversor convierte la corriente continua (CC) en corriente alterna (CA).

Circuito 3. Generación fotovoltaica.
Utilización en CC y CA (dos circuitos independientes).

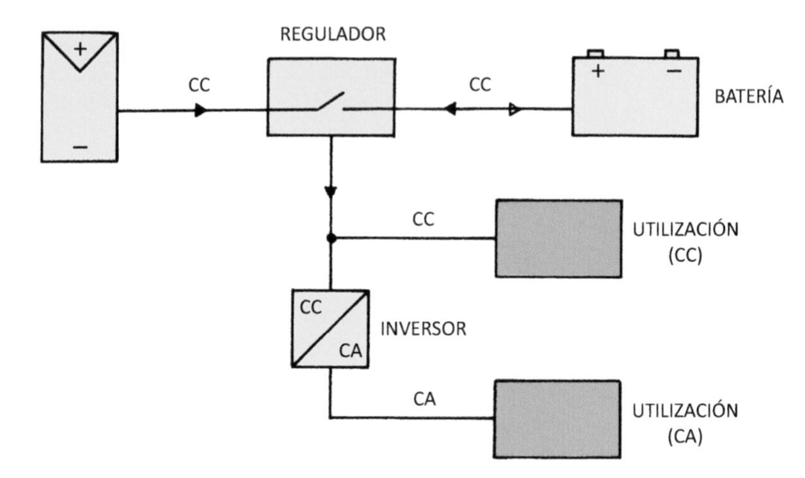

Esquema 7.10. Instalación aislada básica con suministro de corriente continua (CC) y corriente alterna (CA).

① Placa fotovoltaica.
② Regulador.
③ Acumulador (batería).
④ Inversor (CC/CA).
⑤ Consumo (receptor/es) en corriente continua (CC).
⑥ Consumo (receptor/es) en corriente alterna (CA).

Circuito 4. Generación fotovoltaica
Instalación de generación reforzada con un pequeño aerogenerador.
Como complemento de la instalación de generación fotovoltaica o para poner menos paneles, se puede instalar un pequeño aerogenerador, como el que representa el esquema.

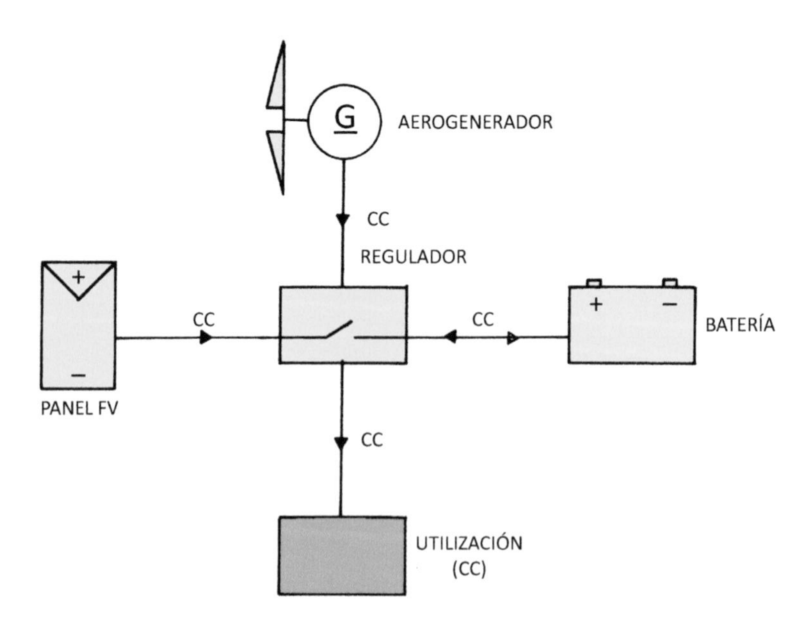

Esquema 7.11. Instalación aislada básica fotovoltaica, apoyada por un aerogenerador con receptores de corriente continua (CC).

Circuito 5. Generación fotovoltaica

Instalación de generación reforzada con grupo electrógeno.

Cuando la energía generada por los paneles fotovoltaicos no es suficiente (más receptores o días cubiertos o de invierno), se puede recurrir a una generación suplementaria como es el caso de un grupo electrógeno.

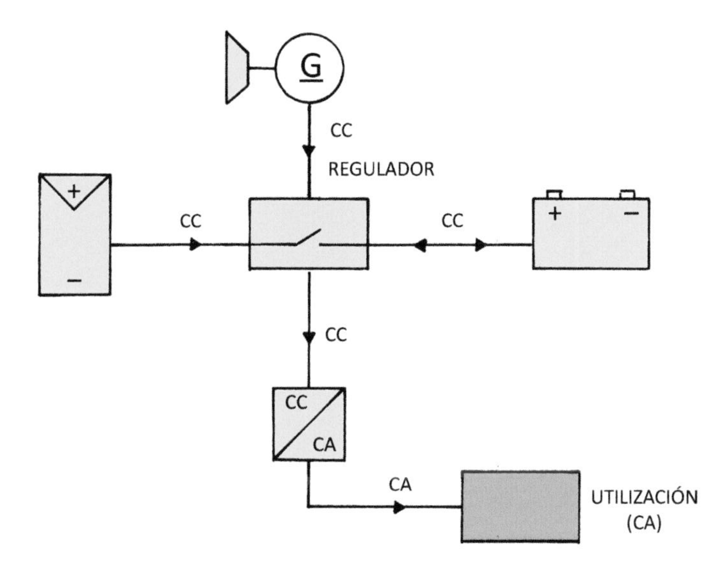

Esquema 7.12. Instalación aislada básica fotovoltaica, apoyada por un grupo electrógeno con receptores de corriente alterna (CA).

Circuito 6. Instalación para alimentación directa de una bomba que extrae agua de un pozo.

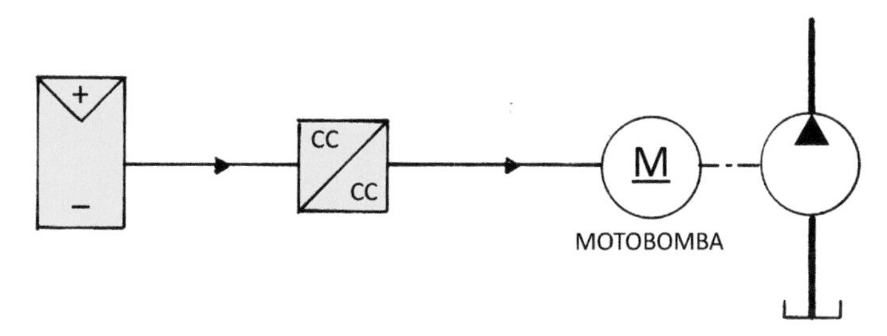

Esquema 7.13. Aplicación de una generación fotovoltaica para alimentar una bomba con motor eléctrico.

Circuito 7. Instalación de generación fotovoltaica con todos sus elementos eléctricos de protección y maniobra

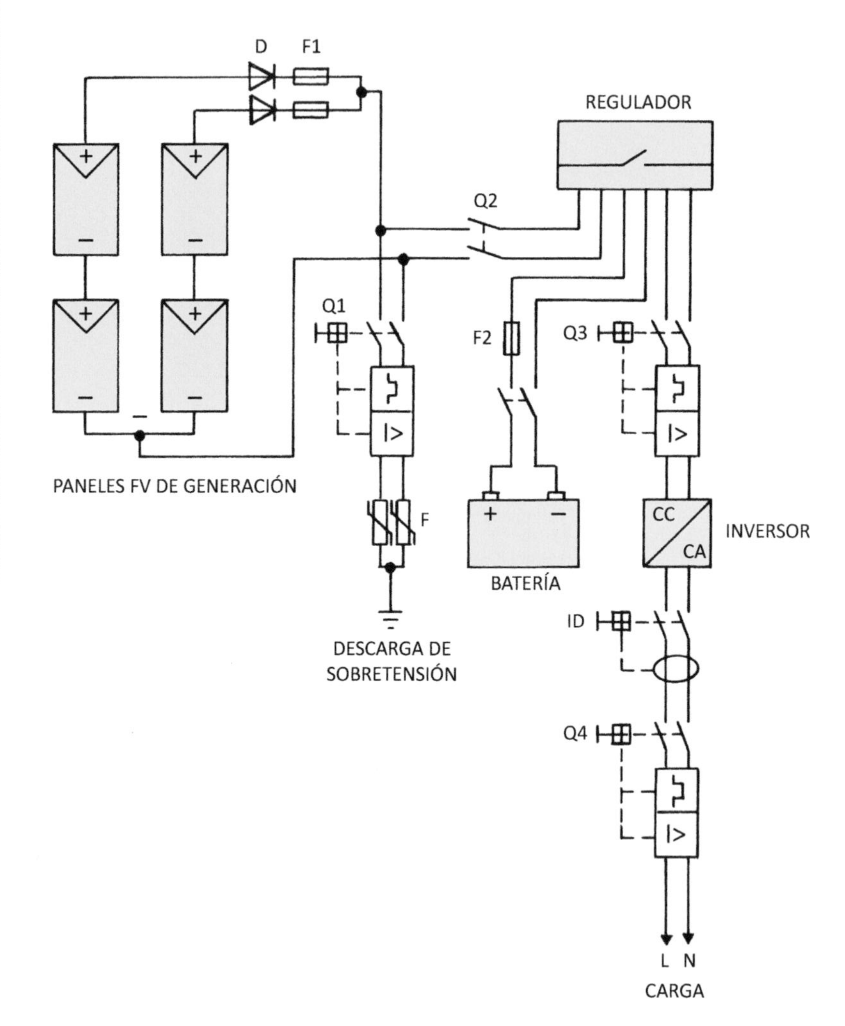

Esquema 7.14. Esquema eléctrico en el que se representan todos sus elementos de protección y maniobra.

6.5. Instalaciones de generación fotovoltaica contectadas a la red

En algunos casos, la electricidad generada por la instalación fotovoltaica es superior a la demandada, permitiendo enviar electricidad a la red eléctrica. En otros casos, se construyen instalaciones de generación fotovoltaica con el único fin de suministrar energía a la red eléctrica.

Circuito 1. Generación fotovoltaica

Instalación que suministra energía a un consumidor privado y a la red eléctrica general.

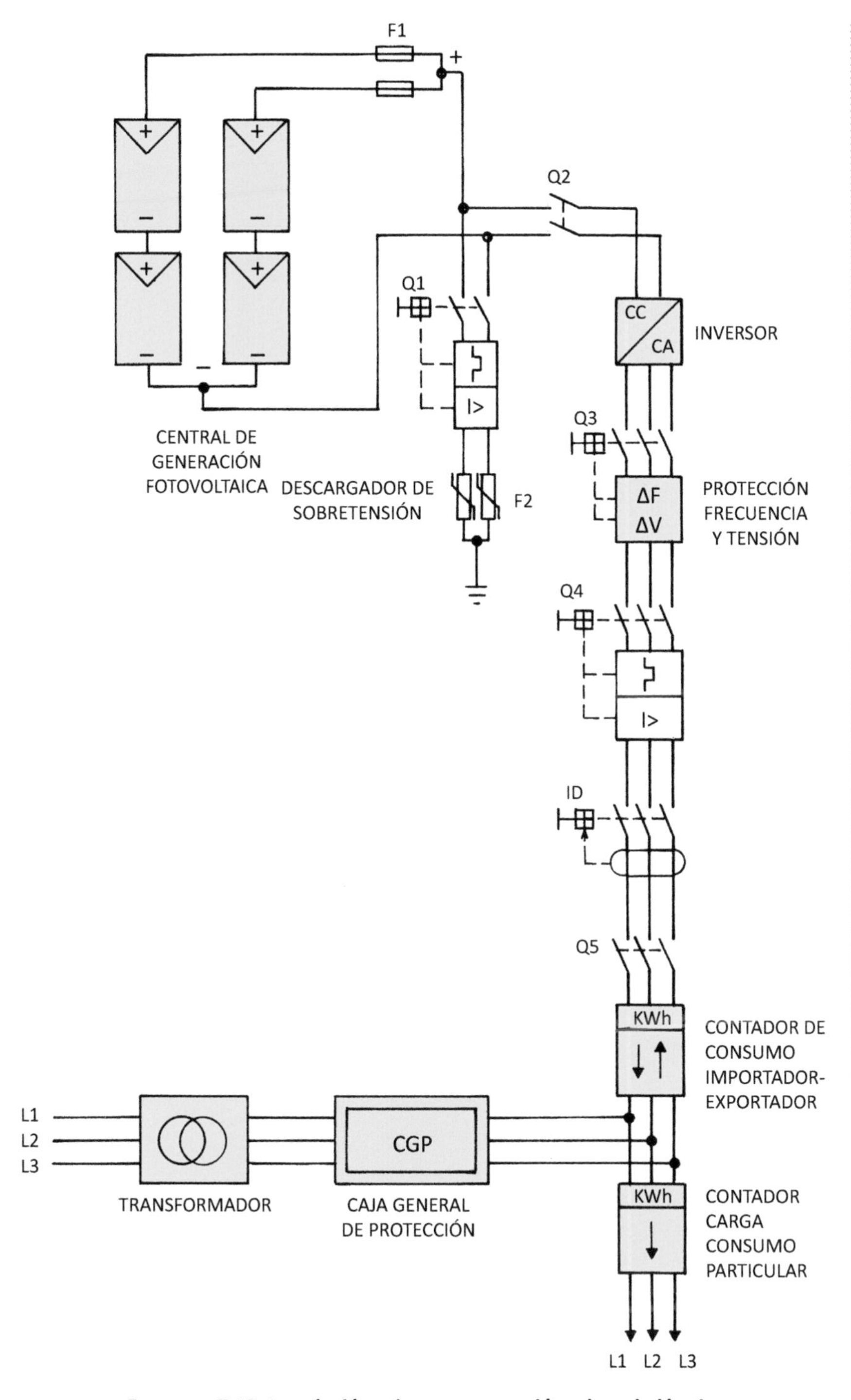

Esquema 7.15. Instalación mixta, con conexión a la red eléctrica.

D – Diodos.
T – Transformador.
CGP – Caja general de protección.
Q – Interruptores.
ID – Interruptor diferencial.

Circuito 2. Instalación de generación fotovoltaica que suministra energía a la red eléctrica.

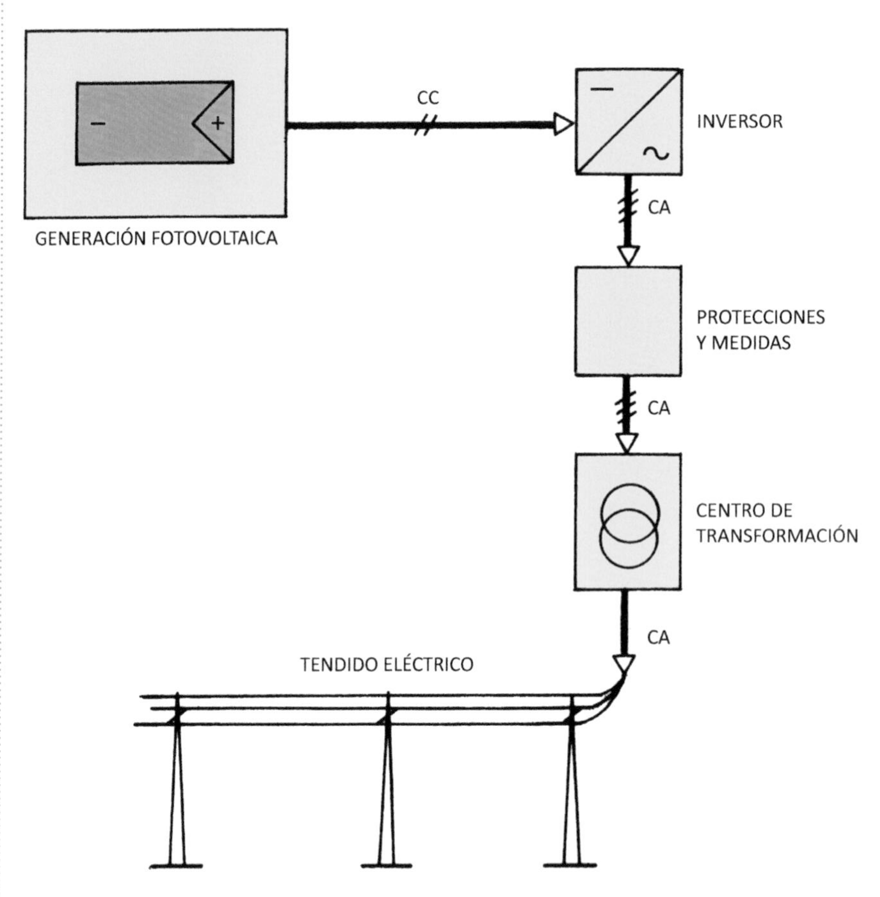

Esquema 7.16. Parque de generación de fotovoltaica, conectado a la red eléctrica.

Figura 7.8. Paneles y estación transformadora en una central eléctrica fotovoltaica.

Estas instalaciones generadoras se conectan a la red pública de electricidad y pertenecen al grupo de "instalaciones conectadas a la red". El otro tipo de generación es el denominado como "instalaciones aisladas", que son más pequeñas.

Los elementos principales de las instalaciones conectadas a la red son: los paneles fotovoltaicos, el inversor de corriente que pasa la corriente continua generada por los paneles, a corriente alterna, que es la forma de corriente utilizada por los receptores eléctricos, y el transformador de tensión.

7.6. Aplicaciones de la generación fotovoltaica

La generación de energía eléctrica fotovoltaica tiene muchas aplicaciones, especialmente en aquellos lugares alejados de las redes y suministros eléctricos, para aplicaciones muy diferentes entre sí.

Las instalaciones fotovoltaicas pueden ser:

a) Aisladas

Cuando no tienen ninguna relación con las redes eléctricas convencionales, y a este grupo o aplicación pertenecen:

- Electrificación rural.
- Aplicaciones agropecuarias.
- Instalaciones de bombeo.
- Telecomunicaciones.
- Señalización y alarma.
- Toma y transmisión de datos.
- Alumbrado autónomo.
- Desalinización.
- Aplicaciones agrícolas, avícolas y ganaderas.
- Aplicaciones aeronáuticas y militares.

Aplicaciones a telecomunicaciones, señalización y control

- Centros remotos de telecomunicaciones.
- Repetidores de radio y televisión.
- Puestos de vigilancia forestal.
- Señalización para tráfico.
- Puestos de toma de muestras y mediciones.
- Estaciones meteorológicas.
- Señalización para ferrocarriles.
- Radiofaros y radiobalizas.
- Control de pasos a nivel.
- Alumbrado de vallas publicitarias.
- Señalización e iluminación de aeropuertos y otras superficies.

Alumbrado en general

- Iluminación de túneles y carreteras.
- Iluminación y control de invernaderos.
- Iluminación de granjas y establos.
- Iluminación de parajes alejados.
- Iluminación de viviendas.
- Iluminación de casetas.

Aplicaciones agrícolas

- Electrificación de cercas para el ganado.
- Riegos por goteo.
- Bombeo de agua.
- Riegos a baja presión.
- Alimentación a centrales frigoríficas apartadas.

Aplicaciones domésticas

- Iluminación.
- Alimentación de electrodomésticos.
- Alimentación de aparatos informáticos y otros.
- Alimentación de bombeo de agua.

Aplicaciones aeronáuticas y militares

- Satélites.
- Comunicaciones diversas.
- Equipos de campaña.
- Radioteléfonos.

b) Conectadas a redes de baja tensión

Cuando la energía generada se suministra a la red general de baja tensión, para su distribución comercial.

Figura 7.9. Instalaciones aisladas. Alimentación de un expendedor de etiquetas y farolas de alumbrado.

Figura 7.10. Generación eléctrica para alimentar un puesto de telecomunicaciones aislado.

Figura 7.11. Subestación en una central de generación fotoeléctrica.

Figura 7.12. Pequeño módulo fotovoltaico.

Figura 7.13. Aplicación de un interruptor crepuscular para el encendido/apagado de un alumbrado exterior.

A este grupo pertenecen las llamadas huertas solares y la energía generada para fines particulares en la que los sobrantes o excedentes puedan ser vendidos a través de la red general de baja tensión.

Suministros de energía a la red eléctrica de baja tensión

- Huertas solares o centrales termoeléctricas.
- Sobrante o excedente de pequeñas instalaciones de generación.
- Instalaciones a las que alude la Exigencia básica HE 5: Contribución fotovoltaica mínima de energía eléctrica.

Otras muchas aplicaciones que todos conocemos

Ejemplo de aplicación para un equipo de medida del nivel de contaminación.

Pequeña placa fotovoltaica instalada para alimentar a un equipo de medida que toma muestras del aire ambiente, para controlar el estado de contaminación de la zona.

De esta forma, no es necesaria una alimentación de corriente de la red eléctrica, especialmente cuando hay dificultad para hacer dicho aprovisionamiento.

7.7. Otras aplicaciones de las células fotovoltaicas

El efecto fotovoltaico tiene muchas aplicaciones, como son las aplicaciones en sensores, detectores e interruptores.

- Sensores de movimiento.
- Detectores de presencia.
- Detectores de paso.
- Interruptores crepusculares.
- Pequeños generadores.

Las células fotovoltaicas modernas reaccionan ante la luz visible, y también ante los rayos infrarrojos, ultravioleta, luz visible roja, luz visible verde y láser, dependiendo de la tecnología que acompaña a la célula fotovoltaica.

Cuando la luz incide en el cátodo de la célula fotovoltaica, se genera una corriente eléctrica que acciona un relé. La combinación de la célula fotoeléctrica con un relé permite construir muchos dispositivos de detección.

Los sensores de luz o interruptores crepusculares permiten conectar alumbrados al anochecer y desconectarlos al almacenar.

El detector fotoeléctrico está constituido básicamente por un emisor de luz asociado a un receptor sensible a la cantidad de luz recibida. Cuando un objeto modifica sensiblemente este haz, provoca el cambio de estado de la salida.

La célula fotoeléctrica con relé conmutado (NC/NA), regulable por ejemplo entre, 2 y 200 lux, con temporización con retardo de encendido (efecto nube).

7.8. Instalaciones híbridas

Paneles solares híbridos son aquellos que además de generar electricidad por medio de sus células fotovoltaicas, también recuperan calor solar para calentar un fluido para utilizarlo en calentar agua para calefacción o para agua caliente sanitaria (ACS).

Los paneles híbridos tienen una construcción diferente a la de los paneles fotovoltaicos que sólo generan electricidad.

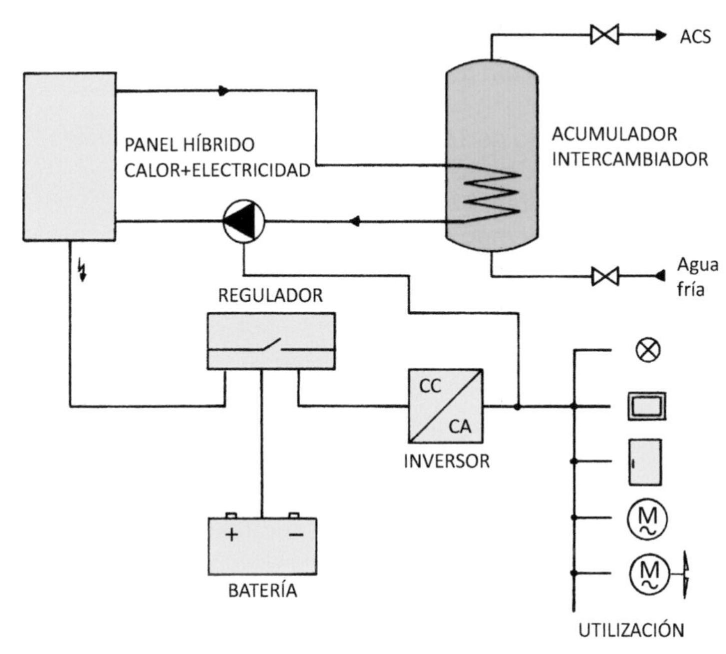

Esquema 7.14. Instalación híbrida (electricidad y calor).

7.9. Mantenimiento de instalaciones fotovoltaicas

Como se ha dicho, los paneles fotoeléctricos tienen una vida útil que supera los 30 años, sin embargo este tipo de instalaciones precisa de un cuidado adecuado para conseguir su máximo rendimiento, como es:

- Limpieza periódica de los paneles, especialmente cuando están expuestos a polvo y a la contaminación ambiental.
- Controlar el estado de los soportes y anclajes, teniendo en cuenta que están expuestos a la intemperie (viento, lluvia, nieve, hielo, etc.).
- Asegurar el correcto aislamiento y estanqueidad de las cajas de conexiones.
- Vigilar el estado de los conductores que están en el exterior, sujetos a las inclemencias del tiempo.
- Verificar el estado de las puestas a tierra de las estructuras y soportes metálicos.
- Verificar la zona de las baterías, especialmente las fugas que puedan darse.
- Conservar las baterías según las instrucciones dadas por el fabricante de las mismas.

7.10. Sencillos datos de cálculo

El cálculo de las instalaciones fotovoltaicas es complejo, sin embrago, en este caso nos limitamos a ejemplos sencillos de cálculo y aplicaciones igualmente sencillas.

7.10.1. Ejemplo de cálculo de la instalación fotovoltaica

Datos de partida:

- Lámpara de sodio de 18 W (bajo consumo, alto rendimiento).
- Batería de 12 V de tensión.
- Tensión del módulo: 12 V.
- Estimación del tiempo de empleo de la energía: 10 horas-día.

Cálculo de los elementos de la instalación:

a) Energía útil, energía perdida y energía total necesaria

$$E_u = \frac{P \cdot t}{V} = \frac{18 \times 10}{12} = 15\ Ah/día$$

Considerando las pérdidas en la instalación de un 20% correspondientes a la reactancia, instalación, batería, regulador y módulo, la energía perdida sería de:

$$E_p = \frac{E_u \cdot \%}{100} = \frac{15 \times 20}{12} = 3\ Ah/día$$

Energía total necesaria:

$$E_T = E_u + E_p = 15 + 3 = 18\ Ah/día$$

b) Horas solar pico (HSP)

Suponiendo que el panel tenga un ángulo de inclinación de 45º

$HSP = Radiación \cdot 0{,}0239 \cdot 0{,}0116 = 6.852 \times 0{,}0239 \times 0{,}0116 = 1{,}8996$

No es lo mismo realizar una instalación en Valladolid, que en Málaga o Tenerife.

Radiación en el mes de enero, en las zonas citadas:

Zona o lugar	Radiación con ángulo de inclinación de 45º
Valladolid	6.852 kJ/m^2
Málaga	13.286 kJ/m^2
Tenerife	12.998 kJ/m^2

Para el cálculo se elige la zona de Valladolid, y paneles de 80 Wp y 4,76 A.

c) Amperios-hora-día (Ah/día)

$Ah/día = I_{módulo} \cdot HSP = 4{,}76 \times 1{,}8996 = 9{,}042096$

d) Número de módulos necesarios (N_m)

$$N_m = \frac{E_T}{Ah / día} = \frac{18}{9{,}042096} = 1{,}99 \text{ paneles} = 2 \text{ paneles de } 80 \text{ Wp}$$

e) Capacidad de carga de la batería (Q_c)

$$Q_C = \frac{100 \cdot E_T \cdot D}{P_c} = \frac{100 \times 18 \times 5}{70} = 128{,}57 \text{ Ah}$$

E – Energía almacenada, en Ah/día.
D – Días de autonomía (para el cálculo, 5 días).
P – Profundidad de la descarga (ejemplo: 70%).

f) Intensidad del regulador (I_R)

$$I_R = N_{pp} \cdot I_p = 2 \times 4{,}76 = 9{,}52 \text{ A}$$

N_{pp} – Número de módulos en paralelo.
I_p – Intensidad máxima que suministra el panel.

g) Elección del módulo o panel fotovoltaico
Características eléctricas principales:

- Potencia pico (Wp)
- Intensidad máxima (A).
- Tensión de suministro (V).

7.10.2. Valores orientativos que proporciona la firma SOLENER para sencillas aplicaciones de instalaciones fotovoltaicas

Aplicación 1

Para una pequeña caseta que tiene los siguientes receptores:

Solamente alumbrado con lámparas de bajo consumo (8 a 20 W) y que se utiliza con poca frecuencia (esporádicamente), con tensión de servicio de 12 V.

Elementos de la instalación eléctrica de generación fotovoltaica:

Aparatos necesarios	Características
1 Panel fotovoltaico	5 Wp a 12 V
1 Regulador	10 A a 12 V
1 Batería monobloc	26 Ah

Aplicación 2

Para una pequeña caseta utilizada los fines de semana en invierno y un mes en verano.

Solamente instalación de alumbrado para hasta 8 puntos de luz de 20 W con tensión de servicio de 12 V.

Elementos de la instalación eléctrica de generación fotovoltaica:

Aparatos necesarios	Características
1 Panel fotovoltaico	55 Wp a 12 V
1 Regulador	30 A a 12 V
1 Batería monobloc	102 Ah

Figura 7.15. Pequeña instalación fotovoltaica con la que se alimenta el alumbrado, un televisor y una bomba.

Aplicación 3

Para una pequeña instalación utilizada los fines de semana de invierno y un mes en verano, con 8 puntos de luz de 20 W (bajo consumo), televisión y pequeños electrodomésticos funcionando en corriente alterna (CA), a 230 V.

Elementos de la instalación eléctrica de generación fotovoltaica:

Aparatos necesarios	Características
2 Paneles fotovoltaicos	55 Wp a 12 V
1 Regulador	30 A a 12 V
1 Batería estacionario	290 Ah
1 Inversor de onda senoidal modificada	700 W a 12 V

Aplicación 4

Instalación para una vivienda utilizada los fines de semana en invierno y un mes en verano, con 8 puntos de luz a 20 W cada uno (bajo consumo), televisión, pequeños aparatos electrodomésticos, y lavadora funcionando en programa frío. Todo en corriente alterna (CA) a 230 V.

Elementos de la instalación eléctrica de generación fotovoltaica:

Aparatos necesarios	Características
4 Paneles fotovoltaicos	55 Wp a 12 V
1 Regulador	30 A a 12 V
1 Batería estacionario	630 Ah
1 Inversor de onda senoidal modificada	1.500 W a 12 V

Aplicación 5

Instalación para una vivienda que se utiliza todos los fines de semana del año, con 12 puntos de luz a 20 W cada uno (bajo consumo), televisión, pequeños electrodomésticos, lavadora funcionando en programa frío y frigorífico de bajo consumo que totalizan 3.000 Wh/día. Todo en corriente alterna (CA) a 230 V.

Elementos de la instalación eléctrica de generación fotovoltaica:

Aparatos necesarios	Características
6 Paneles fotovoltaicos	55 Wp a 12 V
1 Regulador	30 A a 24 V
2 Batería estacionario	436 Ah
1 Inversor de onda senoidal modificada	2.000 W a 24 V

8 Energía eólica

La energía eólica tiene su origen en la fuerza del viento. Es una energía renovable y poco contaminante. Su aportación como energía alternativa es muy importante para la generación de energía eléctrica. En las previsiones de futuro se espera un ascenso progresivo de esta forma de energía, tal como hemos visto al tratar de las energías renovables y alternativas.

El problema de esta forma de generación está en que se trata de una fuente de energía intermitente, dependiendo de si hay o no hay viento con los parámetros adecuados de velocidad.

El 9 de noviembre de 2010 se produjo un máximo histórico de producción instantánea con 14.962 MW, a las 14,46 horas, lo que supuso el 46,65% de la generación eléctrica instantánea. En este caso concreto, la potencia generada fue de casi el doble de la capacidad de generación de las seis centrales nucleares que hay en España (7.742 MW), pero en este caso, generación continuada y regular durante las 24 horas del día.

Aerogeneradores

8.1. Introducción a la energía eólica

Hace 4000 años, los persas ya utilizaban esta energía para la extracción y elevación del agua desde ríos y pozos. En grabados egipcios de 3.000 años (adC), ya aparecen las velas en los barcos que representaban. Después, lo aplicaron los fenicios y los romanos. Todavía hoy, hay embarcaciones que utilizan esta forma de energía.

En el siglo XIV, se desarrolló en Francia el molino de torre (construcción de piedra) que tenía en su parte superior un eje horizontal a cuyo extremo exterior se colocaban 4 u 8 aspas que podían medir de 3 a 9 m. El eje horizontal estaba comunicado con la aplicación por medio de engranajes, y su finalidad principal era, moler granos de cereales.

En el siglo XII había molinos de viento en Francia e Inglaterra.

El "Quijote" cita los molinos de viento de la Mancha, en los que se molían los granos.

En los siglos XV al XIX, los molinos también se aplicaron para el bombeo de agua, aserraderos de madera, fabricación de papel, prensado de materias, etc. Como dato anecdótico señalaremos, que a lo largo del siglo XIX se construyeron en Holanda alrededor de 9.000 molinos, algunos de los cuales perduran hasta la fecha y dan al paisaje la fisonomía que todos conocemos.

En 1745 se inventó el molino de aspas en forma de abanico.

En 1772 se le introdujo una variante que consistía en la posibilidad de regular sus aspas y su orientación, con lo que se conseguía un mejor aprovechamiento de la energía del viento y una velocidad mucho más constante.

Los "molinos" también fueron aplicados a la generación de electricidad, pero a pequeña escala. Fue a finales de los años 70, y tras el encarecimiento de las materias primas energéticas, cuando comenzó la aplicación de los aerogeneradores para la producción de energía eléctrica.

La energía eléctrica obtenida por medio de la fuerza del aire, puede llegar a representar en un futuro próximo, el 10% del total de energía eléctrica consumida.

En 1986 se inauguró en España un parque eólico de gran potencia en Tenerife (Islas Canarias).

La fuerza del viento se viene empleando desde la antigüedad, por tanto no se trata de una aplicación reciente como podemos ver con un pequeño repaso a la historia.

La Tierra recibe del Sol en torno a 2.000 kWh/m^2/año, de la que el 2% de la energía que nos llega se convierte en energía eólica.

La energía contenida en el viento y que se transforma en energía mecánica está comprendida entre el 20 y el 40% dependiendo de la geometría de las palas (rotor).

Se tienen referencias de que hace 1700 años a. C. en Babilonia se utilizaba la fuerza del viento para elevar agua. Los barcos en la antigüedad eran empujados por la fuerza del viento, y todavía se sigue empleando esta energía eólica.

El molino de viento inicia su desarrollo en Francia el año 1180, para poco a poco ir generalizándose.

Los molinos de la Mancha son famosos desde que el año 1605 fuera publicado el Quijote, escrito por Miguel de Cervantes.

Los molinos accionados por la fuerza del viento tuvieron una evolución en su mecánica y palas en los siglos XVIII y XIX, con lo que se ganó eficiencia.

Las primeras bombas de pistón accionadas por un rotor multipala de accionamiento eólico, aparecen en Estados Unidos en 1854, desarrolladas por Daniel Halladay.

En 1880, Charles F Brush construyó en EE.UU. una turbina eólica para producir energía eléctrica (corriente continua), que luego se almacenaba en baterías.

En 1892, el profesor danés Lacour diseñó el primer prototipo de un aerogenerador eléctrico. Tenía cuatro palas y una capacidad de generación entre 5 y 25 kW.

La crisis energética de 1973 dio un fuerte impulso a las energías alternativas, y de entre ellas la energía eólica.

En 1975 se construye en EE.UU. un aerogenerador bipala de 35 metros de diámetro, palas de metal, que daba una potencia de 100 kW. En 1987 y siguiendo con las pruebas se instala en Haway un aerogenerador de 100 metros de diámetro y 3,2 MW de potencia.

En 1979 se promueve un estudio sobre el aprovechamiento de la energía eólica, que se materializa a principio de los años 80 en el parque eólico experimental de Punta Tarifa (Cádiz) en dos fases.

En los años 90 se dio un gran impulso a esta tecnología en España

En 1986 había en España 0,5 MW instalados, en 1992 eran 45 MW y en 1996, 210 MW.

El año 2008 se instaló la mayor turbina española sobre una torre de 140 metros de altura, 100 m de diámetro y 3 MW de potencia.

Observación:

A principios del año 2011, Navarra tenía instalados en torno a 1.200 aerogeneradores en 40 parques eólicos que sumaban una potencia total instalada de 950 MW. Se estima que el tiempo real de generación está en torno a 8 horas/día (1/3 de las 24 horas del día), aunque puede pasar un día entero, e incluso más, sin que se genere electricidad por falta de viento.

Sta. María de Garoña (generación media/pequeña) genera 460 MW en continuo. Las centrales más modernas superan los 1.000 MW por generador.

Harían falta 3.000 MW de generación eólica para proporcionar la misma energía que una sola central nuclear de un generador de 1.000 MW.

Hay que pensar que si con 1.200 aerogeneradores se ven casi todas las cumbres plantadas de torres, ¿qué será con 3.000 aerogeneradores? Para pensarlo.

8.2. Conceptos generales

8.2.1. La atmósfera terrestre

La capa atmosférica se comporta como un escudo protector contra la penetración de radiaciones nocivas, contra los meteoritos y en concreto, contra el ozono, protegiendo la atmósfera que respiramos y la superficie del planeta.

8.2.2. El viento

Corrientes de aire producidas sobre la superficie terrestre.

El aire se desplaza de unas zonas o áreas de más presión a otras con menos presión, en forma de viento. Las diferencias de temperatura colaboran a las variaciones de presión.

8.2.3. Isobaras

Las isobaras son líneas que unen puntos de igual presión. Los mapas de isobaras son usados por los meteorólogos para los mapas de predicción del tiempo.

8.2.3. Presión atmosférica

Presión ejercida por el aire sobre la superficie de la Tierra.

La presión atmosférica media a nivel del mar es de 1.013 milibares (mb) o hectopascales (hPa) o 102.325 N/m^2 o pascal (Pa).

En el Sistema Internacional de unidades (SI), la unidad de presión es el newton por metro cuadrado (N/m^2).

$$1\ N/m^2 = 1\ \text{Pa}$$

A medida que se sube en altura sobre el nivel del mar, baja la presión y decrece la masa de aire. Por ejemplo, a 10.000 m de altitud, la presión es cuatro veces inferior a la del nivel del mar y la temperatura es de –55 °C.

8.2.4. Altas y bajas presiones

Se denomina zona con ***altas presiones***, cuando la presión reducida al nivel del mar y a 0 °C, es mayor de 1.013 mb (milibares) y ***bajas presiones,*** si el valor es menor que ese valor.

8.2.5. Presión absoluta y relativa

a) Presión absoluta

Corresponde a la presión que realmente soportan las moléculas de un determinado fluido o las paredes del recipiente que las contiene.

b) Presión relativa

Es la diferencia existente entre la presión absoluta y la presión atmosférica.

c) Aparatos para medir la presión

- ***Manómetros:*** miden la presión relativa superior a la atmosférica.
- ***Vacuómetros:*** miden la presión relativa inferior a la atmosférica.
- ***Barómetros:*** miden la presión atmosférica.

d) Fórmula con la que se calcula la presión:

$$P_{red} = P_{abs} + 1.013,25 \left[1 - \left(\frac{288 - 0,0065\,h}{288} \right)^{5,255} \right] \quad \text{(hPa)}$$

P_{red} – Presión atmosférica reducida al nivel del mar, en hPa.
P_{abs} – Presión atmosférica absoluta, en hPa.
h – Altura a nivel del mar, en metros.
Δ_p – Diferencia de presión, en hPa.

Tabla 8.1. Determinación de la altura sobre el nivel del mar en función de la diferencia de presión entre dos puntos.

Altura (h) sobre nivel del mar (m)	Diferencia de nivel (Δ_p) (hPa)	Altura (h) sobre nivel del mar (m)	Diferencia de nivel (Δ_p) (hPa)
100	12,0	1.100	125,4
200	23,8	1.200	136,1
300	35,5	1.300	146,8
400	47,2	1.400	157,3
500	58,7	1.500	167,7
600	70,1	1.600	178,1
700	81,3	1.700	188,3
800	92,5	1.800	198,4
900	103,6	1.900	208,4
1.000	114,5	2.000	218,4

Tabla 8.2. Características de la atmósfera en función de la altura sobre el nivel del mar.

Altura m	Presión mb	Densidad G · dm^{-3}	Temperatura °C
0	1013	1,226	15
1.000	898,6	1,112	8,5
2.000	794,8	1,007	2
3.000	700,9	0,910	-4,5
4.000	616,2	0,820	-11
5.000	540,0	0,736	-17,5
10.000	264,1	0,413	-50
15.000	120,3	0,194	-56,5

[1] Para latitudes templadas.

8.2.6. Humedad

a) Humedad absoluta

Corresponde a la cantidad de vapor acuoso contenido en el volumen de un metro cúbico (m^3) de aire.

b) Humedad relativa

Relación entre la masa de vapor acuoso contenido en un determinado volumen de aire y la que existiría en el mismo volumen, si el aire estuviera saturado. La humedad se da en tanto por ciento (%).

$$Humedad = 100 \cdot \frac{presión\ parcial\ del\ vapor\ de\ agua}{presión\ de\ vapor\ a\ la\ misma\ temperatura}\ (\%)$$

Nota:

Una humedad relativa normal junto al mar puede ser del 90%, lo que significa que el aire contiene el 90% del vapor de agua que puede admitir, mientras un valor normal en una zona seca puede ser de 30%.

Tabla 8.3. Peso de gases y vapores industriales a 0 °C y 1 atm

Gases y vapores	g/dm^3	Gases y vapores	g/dm^3
Acetileno	1,1709	Etileno	1,2605
Ácido clorhídrico	1,6391	Flúor	1,6950
Aire seco	1,2928	Freón 11 (fluoruro clorofórmico)	-
Alcohol etílico	2,0430	Freón 12 (difluordiclormetano)	5,1100
Amoníaco	0,7714	Freón 13 (trifluorclormetano)	-
Anhídrido carbónico	1,9768	Gas del alumbrado	0,5600
Anhídrido sulfuroso	2,9263	Helio	0,1785
Butano-n	2,7030	Hidrógeno	0,0898
Argón	1,7839	Metano	0,7168
Cloro	3,2140	Nitrógeno	1,2505
Cloroformo	5,2830	Óxido de carbono	1,2500
Cloruro metálico	2,3070	Oxígeno	1,289
Etano	1,3560	Ozono	2,1440
Éter	-	Propano	2,0037
Éter metílico	2,1097	Vapor de agua	0,7680

Tabla 8.4. Saturación de agua en el aire (humedad) en función de la temperatura.

Temperatura (°C)	Saturación (g/m³)
-20	0,89
-10	2,16
0	4,85
10	9,40
20	17,30
30	30,37
40	51,17

a) Agua en la atmósfera

El aire puede contener vapor de agua en suspensión.

- La atmósfera contiene en torno a 12.000 km³ de agua.
- Entre 0 y 1.800 metros sobre el nivel del mar, está la mitad del agua contenida en la atmósfera.
- Se evaporan y más tarde se licuan en torno a los 500.000 km³/año.
- La evaporación potencial es de: 940 litros por m² y año en los océanos, y entre 200 a 6.000 litros por m² en los continentes.

b) Humedad del aire

La cantidad de agua en el aire tiene un límite y depende fundamentalmente de su temperatura.

- Cuando el aire está saturado de agua da lugar a que se licue en forma de gotas.
- El aire frío admite menos agua que el aire caliente.
- 1 m³ de aire a 0 °C puede contener como máximo: 4,85 g de vapor de agua.
- 1 m³ de aire a 25 °C puede contener 23,5 g de vapor de agua.

c) Humedad de saturación

Es la cantidad máxima de vapor de agua que contiene un metro cúbico de aire en unas condiciones determinadas de presión y temperatura.

8.2.7. Densidad del aire

La densidad del aire es un factor muy importante para determinar la fuerza del viento y su transformación en energía mecánica a través de los aerogeneradores.

La densidad del aire decrece con el incremento de temperatura y aumenta con la humedad.

a) Densidad absoluta

Es la magnitud que expresa la relación entre la masa y el volumen de un cuerpo. La unidad en el SI es el kilogramo por metro cúbico (kg/m³).

$$\rho = \frac{m}{V}$$

m – Masa.
V – Volumen.

b) Densidad relativa

Respecto a una sustancia, es la relación existente entre su densidad y la de otra sustancia de referencia, no tiene unidad, por lo que es una magnitud adimensional.

$$\rho_r = \frac{\rho}{\rho_0}$$

ρ_r – Densidad relativa.
ρ – Densidad de la sustancia.
ρ_0 – Densidad de referencia o absoluta.

Para los gases, la densidad de referencia habitual es la del aire a la presión de 1 atm y la temperatura de 0 °C.

Unidades de densidad:

- Kilogramo por metro cúbico (kg/m).
- Gramo por centímetro cúbico (g/cm^3).

8.2.8. Cálculo de la densidad del aire

Valores fundamentales:

- Presión atmosférica (hPa).
- Temperatura ambiente (°C).
- Humedad relativa (%).
- Densidad del aire (kg/m^3).

Fórmula para calcular la densidad del aire:

$$\rho = \frac{p \cdot Ma}{Z \cdot R \cdot T} \cdot \left[1 - x_v \cdot \left(1 - \frac{Mv}{Ma} \right) \right] \; (kg/m^3)$$

ρ – Densidad del aire, en kg/m^3.
p – Presión atmosféricas, en Pa.
Ma – Masa molecular del aire seco: 0,0289635 kg/mol
Para aire con una fracción molar de CO_2: $x_{co2} = 0,0004$.
Z – Factor de compresibilidad adimensional.
R – Constante molar de los gases: 8,31451 J/K (mol).
T – Temperatura termodinámica, en K, T = t + 273,15.
x_v – Fracción molecular de vapor de agua, adimensional.
Mv – Masa molar del vapor de agua: 0,0180154 kg/mol.

Tabla 8.5. Densidad del aire a presión atmosférica estándar

Temperatura °C (Celsius)	Temperatura °F (Fahrenheit)	Densidad del aire Kg/m³	Contenido de agua Kg/m³
-25	-13	1,423	
-20	-4	1,393	
-15	5	1,368	
-10	14	1,342	
-5	23	1,317	
0	32	1,292	0,005
5	41	1,269	0,007
10	50	1, 247	0,009
15	59	1,225 (1)	0,013
20	68	1,204	0,017
25	77	1,184	0,023
30	86	1,165	0,030
35	95	1,146	0,039
40	104	1,127	0,051

(1) La densidad del aire seco a la presión atmosférica estándar sobre el nivel del mar a 15 °C se utiliza como estándar en los cálculos eólicos (industria eólica).

8.2.9. Composición del aire atmosférico

De los componentes del aire atmosférico, el 99,02% de su volumen pertenece a solo dos gases, nitrógeno y oxígeno.

Tabla 8.6. Composición del aire.

Gas	Símbolo	Volumen (%)
1 Nitrógeno	N	78,03
2 Oxígeno	O	20,99
3 Dióxido de carbono	CO_2	0,03
4 Argón	Ar	0,94
5 Neón	Ne	0,00123
6 Helio	He	0,0004
7 Criptón	Kr	0,00005
8 Xenón	Xe	0,000006
9 Hidrógeno	H	0,01
10 Metano	CH_4	0,0002
11 Óxido nitroso	N_2O	0,00005
12 Vapor de agua	H_2O	Variable
13 Ozono	O_3	Variable
14 Partículas		Variable

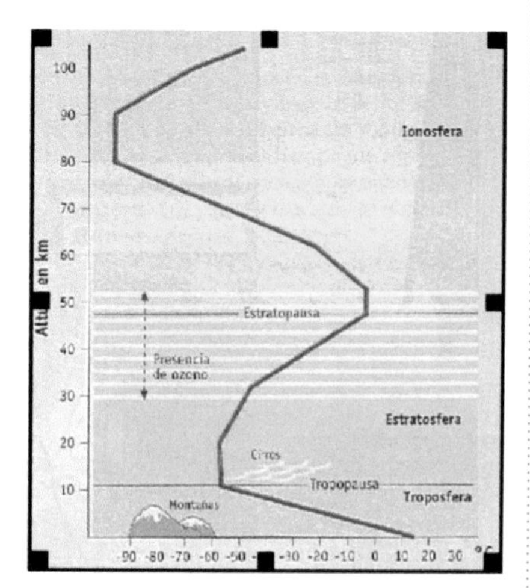

Figura 8.1. Niveles atmosféricos.

8.2.10. Zonas atmosféricas

Disponemos de aire en los 10 primeros kilómetros sobre el nivel del mar.

8.3. Fórmulas relacionadas con el aire

A continuación se presentan las principales magnitudes, unidades y fórmulas relacionadas con el aire.

8.3.1. Principales unidades de presión para fluidos gaseosos

Tabla 8.7. Relación entre diferentes unidades de presión.

	Pascal (Pa)	bar	N/mm^2	kp/m^2	kp/cm^2	atm	Torr
1 Pa (N/m^2) =	1	10^{-5}	10^{-6}	0,102	$0,102x10^{-4}$	$0,987x10^{-5}$	0,0075
1 bar (daN/cm^2) =	100.000	1	0,1	10.200	1,02	0,987	750
1 N/mm^2 =	10^6	10	1	$1,02x10^5$	10,2	9,87	7.500
1 kp/m^2 =	9,81	$9,81x10^{-5}$	$9,81x10^{-6}$	1	10^{-4}	$0,968x10^{-4}$	0,0736
1 kp/cm^2 =	98.100	0,981	0,0981	10.000	1	0,968	736
1 atm (760 Torr) =	101.325	1,013	0,1013	10.330	1,033	1	760
1 Torr (mmHg) =	133	0,00133	$1,33x10^{-4}$	13,6	0,00132	0,00132	1

8.3.2. Fórmulas relacionadas con fuerza y presión

a) Presión (P)

$$P = \frac{F}{S}$$

b) Fuerza (F)

$$F = m \cdot a$$

c) Presión en función de la fórmula anterior

$$P = \frac{m \cdot a}{S}$$

8.3.3. Dilatación de los gases:

- Cuando un gas se calienta, aumenta su volumen si la presión se mantiene constante o aumenta su presión si el volumen no varía.
- Se llama coeficiente de dilatación (α) a presión constante, el aumento que experimenta la unidad de volumen del gas, cuando éste eleva su temperatura un grado.

$$Vt = Vo\,(1 + \alpha \cdot t)$$

Vt – Volumen a temperatura (t).
Vo – Volumen inicial.
α – Coeficiente de dilatación.
t – Temperatura en que se ha incrementado el gas.

8.3.4. Leyes relacionadas con el aire

8.3.4.1. Ley de Boyle-Mariotte

A temperatura constante, los volúmenes ocupados por una masa gaseosa son inversamente proporcionales a las presiones a que se les somete.

$$\frac{P_1}{P_2} = \frac{V_1}{V_2} \; ; \; P_1 \cdot V_1 = P_2 \cdot V_2$$

A la presión P_1, el gas ocupa el volumen V_1.
A la presión P_2, el gas ocupa el volumen V_2.
El producto de la presión de un gas por su volumen, es una cantidad constante para ese gas, si la temperatura no varía.

8.3.4.2. Leyes de Gay-Lussac

1ª Ley. El valor del coeficiente (α) de dilatación de todos los gases es el mismo a presión constante y su valor es de:

$$\alpha = \frac{1}{273} = 0{,}00366$$

El volumen ocupado por un gas a 0 K, sería igual a cero.

2ª Ley. A presión constante, los volúmenes ocupados por una masa gaseosa son proporcionales a sus temperaturas absolutas.

$$\frac{V_1}{V_2} = \frac{T}{T'}$$

Vt – Volumen a temperatura t.
Vt' – Volumen a temperatura t'.
T y T' – Temperaturas absolutas.

3ª Ley. Las presiones a que está sometido un gas a volumen constante, son proporcionales a las temperaturas absolutas.

8.3.5. Ecuación de los gases perfectos

Son gases perfectos, aquellos que cumplen las leyes de Boyle-Mariotte y Gay-Lussac.

$$P \cdot V = Po \cdot Vo\,(1 + \alpha \cdot t)$$

Po – Presión inicial del gas.
P – Presión final del gas.
Vo – Volumen inicial del gas.
V – Volumen final del gas.
α – Coeficiente de dilatación del gas (0,00366).
T – Temperatura final del gas.

8.4. Velocidad del viento y presión que ejerce

El viento es una forma de energía que a través de un aerogenerador se transforma en energía mecánica rotativa y en energía eléctrica por medio de la dinamo o un alternador eléctrico.

8.4.1. Energía eólica

La potencia de los generadores eólicos se fundamenta básicamente en tres factores, que son:

- Superficie o área barrida por las palas del rotor, en m^2.
- Densidad del aire, en kg/m^3.
- Velocidad del viento, en m/s.

8.4.2. Inicios de la generación eólica en España

En 1987 se instala el primer aerogenerador en Tarifa (Cádiz). Su potencia era de 100 kW.

- En 1984 se monta el primer parque eólico en el Ampurdán (Gerona), con cinco generadores con 24 kW de potencia.
- En 15 años hemos visto como han proliferado los parques eólicos por toda España y se ha llegado a generar 5.000 MkW.

Referencias:

El consumo de energía de una vivienda media es de 3.600 kWh/año.

Un aerogenerador de 150 kW de potencia, con un rendimiento entre el 85 y 95% produce una energía de 373.000 kWh/año, es decir, que podría suministrar energía a 100 viviendas.

8.4.3. Identificación del viento

La fuerza del viento es el elemento energético que por medio de aerogeneradores se transforma en energía eléctrica. Habrá generación de electricidad, cuando se tenga viento dentro de un campo de velocidades. No se puede sobrepasar la velocidad máxima fijada, porque se pueden dar graves problemas en el conjunto aerogenerador.

Tabla 8.8. Velocidad del viento y equivalencias.

Velocidad (m/s)	Velocidad (km/h)	Velocidad (nudos)[1]	Clasificación del viento
0,0 – 1,8	0 a 6,48	0 a 3,5	Calma
1,8 a 5,8	6,48 a 20,88	3,5 a11	Ligero
5,8 a 8,5	20,88 a 30,60	11 a 17	Moderado
8,5 a 14	30,60 a 50,40	17 a 28	Fresco
14 a 21	50,40 a 75,6	28 a 41	Fuerte
21 a 25	75,6 a 90,00	41 a 48	Temporal
25 a 29	90,00 a 104,4	48 a 56	Fuerte
29 a 34	104,4 a 122,4	56 a 65	Temporal
>34	>122,4	>65	Huracán

[1] Un nudo equivales a una milla marina, a una velocidad de 0,5144 m/s o 1,85184 km/h. Los valores de la tabla están dados en nudos/s.

Tabla 8.9. Potencia del viento [1] en función de su velocidad

Velocidad m/s	Potencia W/m²	Velocidad m/s	Potencia W/m²	Velocidad m/s	Potencia W/m²
0	0	8	313,6	16	2.508,8
1	0,6	9	446,5	17	3.009,2
2	4,9	10	612,5	18	3,572,1
3	16,5	11	815,2	19	4.201,1
4	39,2	12	1.058,4	20	4.900,0
5	76,5	13	1.345,7	21	5,672,4
6	132,3	14	1.680,7	22	6.521,9
7	210,1	15	2.067,2	23	7.452,3

[1] Para la densidad del aire seco (1,225 kg/m³) que corresponde a la presión atmosférica estándar a 15 °C sobre el nivel del mar.

Tabla 8.10. Escala de Beaufort sobre velocidad del viento y sus efectos.

Escala Beaufort	nudos	km/h	m/h	m/s	Definición	Efectos del viento		
						En tierra	En la mar cerca de la costa(referidos a botes de vela)	En alta mar
0	0	0	0	0	calma	Bonanza, el humo sube verticalmente	bonanza, los barcos no gobiernan	La mar está como un espejo
1	1-3	1-5	1-3	<2	ventolina	La dirección del viento es indicada por el humo pero no por las banderas	Los barcos empiezan a moverse	Se forman rizos como escamas de pescado pero sin espuma
2	4-6	6-11	4-7	2-3	Suave (brisa muy débil)	El viento se siente en la cara y la banderola se mueve	El viento infla las velas de los barcos que van a aproximadamente 1-2 nudos	Pequeñas olas, crestas de apariencia vítrea, sin romperse
3	7-10	12-19	8-12	4-5	Leve (brisa débil)	Las hojas y las pequeñas ramas se mueven	Los barcos empiezan a inclinarse y van a aproximadamente 3-4 nudos	Pequeñas olas cuyas crestas empiezan a romper, espuma de aspecto vítreo aislados vellones de espuma
4	11-16	20-28	13-18	6-7	Moderado (brisa moderada)	Se levantan papel y polvo, las ramas más finas se mueven	Viento manejable: los barcos tienen las velas bien inclinadas	Las olas se hacen más largas. Borreguillos numerosos.
5	17-21	29-38	19-24	8-10	regular (brisa fresca)	Empiezan a moverse los árboles pequeños	Los barcos arrían las velas	Olas moderadas alargadas. Gran abundancia de borreguillos, eventualmente algunos rociones
6	22-27	39-49	25-31	11-13	fuerte (brisa fuerte)	Se mueven las ramas grandes. Se utilizan con dificultad los paraguas.	Los barcos ponen dos manos de rizos a la vela maestra	Comienzan a formarse olas grandes. Las crestas de espuma blanca se extienden por todas partes.
7	28-33	50-61	32-38	14-16	Muy fuerte (viento fuerte)	Se mueven los árboles más grandes es difícil andar contra el viento	Los barcos se quedan en el puerto, las que están en el mar echan las anclas, si es posible alcanzan una zona de refugio	La mar engruesa; la espuma blanca que proviene de las olas es arrastrada por el viento
8	34-40	62-74	39-46	17-20	temporal (duro)	Se rompen las ramas de los árboles, es muy difícil andar al aire libre	Todos los barcos se dirigen al puerto más cercano	Olas de altura media y más alargadas. De las crestas se desprenden algunos rociones en forma de torbellinos. La espuma es arrastrada en nubes blancas.
9	41-47	75-88	47-55	21-24	temporal fuerte (muy duro)	Se levantan los tejados	-	Grandes olas, espesas estelas de espuma a lo largo del viento, las crestas de las olas se rompen en rollos, las salpicaduras pueden reducir la visibilidad
10	48-55	89-102	56-64	25-28	temporal muy fuerte	Se observa rara vez en tierra. Arranca árboles y ocasiona daños de consideración en los edificios	-	Olas muy grandes con largas crestas en penachos, la espuma se aglomera en grandes bancos y es llevada por el viento en espesas estelas blancas en conjunto la superficie esta blanca, la visibilidad esta reducida
11	56-63	103-117	65-73	29-32	tempestad	Daños graves a los edificios y destrozos	-	Olas excepcionalmente grandes (los buques de pequeño y mediano tonelaje pueden perderse de vista). La mar está completamente cubierta de bancos de espuma blanca extendida en la dirección del viento. Se reduce aún más la visibilidad
12	>64	>118	>74	>33	temporal huracanado (huracán)	Daños muy graves	-	El aire está lleno de espuma y de rociones. La mar está completamente blanca debido a los bancos de espuma. La visibilidad es muy reducida.

10 hPa – Hectopascal.

Presión del viento a 100 km/h = 27,777 m/s, le corresponde una presión de:

- 10 hPa = 10 mbar

8.4.4. Unidades y equivalencias sobre velocidad del viento

1 km = 1.000 m
1 km/h = 0,278 m/s
1 m/s = 3,6 km/h
1 nudo = 1 milla náutica por hora = 1.852 m/h = 1,852 km/h
1 nudo = 0,514444 m/s
1 nudo = 30,86666 m/min
1 milla terrestre = 1.609 m
1 milla marina = 1 milla náutica = 1.852 m
1 m/s = 3,6 km/h = 2,237 millas/h = 1,944 nudos
1 nudo = 1 milla náutica/h = 0,5144 m/s = 1,852 km/h = 1,125 millas/h

Tabla 8.11. Estimación de la velocidad del viento por la apreciación sobre el terreno.

Escala	Denominación	Velocidad en m/s	Evaluación del viento en tierra
0	Calma	0 a 0,4	No se nota ningún movimiento en las ramas de los árboles.
1	Casi calma	0,5 a 1,5	La dirección del humo sufre un pequeño desvío.
2	Brisa leve	1,6 a 3,4	Las hojas son levemente agitadas.
3	Viento fresco	3,5 a 5,5	Las hojas quedan en agitación continua.
4	Viento moderado	5,6 a 8	Polvo y pedazos de madera son levantados.
5	Viento regular	8,1 a 10,9	Los árboles pequeños comienzan a oscilar.
6	Viento medio forte	11,4 a 13,9	Gajos mayores quedan agitados.
7	Viento fuerte	14,1 a 16,9	Es difícil andar contra el viento.
8	Viento muy fuerte	17,4 a 20,4	Es imposible andar contra el viento.
9	Ventarrón	20,5 a 23,9	Tejas pueden ser arrancadas.
10	Vendaval	24,4 a 28	Árboles son derribadas.
11	Huracán	83,0 a 125	Producen efectos devastadores.

8.4.5. Condiciones del viento para aerogeneradores

Las condiciones del viento son básicas cuando se estudia una instalación de aerogeneradores, de acuerdo con los parámetros siguientes:

- Densidad del aire.
- Intensidad de turbulencia.

Si la velocidad del viento y las turbulencias son altas, la vida útil del aerogenerador se acorta. Si hay turbulencias en la zona pero la velocidad del viento es baja, se puede mitigar el efecto negativo.

La turbulencia es el parámetro que describe las variaciones o fluctuaciones a corto plazo del viento.

- Velocidad media del viento en la zona.
- El parámetro k.
- La topografía del terreno es un factor que puede incidir sobre la velocidad y perfil del viento y las turbulencias. También incide sobre la incidencia del viento (inclinación) sobre las palas del aerogenerador.

Límites de velocidad de funcionamiento: entre 19 km/h y 100 km/h

Velocidad optima de funcionamiento: entre 40 y 50 km/h

8.5. Aplicaciones de la fuerza del viento

La energía eólica proviene de la fuerza del viento.

Muchas son las aplicaciones del aire desde la antigüedad

Entre las más importantes están:

- Navegación marítima.
- Molinos de viento para la molienda de granos.
- Molinos de viento para la extracción de agua.
- Molinos eólicos para producir electricidad

En la actualidad, esta es la más importante aplicación de la fuerza del viento.

El inconveniente principal que tiene este tipo de energía es su discontinuidad, con momentos de calma, variación de su velocidad y cambios en la dirección e intensidad.

Para el aprovechamiento de la energía eólica y su transformación en energía eléctrica se tendrá en cuenta lo siguiente:

a) Horas anuales con viento útil

Media anual de funcionamiento: 1.500 a 2.500 horas año.

El año tiene: 8.760 h

Porcentaje medio de viento: entre 17 y 25 % del día (4 a 6 h/día).

Pueden darse hasta varios días sin viento capaz de accionar los aerogeneradores.

b) Potencia generada en un año

$$Pg_{anual} = P_{instan.} \cdot h_{fun.\ anual}$$

Pg_{anual} – Potencia generado anual.
$P_{instan.}$ – Potencia total instalada.
$h_{fun.\ anual}$ – Horas de funcionamiento anual

8.6. Aerogeneradores

Los aerogeneradores son máquinas que transforman la energía contenida en la fuerza del viento, en energía mecánica rotativa (palas y multiplicador mecánico), y ésta en energía eléctrica, por medio de un generador eléctrico.

8.6.1. Datos generales

- La energía contenida en el viento es 80 veces más grande que el consumo energético de toda la humanidad.
- En 2009, la energía eólica supuso el 2% del consumo mundial de energía.
- El 8 de noviembre de 2009, el 50% de la electricidad generada ese día procedía de la energía eólica.
- En el año 2009, la energía eólica supuso el 13,8% de la energía eléctrica generada.
- En España, la media de funcionamiento anual de los aerogeneradores es de 2.530 horas. En Galicia 2.830 horas. En algunos parques se llega a las 3.000 horas.

- Actualmente se trabaja en la construcción de aerogeneradores entorno a 3 MW de potencia.
- Las potencias más normales de los aerogeneradores en los parques eólicos están comprendidas entre 500 kW y 1.500 kW.
- El rendimiento de los actuales aerogeneradores está en torno al 50%.
- El eje lento (rotor con palas) gira entre 16 y 30 rpm. El eje rápido (eje del alternador) gira a 1.500 rpm, después de pasar por un multiplicador mecánico de velocidad.
- Un aerogenerador de 600 kW, tiene palas de 20 metros de longitud.
- Un aerogenerador empieza su funcionamiento cuando la velocidad del viento esta en torno a 19 km/hora.
- El máximo rendimiento de un aerogenerador corresponde cuando la velocidad del viento está entre 40 y 48 km/hora.
- Cuando la velocidad del viento llega a 100 km/h, los aerogeneradores dejan de funcionar.

8.6.2. Tipos de aerogeneradores

Los tipos de aerogeneradores se clasifican en función a la posición de su eje, respecto a la posición en la que reciben el viento y al tipo de palas.

8.6.2.1. Respecto a la posición de su eje

a) De eje horizontal

Son los más utilizados y corresponden a los aerogeneradores que vemos instalados en los parques eólicos y en instalaciones aisladas con pequeños aerogeneradores.

b) De eje vertical

Se distinguen tres tipos de aparatos:

- ***Darrieus.*** Se basan en dos o tres arcos que giran alrededor de su eje.
- ***Panemonas.*** Se basan en cuatro o más semicírculos unidos al eje. Bajo rendimiento.
- ***Sabonius.*** Se basan en dos o más filas de semicilindros colocados opuestamente.

8.6.2.2. Respecto a la posición del equipo con referencia al viento

a) A barlovento

Cuando el rotor recibe el aire de frente. Es el caso de la mayoría de los aerogeneradores instalados en los parques eólicos.

El aerogenerador necesita dispositivo orientador para encararse al viento. Hay que alejar el rotor de la torre para minimizar la pérdida de empuje que supone cuando la pala está en línea con la torre.

b) A sotavento

En este caso, el viento ataca por el lado de la góndola que tiene un diseño especial. El aerogenerador no necesita mecanismo de orientación ya que el aire lo posiciona de la forma correcta. La construcción de las palas es especial para este caso.

Tiene las pérdidas que suponen la góndola y el soporte.

8.6.2.3. Respecto a las palas

a) Una pala

No es normal esta construcción por que afecta a la mecánica por el desequilibrio que supone una carga giratoria no distribuida.

b) Dos palas

Equipan pequeños aerogeneradores.

c) Tres palas

Es el tipo de rotor más utilizado.

d) Multipalas

Se denomina modelo americano, ya que fue en EE.UU. donde empezó a utilizarse este sistema para la extracción de agua de pozos.

8.6.3. Principales elementos de un aerogenerador

Constitución básica de un aerogenerador:

- Base de obra civil.
- Soporte o torre.
- Rotor formado por las palas y buje.
- Góndola, conteniendo en su interior:
  - Eje de baja velocidad.
  - Variador mecánico de velocidad (multiplicador).
  - Acoplamientos.
  - Eje de gran velocidad.
  - Generador eléctrico (entre 500 y 2.000 kW). Los hay de mayor potencia.
  - Mecanismo de orientación.
  - Controlador electrónico.
  - Equipo regulador (orientación y maniobra).
  - Unidad de refrigeración.
  - Mecanismo de orientación.
  - Sistema de frenado.
- Anemómetro y veleta.
- Equipo de comunicaciones a través del cual se reciben las órdenes.
- Cableado eléctrico.

En la figura 8.2 se muestran los principales elementos de un aerogenerador.

8.6.4. Condiciones climáticas de la zona

Las condiciones climáticas de la zona determinan el rendimiento del aerogenerador y entre las principales están:

- Temperaturas que se dan en la zona a lo largo del año.
- Humedad relativa del aire.
- Velocidad media del viento.
- Dirección, frecuencia y vientos predominantes
- Meteorología general de la zona.

Figura 8.2. Elementos principales de un aerogenerador.

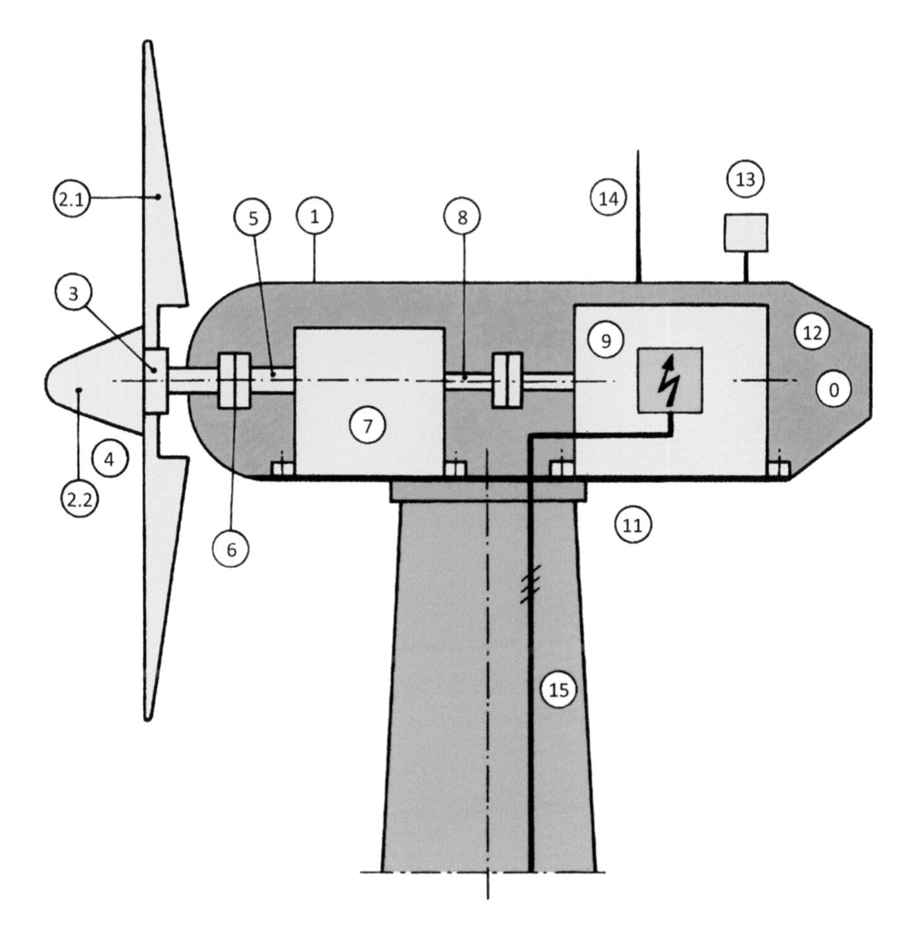

1. Góndola.
2. Conjunto rotor.
   2.1. Palas.
   2.2. Buje.
3. Mecanismos para la orientación de palas.
4. Freno.
5. Eje de baja velocidad.
6. Acoplamientos.
7. Caja mecánica de multiplicación de velocidad.
8. Eje de gran velocidad.
9. Generador eléctrico.
10. Unidad de refrigeración.
11. Mecanismo para orientación de la góndola.
12. Equipo electrónico de regulación y control.
13. Anemómetro y veleta.
14. Equipo de comunicaciones.
15. Cableado eléctrico.

Tabla 8.12. Velocidades de funcionamiento de aerogeneradores.

Velocidad del viento	km/h	m/s
· Velocidad mínima para el inicio de giro	19	5
· Velocidad que proporciona máximo rendimiento.	40 a 48	11 a 13
· Velocidad máxima de funcionamiento	100	28

Figura 8.3. Aerogeneradores.

8.7. Principales características de algunos elementos del aerogenerador

En este apartado se hace un pequeño descriptivo de los principales componentes de una torre de generación eléctrica a partir de energía eólica.

8.7.1. Torre soporte

Además de soportar la carga del conjunto aerogenerador, también soportan el empuje del aire por lo que su construcción ha de ser robusta.

Las características principales de una torre son:

Lugares de instalación ideal:

Donde el viento sople regularmente a una velocidad media de 21 km/h.

- Tipo de estructura (los hay de perfilaría, pero los más empleados son los tubulares).
- Materiales. El material empleado en su construcción suele ser acero al carbono.
- Especificaciones de las virolas.
- Medias y pesos de los tramos.
- Forma de unión entre virolas (bridas o soldadura).
- Tratamiento superficial exterior e interior.
- Diámetro superior e inferior de la torre.
- Altura de la torre.
- Equipamiento interior de la torre.
- Accesos interiores.
- Plataforma superior y sistema de giro de la góndola)
- Peso total del conjunto.

8.7.2. Características del rotor

El conjunto del rotor con sus palas recibe la fuerza del viento para transformar su energía en energía mecánica rotativa.

- Diámetro (m).
- Área de barrido (m^2).
- Velocidad de rotación (rpm).
- Sentido de rotación (derecha o izquierda).
- Orientación.
- Ángulo de inclinación (grados).
- Conicidad del rotor.
- Número de palas.
- Tipo de frenado aerodinámico.

La figura 8.4 muestra algunos de los tipos de rotores y sus palas. Como puede apreciarse, hay una gran variedad de ellos, aunque al final son pocos los tipos empleados, como podemos apreciar en la realidad.

Los aerogeneradores de mediana y gran potencia utilizan el rotor de tres palas.

8.7.3. Características de las palas

Las palas son el elemento que recibe el impulso de la fuerza del viento en situaciones muy diferentes, lo que obliga a que tengan una construcción especial en diseño, material y resistencia unido a su flexibilidad.

Las palas se construyen con fibra de vidrio que les permiten tener poco peso, y ser flexibles y resistentes.

Principales características:

- Longitud de la pala (m).
- Material del que está construido la pala.
- Forma constructiva de las palas (geometría).
- Perfiles aerodinámicos.
- Torsión y flexión admisible.
- Cuerda de las palas.
- Distancia entre raíz de las palas hasta el centro del buje (m).
- Fijación de las palas al rotor.
- Sistema de orientación de las palas.
- Peso de la pala (kg).

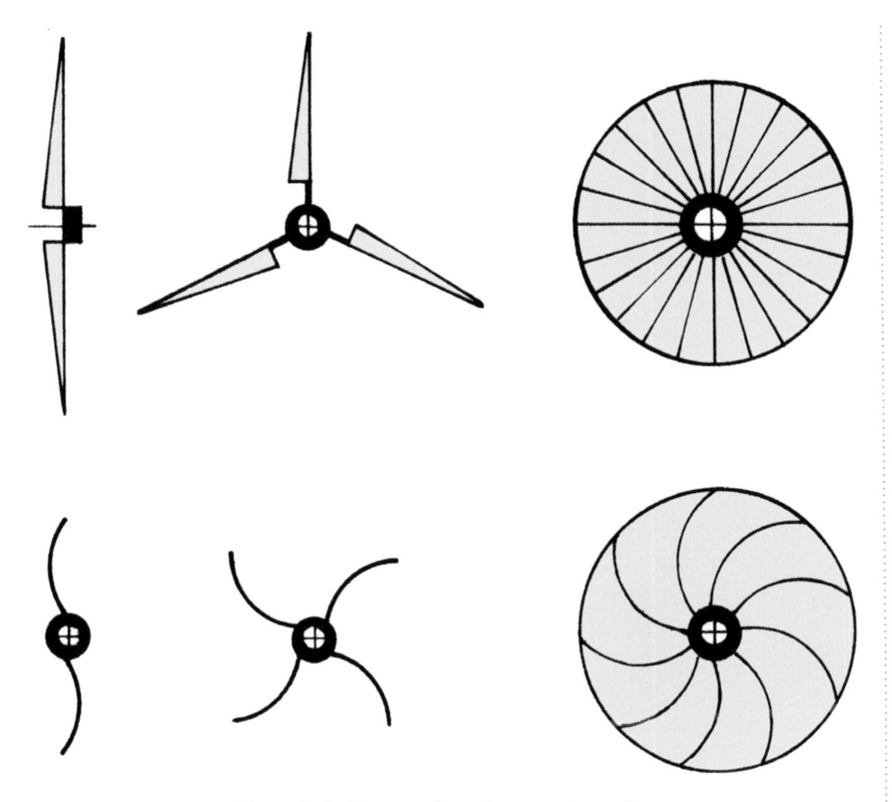

Figura 8.4. Algunos tipos de rotores y palas.

8.7.4. Ejes de transmisión

La transmisión se divide en dos partes:

a) Eje lento es el que está unido al rotor (palas).
b) Eje rápido es el que sale de la caja multiplicadora y mueve el generador eléctrico.

Los ejes lentos tienen mayor diámetro que los ejes rápidos, También los acoplamientos son de mayor dimensión en los ejes lentos que en los ejes rápidos.

De los ejes hay que conocer:

- Diámetros.
- Rodamientos. Los del eje lento soportan grandes cargas debido al diámetro de las palas y los esfuerzos que tienen que soportar.
- Acoplamientos.

8.7.5. Caja multiplicadora

La misión de la caja multiplicadora es la de pasar la velocidad lenta (≈ 20 rpm) proporcionada por el rotor a palas, a la alta velocidad que interesa al alternador (≈ 1.500 rpm).

- Tipo de multiplicador mecánico.
- Relación de transformación (Ejemplo: 20/1.500 rpm).
- Procedimiento de refrigeración (aire, aceite, etc.).
- Si dispone de sistema de calentamiento del aceite para épocas frías en las que se puede enfriar por falta de trabajo y soportar bajas temperaturas.

- Dimensiones de la caja multiplicadora de velocidad.
- Noticia de mantenimiento.
- Peso del elemento (kg).

8.7.6. Generador

Los aerogeneradores destinados a la producción de energía eléctrica están equipados con alternadores eléctricos que generan por lo general corriente alterna trifásica.

Características principales del generador:

- Tipo de generador (alternador o dinamo).
- Los pequeños aerogeneradores pueden ser para corriente continua (dinamos) o para corriente alterna (alternadores monofásicos).
- Potencia nominal ((W, kW, MW).
- Tensión de salida (V, kV).
- Si se trata de un alternador.
    - Frecuencia (Hz).
    - Número de polos (2p).
    - Velocidad nominal de rotación (rpm).
    - Factor de potencia (cos φ)
    - Dimensiones
    - Clase de aislamiento (IP--).
    - Peso (kg)

8.7.7. Góndola

La góndola almacena en su interior los elementos de transmisión, control, regulación y generación de energía eléctrica. Deben tener acceso para que pueda efectuarse el mantenimiento de los diferentes componentes, todo ello con las debidas garantías de seguridad para el personal.

Su construcción es aerodinámica para reducir pérdidas por frenado de la fuerza del viento en su estructura.

Características principales de la góndola:

- Material de fabricación, generalmente liviano y resistente.
- Aislamiento acústico del recinto.
- Espacio suficiente.
- Iluminación (claraboyas y alumbrado eléctrico).
- Ventilación adecuada.
- Contiene instrumentos y electrónica de control y regulación.
- Sistema de orientación y giro.

8.7.8. Sistemas de control

El sistema de control del aerogenerador constituye el cerebro del mismo, ya que se autocontrola de acuerdo con los parámetros que se le han fijado. El sistema de detección y control, así como los motores de posicionamiento tienen su alimentación independiente de la que pueda generar el alternador.

- Medida y control de la velocidad del viento.
- Sistema del mecanismo de giro del conjunto góndola.
- Sistema de control que coordina el funcionamiento de todos los elementos de la instalación a través de diferentes sensores, ordenando el funcionamiento, reactivando las anomalías detectadas y

arrancando y parando la actividad cuando las condiciones son favorables o dejan de serlo.

- Cada aerogenerador tiene comunicación con el edificio central del parque, pudiendo recibir órdenes en uno u otro sentido. El control de la instalación también puede estar centralizado en un lugar lejano, desde el cual se controlan varios parques eólicos.

8.7.9. Cimentación para la torre

El peso total de la torre al que se suma el contenido de la góndola con su rotor y la fuerza del viento, está soportado por la cimentación en obra civil que se realiza con este fin. Las cargas a soportar son muy importantes por lo que el anclaje y cimentación deberán soportar el momento de vuelco que se crea en el conjunto, y de forma especial, cuando el aerogenerador está funcionando con el empuje del viento.

Respecto a la base de fijación podrá ser una base de cimentación o zapatas de anclaje.

Los principales elementos de cálculo son las cargas:

- Cargas gravitatorias.
- Cargas producidas por la acción del viento.
- Tensión admisible en el terreno.
- Hormigón a emplear.
- Tipo de acero de la torre.
- Tipo de anclaje.

Figura 8.5. Torre soporte de un aerogenerador.

8.8. Instalación eléctrica del aerogenerador

Se distinguen tres tipos de instalaciones:

a) Instalación informática, con sus sensores y captores de información.

b) Instalación de los elementos de accionamiento de posicionamiento de palas, de la góndola y de otros elementos auxiliares que pudiera tener la torre, como es el caso de ascensor interior.

c) Instalación eléctrica desde el alternador hasta la subestación.

La tensión de la instalación de salida corresponde a la tensión de generación. En la subestación se eleva la tensión al valor que corresponda, a media tensión.

Todas las instalaciones tienen sus elementos de control, accionamiento y seguridad que les son propios.

Cada aerogenerador tiene su sistema de control y accionamiento automatizado y coordinado con el resto de aerogeneradores del parque eólico.

8.9. Ejemplos de características y producción de un aerogenerador

Para ayudarnos a comprender mejor el funcionamiento de un aerogenerador se presentan dos ejemplos de instalación.

8.9.1. Ejemplo de las principales características de un aerogenerador

Características de un aerogenerador de grande medidas y potencia.

Potencia..............................2.000 kW = 2 MW
Diámetro rotor...................80 m
Número de palas 3
Longitud de las palas..........39 m
Peso de las palas ....6.500 kg
Velocidad del viento ....4 m/s para el ***arranque***.
25 m/s para el ***corte*** o ***parada*** por exceso de velocidad.
Altura de la torre ....Entre 60, 67, 78 y 100 m
4 Motores de orientación del rotor y palas.

8.9.2. Ejemplo de producción de un aerogenerador

Tabla 8.13. Relación de las características de un aerogenerador y su entorno.

Características	Valores
Potencia eólica	1,5 MW
Capacidad nominal	1,5 MW equivalente a 1.500 kW
Velocidad nominal del viento	12 m/s (43,2 km/h)
Horas año de producción	3.500 h/año [2]
Producción máxima de electricidad/año	3.500 x 1.500 = 5.250.000 kWh/año
Capacidad total de producción/año	1.500 x 8.760 = 13.125.000 kWh/año [1]
Número de palas	3
Diámetro del rotor	70,5 – 77 m
Velocidad de rotación	10,1 – 22,2 rpm

[1] Un año tiene 24 x 365 = 8.760 horas.
[2] De las 8.760 h de un año, sólo 3.500 h son hábiles para la generación de electricidad, en emplazamientos muy favorables.

Figura 8.6. Presentación comercial de un higrómetro digital que también incorpora termómetro.

8.10. Aparatos de medida

Además de los aparatos de medida y control que incorporan los aerogeneradores, se emplean aparatos para la medida de la velocidad del viento (anemómetros), la humedad del aire (higrómetros) y la temperatura ambiente o de parte de los elementos de la instalación (termómetros).

8.10.1. Higrómetro

Aparato de mide la humedad del aire.

8.10.2. Anemómetros

El anemómetro es un aparato destinado a medir la velocidad del viento. Los hay portátiles y fijos. Los aerogeneradores incorporan anemómetros fijos y son su elemento principal de funcionamiento que determinará su entrada en servicio o su paro por baja velocidad o excesiva velocidad del viento.

8.10.3. Termómetros

Aparatos destinados a la medida de la temperatura. Se puede medir la temperatura ambiente y la temperatura de diferentes elementos de situados en el interior de la góndola (rodamientos, variador mecánico de velocidad, alternador, cables, temperatura interior de la góndola y torre, etc.).

8.11. Ventajas e inconvenientes de la generación eólica

La generación eólica consiste en el aprovechamiento de la energía o fuerza del viento (energía cinética del viento) para generar una corriente eléctrica a través de dinamos o alternadores. Este tipo de aprovechamiento de la energía del viento tiene sus ventajas e inconvenientes.

a) Ventajas:

- La energía base (viento) es gratuita y renovable.
- El proceso no produce contaminación ambiental ni efecto invernadero.
- La instalación de un parque eólico es muy rápido de montar (6 meses a un año).
- Ahorra otras materias primas no renovables.
- En el caso de España, se trata de un sistema viable y con futuro, dada su orografía, y con el que se puede cubrir un porcentaje importante de las necesidades del consumo total (superior a 25%).

b) Inconvenientes:

- No es una forma de producción continua. Está sujeta a si hay o no viento.
- Se trata de una energía muy subvencionada.
- Los parques eólicos tienen un impacto ambiental (fauna y paisaje)
- Para el acceso a los parques se necesita abrir pistas, que en algunos casos causan un impacto ambiental.
- Generan ruido que puede ser molesto, si hay viviendas próximas en la zona.
- Pueden ser causa de la muerte de aves, especialmente de paso, ya que los parques se instalan en zonas con corrientes de aire que son las que aprovechan las aves migratorias.

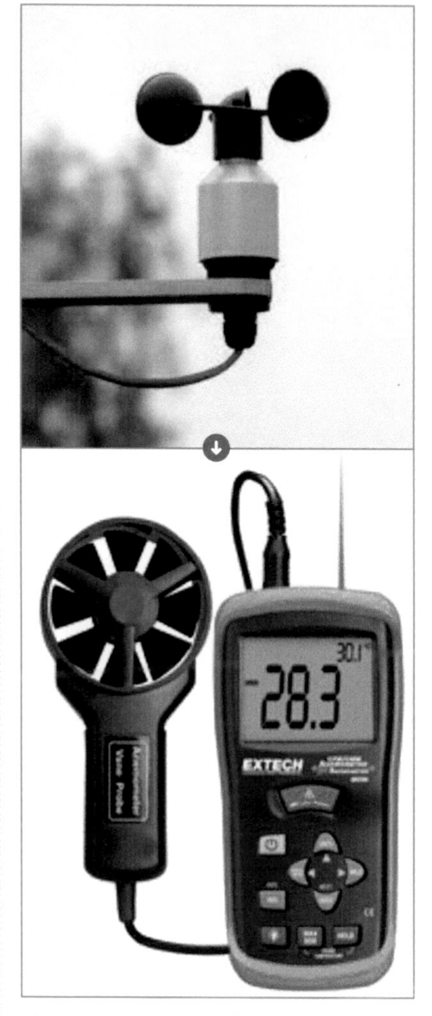

Figura 8.7. Presentación comercial de anemómetros. Anemómetro fijo y anemómetro digital con indicación de la temperatura.

8.12. Instalaciones eólicas

EL objetivo para el año 2012 es de llegar a una capacidad de producción de 20.000 MW.

La capacidad actual de generación es de 11.615 MW.

El 19 de marzo de 2007, con una situación favorable de viento sobre todo el territorio nacional, se consiguió una producción puntual de 8.375 MW, producción que se ha mejorado en marzo del 2008.

Los 8 reactores nucleares en funcionamiento en España generan 7.742,32 MW de forma continua, sin intermitencias, como sucede con la energía eólica.

Este sistema empezó a generalizarse en los años 80.

Nota:

La generación eléctrica a partir de la fuerza del viento, es complementaria de otras formas de generación continua de electricidad (hidráulica, térmica o nuclear). La energía eléctrica de generación eólica no puede asegurar por sí misma, un suministro continuo, por su funcionamiento intermitente.

Figura 8.8. Molino (rueda de aletas) movido por la fuerza del viento. Puede aplicarse a mover una bomba de extracción de agua o un generador eléctrico.

Figura 8.9. Pequeños aerogeneradores.

La energía base, el viento, se emplea en las condiciones siguientes:

- Velocidad mínima del viento para el arranque .... 5 m/s
- Velocidad media de funcionamiento .... 15 m/s
- Velocidad de parada por exceso de velocidad .... 28 m/s

Las turbinas generadoras (alternadores) pueden ser de 4 polos y potencia de 500, 650, 1.000, 1.500 kW y más.

El número de revoluciones del rotor viene a ser de entre 16,33 y 30 rpm.

La orientación de las paletas y del propio molino se hace de forma automática, en función de la dirección y velocidad del viento, con el fin de conseguir siempre la misma velocidad y el mayor rendimiento mecánico.

Ejemplos de generación de energía eléctrica, a partir de aerogeneradores accionados por la acción del viento se presentan en el apartado 8.14.

a) Otras aplicaciones de la fuerza del viento:

La agricultura aplica la fuerza del viento para extraer agua de pozos y con molinos de viento se accionaban las ruedas con las que se molían los cereales (cebada, trigo, maiz, etc.).

Aplicación del aire para accionar una bomba con la que se extrae agua de un pozo destinada al riego de huertas. Se trata de una vieja aplicación del viento desde hace muchos años. También, para generar electricidad si el movimiento se aplica a una dinamo (corriente continua) o a un alternador (corriente alterna).

b) Torre eólica

En las cercanías de Manzanares (Ciudad Real) se está construyendo una torre o chimenea eólica de 750 m de altura, para la generación de energía eléctrica y que en su interior tiene 32 turbinas. El sistema consiste en el calentamiento del aire, por medio de un colector solar que aprovecha el calor proporcionado por el Sol. El aire caliente asciende por la chimenea a una temperatura de 70 °C que adquiere una velocidad de 60 km/hora y que moverá las turbinas y sus generadores correspondientes, para producir en torno a 40 MW.

En Australia se está construyendo una torre eólica similar de 1.000 m de altura que generará 200 MW de potencia.

8.13. Cálculo de la potencia generada por un aerogenerador

Como se ha indicado la potencia de un aerogenerador (motor accionado por la fuerza del viento) depende de tres conceptos fundamentales, a saber: Superficie o área barrida por las palas del rotor, densidad del aire y velocidad del viento.

a) Fórmula general

$$P = \frac{1}{2} \cdot S \cdot d \cdot v^3$$

S – Superficie o área barrida por las palas del rotor, en m^2.
d – Densidad del aire, en kg/m^3.
v – Velocidad del viento, en m/s.

b) Fórmula para calcular la potencia eólica aprovechable

$$P = \frac{1}{2} \cdot S \cdot G \cdot v^3 \cdot Cp$$

Cp – Coeficiente de potencia, que depende del tipo de máquina y de la relación entre velocidad periférica de las palas y la velocidad del viento.

c) Otra fórmula de cálculo de la potencia eólica

$$P = \frac{1}{2} \cdot d \cdot Cp \cdot \eta \cdot S \cdot v \text{ (W)}$$

P – Potencia, en W.
d – Densidad de la masa de aire, en kg/m^3.
Cp – Coeficiente de potencia máxima de una turbina ideal de eje horizontal, igual a: 16/27 = 0,593
η – Rendimiento o eficiencia mecánica de la turbina.
S – Área circular de movimiento de las palas del rotor (área barrida), medida en m^2.
v – Velocidad de la masa de aire, antes de pasar por las palas, en m/s.

8.14. Parques eólicos para generar electricidad

Los parques eólicos son centrales en las que se genera electricidad transformando la fuerza del viento en energía mecánica, y ésta, en energía eléctrica.

8.14.1. Puestos de generación eléctrica

Las centrales o parques eólicos están constituidos por un número determinado de generadores, que están situados en lo alto de una torre, generalmente metálica. Las hay de diversos tipos, pero las más generalizadas son las de tubo cónico. En la figura 8.10. se muestran dos ejemplos de torres. Para llegar a disponer de un parque eólico de generación eléctrica, hay que recorrer un largo camino, como es el que de forma resumida se indica a continuación.

a) Elementos que integran la instalación

- Parque de aerogeneradores.
- Edificio principal con el siguiente cometido:
  - Centro de control y maniobra.
  - Almacén de materiales de mantenimiento y conservación.
  - Oficina de control, mando y telemando del parque.
  - Anexa, estación de subestación con transformadores y equipos de protección.
- Red interna de baja tensión (BT/MT) a la tensión de 400 o 20.000 V, que conecte cada generador con el centro de transformación correspondiente.
- Red externa de media tensión (MT) a la tensión de 20.000 V, que conecte el parque con la subestación del centro de transformación de la red de distribución pública (parques conectados a la red general).

Figura 8.10. Torres de tubo metálico y estructura de perfilaría.

Figura 8.11. Parques eólicos para generación eléctrica.

b) Proyecto

- Elección de la zona.
- Estudio del potencial eólico de la zona.
- Estudio del terreno.
- Elección de la tecnología de generación que mejor se adapte a la zona.
- Estudiar el potencial eólico de la zona.
- Estudio del impacto medioambiental de la zona.
- Estudio económico y rentabilidad del parque.
- Efectuar el diseño de la zona (acceso y ubicación de aerogeneradores y centro de control y elementos auxiliares).
- Estudio de la obra civil a realizar, para el montaje de las torres.
- Estudio de la línea eléctrica entre aerogeneradores y la subestación del parque.
- Estudio de la línea que conecta con la red eléctrica general.
- Gestión de permisos ante las administraciones.

c) Normativa

En la construcción de un parque eólico se tendrán en cuenta la normativa en vigor a nivel general, así como las normas emitidas por las administraciones locales, autonómica y del estado.

8.14.2. Otros elementos del parque de generación eléctrica eólica

Algunos de los elementos que constituyen los parques eólicos.

Además de las torres que contienen el elemento principal de generación como son:

- Interior de la torre.
- Caseta eléctrica de recepción.
- Canalización para la conducción de cables a subestación..
- Subestación con transformadores y local de control.
- Línea de enlace con la red general de electricidad.

La elección de los aerogeneradores es muy importante para el proyecto, dado el gran número de tipos de aerogeneradores (tecnologías, potencias, curvas de funcionamiento, vientos predominantes y su velocidad, etc.).

8.14.3. Ejemplo de las características de transformadores y celdas

A continuación se presentan a modo de ejemplos las características de un transformador y celda para un parque eólico de generación de electricidad.

Ejemplo de transformación BT/MT que podrá ser de tipo seco y aislado con materiales auto-extinguibles. Suponiendo un transformador de 900 kVA.

Tabla 8.14. Características del transformador.

Características	Valores
Tipo de transformador	Trifásico, seco encapsulado.
Relación de transformación	20 kV/690 V
Potencia nominal	900 kVA

Características	Valores
Frecuencia	50 Hz
Grupo de conexión	Dyn11n11
Tensión de cortocircuito	≤ 6%
Clase de aislamiento	F
Nivel de aislamiento del primario	Frecuencia industrial: 24 kV Impulso tipo rayo: 125 kV
Nivel de aislamiento del secundario	Frecuencia industrial: 3 kV
Dimensiones aproximadas	860 x 1.720 x 1.660 (alto) (mm)
Peso aproximado	2.900 a 3.000 kg
Norma UNE	UNE 21538

Las celdas que protegen al transformador BT/MT y la conexión a los cables de la red de MT, podrán ser modulares o compactas, y sus características más comunes son las recogidas en la tabla 8.15.

Tabla 8.15. Características de las celdas.

Características	Valores
Tipo	Aparamenta aislada SF6
Servicio	Continuo
Instalación	Interior
Nº de fases	3
Nº de embarrados	1
Tensión nominal asignada	24 kV
Tensión del servicio	20 kV
Frecuencia nominal	50 Hz
Intensidad nominal: Función protección (P) Función de conexión a la red (L)	200 A 400 A
Nivel de aislamiento: Sobre distancia de seccionamiento (frecuencia industrial/tipo rayo) A tierra, entre polos y entre bornas (frecuencia industrial/tipo rayo)	 60 kV/145 kV 50 kV/125 kV
Intensidad de cortocircuito: Admisible de corta duración (1 s Nominal Voltaje	 16 kA 40 kA 24 kV
Medidas: dimensiones aproximadas	1.200 x 800 x 2.090 (alto) (mm)

Figura 8.12. Interior de una torre y caseta eléctrica a pie de torre.

Figura 8.13. Subestación y armario de control.

Figura 8.14. Los aerogeneradores, su estructura e instalación eléctrica deben mantenerse adecuadamente.

8.15. Mantenimientos de parques eólicos

El mantenimiento de una instalación del tipo que sea, puede ser:

a) Mantenimiento ***"correctivo"*** es el que se hace cuando surge una avería, y por tanto, no está programado.

b) Mantenimiento ***"predictivo"*** es el que se hace de forma programada. No se espera a que se produzca la avería, se interviene antes de que pueda darse de forma intempestiva, causando problemas productivos.

El mantenimiento debe hacerse por personal cualificado, siguiendo el *"cuaderno de instrucciones"* y las recomendaciones del fabricante respecto a los materiales y su reposición, de acuerdo con su vida útil.

Respecto a un parque eólico, estas son las principales acciones de mantenimiento a realizar:

a) En el rotor (buje y palas)

- Inspección visual del conjunto.
- Detección de roturas y fisuras en el buje y palas.
- Estado del apriete de los tornillos.
- Inspección del estado superficial de las palas (conservación, limpieza, etc.).
- Inspección del sistema de regulación de las palas.

b) Transmisión mecánica (ejes, acoplamientos)

- Situación de los rodamientos.
- Situación de los acoplamientos.
- Situación de anclajes.
- Lubricación.
- Estado del sistema anti-rotación.

c) Multiplicadora de velocidad

- Control de los niveles de aceite.
- Cambio de aceite cuando proceda.
- Revisión de acoplamientos.
- Revisión de la tornillería de fijación.
- Otras revisiones que aconseje el fabricante.

d) Sistema de amortiguación

- Lubricación disco de muelles derecho / izquierdo.
- Verificar el par de apriete de los tornillos.

e) Sistema de frenado

- Verificar el par de apriete de los tornillos.
- Comprobar estado de pinzas y pastillas de frenos.
- Verificar el disco de freno.
- Lubricación general.
- Verificar el eje de transmisión.

f) Generador eléctrico

- Verificar los parámetros del alternador señalados por el fabricante en las instrucciones de mantenimiento, tanto de tipo eléctrico como mecánico.
- Verificar el aspecto general del exterior del alternador.
- Verificación de anclajes.

- Atención al ventilador de refrigeración.

g) Mecanismo de orientación

- Verificar motor de accionamiento.
- Revisión de de los niveles de engrase.
- Revisión de posibles fugas de aceite.
- Efectuar engrases y cambios de filtro según lo señalado en el libro de instrucciones.
- Revisión de rodamientos, piñones y engranajes.
- Revisión de la tornillería en general.
- En el sistema hidráulico, revisión de niveles, fugas y cambio de filtro cuando proceda.

h) Góndola, carcasa y corona

- Comprobación de su estado general.
- Limpieza.
- Limpiar posibles fugas de aceite.
- Control del material y apriete de tornillos.

i) Torre

- Verificación de la fijación de la góndola a la torre
- Aspecto exterior (pintura protectora).
- Anclaje a la base.

j) Aparatos electrónicos, sensores y detectores

- Verificación general de los mismos.

k) Cableado general

- Cableado de la red de salida de la electricidad generada.
- Cableado informático y de sensores y detectores.
- Cableado de los aparatos de accionamiento.

j) Consumibles

En el almacén se dispondrá de materiales de utilización habitual, tales como: fusibles, interruptores, tornillería, aceites, lubricantes, juntas de estanqueidad, etc. Y las herramientas y aparatos de comprobación que requieran las tareas a realizar.

k) Condiciones del viento

Las condiciones del viento son básicas cuando se estudia una instalación de aerogeneradores, de acuerdo con los parámetros siguientes:

- Densidad del aire.
- Intensidad de turbulencia.

Si la velocidad del viento y las turbulencias son altas, la vida útil del aerogenerador se acorta. Si hay turbulencias en la zona pero la velocidad del viento es baja, se puede mitigar el efecto negativo.

La turbulencia es el parámetro que describe las variaciones o fluctuaciones a corto plazo del viento.

- Velocidad media del viento en la zona.
- El parámetro k.
- La topografía del terreno es un factor que puede incidir sobre la velocidad y perfil del viento y las turbulencias. También tiene influencia el ángulo con que llega el viento a incidir sobre las palas del aerogenerador.

Figura 8.15. Parque eólico y armarios eléctricos situados en el edificio de la subestación.

8.16. Pequeñas instalaciones de generación eólica

Desde hace bastante tiempo, se viene utilizando la generación eólica de electricidad (antes que la fotovoltaica), para suministrar de esta energía a instalaciones aisladas. También se ha empleado la energía eólica para sacar agua de pozos accionando bombas hidráulicas.

A continuación se estudian algunos ejemplos de pequeñas instalaciones para generar electricidad u extraer agua de pozos.

Esquema 1. Puesto aislado de generación eléctrica para suministrar energía en corriente alterna a una vivienda aislada.

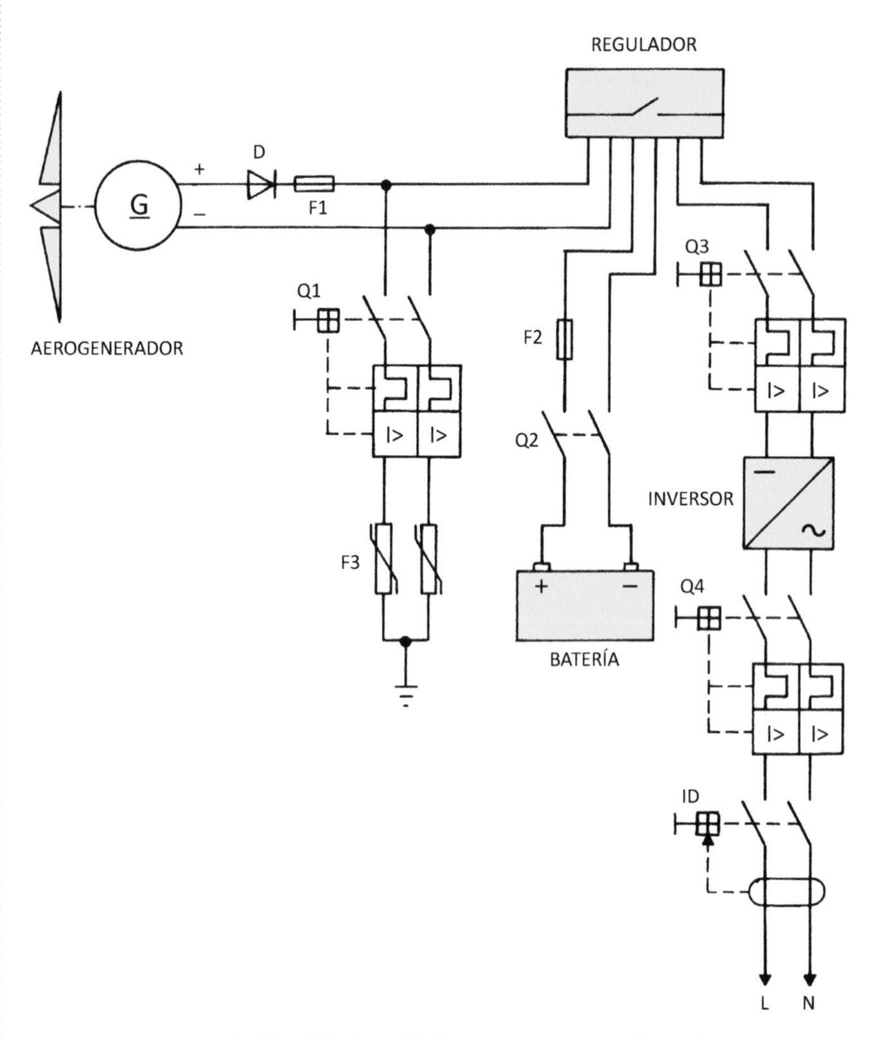

Esquema 8.1. Instalación eléctrica aislada para generar corriente eléctrica a una vivienda.

Esquema 2. Generación eólica para alimentar una motobomba.

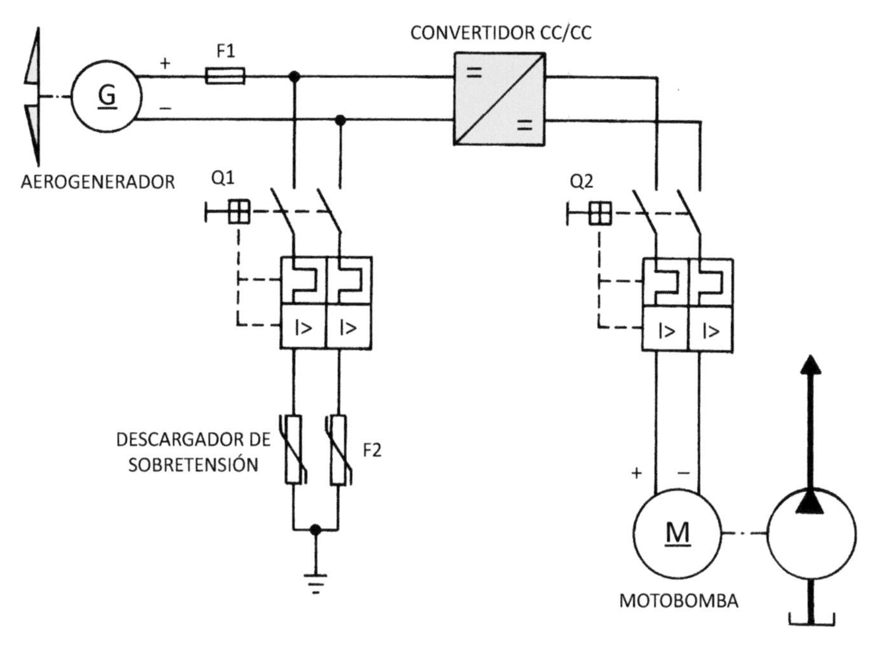

Esquema 8.2. Instalación eléctrica para alimentar una motobomba.

Esquema 3. Instalación mixta de generación eléctrica eólica y fotovoltaica.

Cuando una instalación de generación eléctrica resulta escasa o no se quiere hacer mayor la que se tenía, por ejemplo la fotovoltaica, y se necesita más energía se puede suplir la deficiencia instalando un aerogenerador, como el que se presenta en este esquema.

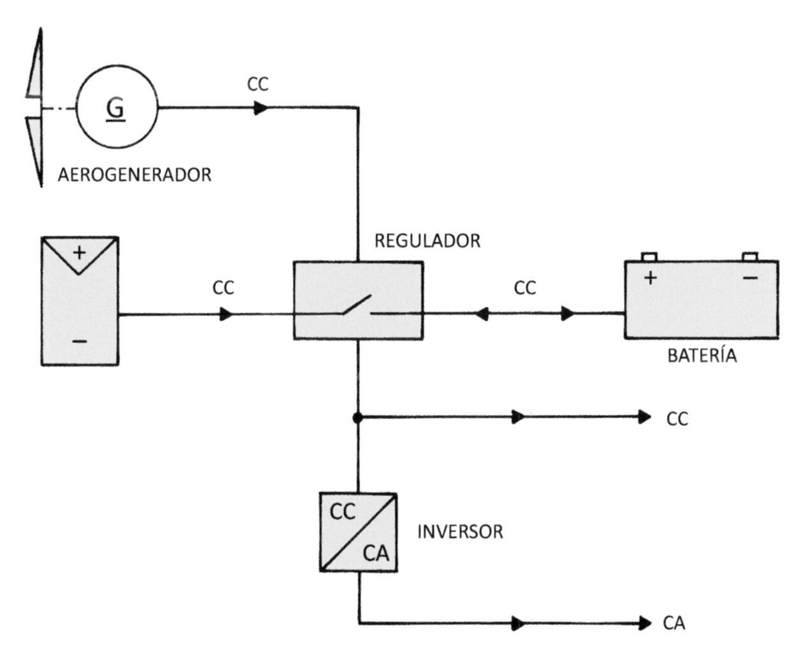

Esquema 8.3. Instalación eléctrica aislada de generación fotovoltaica y eólica, con acumulación de energía.

Figura 8.16. Vista parcial de parques eólicos de generación eléctrica.

8.17. Grandes instalaciones de generación eólica

Las grandes instalaciones o parques de generación eléctrica por transformación de la energía eólica, constan de los siguientes elementos básicos.

a) Aerogeneradores.
b) Equipo eléctrico de recepción de la energía a pie de torre, bien sea en el interior o en caseta junto a la torre.
c) Canalización eléctrica hasta la subestación.
c) Equipo transformador en la subestación. Eleva la tensión generada a la tensión que corresponda en media tensión para transportar la energía hasta la red eléctrica general.
d) Red de transporte de energía eléctrica generada en media tensión, hasta la red eléctrica general.

8.18. Producción de energía eléctrica eólica y previsiones de futuro

Los últimos años han sido muy positivos en la evolución de esta tecnología, incentivados por ayudas de la administración que ahora se han reducido sensiblemente, lo que ha llevado a una ralentización respecto a la evolución ascendente de los últimos años, a nivel de España.

8.18.1. Producción energética a nivel mundial

España está bien situada en el desarrollo del aprovechamiento eólico. Poco a poco, la energía eólica va teniendo un peso significativo en la generación de energía eléctrica.

Tabla 8.16. Producción de energía eléctrica por países

Capacidad total de energía eólica instalada					
Posición	País	Capacidad (MW)			
		2004	2005	2006	2008
1	USA	6.725	9.149	11.603	25.170
2	Alemania	16.628	18.428	20.622	23.903
3	España	8.504	10.028	11.730	16.754
4	China	764	1.260	2.405	12.210
5	India	3.000	4.430	6.270	9.654
6	Italia	1.265	1.717	2.123	3.736
7	Francia	386	757	1.567	3.404
8	Reino Unido	888	1.353	1.963	3.241
9	Dinamarca	3.124	3.128	3.136	3.180
10	Portugal	522	1.022	1.716	2.862
	Total mundial	**47.671**	**58.982**	**73.904**	**120.791**

8.18.2. Previsión de evolución futura

Se tienen puestas muchas esperanzas en esta energía limpia y no contaminante, que está en todas las partes. Hay que buscar las zonas más rentables, así como la instalación de grandes aerogeneradores que den lugar a producciones significativas.

Las previsiones futuras son muy halagüeñas y su evolución en los últimos años ha sido muy positiva.

8.18.3. Ejemplo real de evolución de esta energía en España

En el mes de marzo de 2011, la energía eléctrica proveniente de generación eólica ocupó el primer puesto en España con 4.738 MWh, supuso el 21% del total de energía eléctrica generada. Le siguió la energía nuclear con el 19%.

La instalación eólica en este momento en España, suponía 20.676 MW.

El total de energía eléctrica producida en este mes de marzo fue de 22.799 GWh.

9 Energía hidráulica

La energía hidráulica es una energía renovable que se viene utilizando desde la antigüedad. Es limpia y no contaminante.
Se denomina energía hidroeléctrica, aquella energía eléctrica que es transformada desde una fuente de energía hidráulica.
La energía hidráulica tiene la ventaja de que puede ser reutilizada varias veces en el curso de un río, bien por retención en grandes pantanos, en pequeños desniveles o por el impulso de la propia corriente.
Se trata de una energía renovable que está sujeta a las precipitaciones que se den a lo largo del año, escasea más en los meses de estío, cuando se dan períodos secos, en los que se reduce este aprovechamiento energético.

Masa de agua almacenada en pantanos.

9.1. Pequeña reseña histórica

En la antigüedad, el agua estaba considerada como un elemento básico, junto con el aire, la tierra y el fuego. Las aglomeraciones de personas se formaban en torno a los ríos, para asegurarse este líquido de primera necesidad.

Los griegos, los romanos y otras culturas utilizaron la energía del agua (hidráulica) por medio de norias, para mover ruedas de molinos y elevar el agua.

No fue un método muy empleado porque en esta época se disponía de animales de carga y esclavos con los que se realizaban el movimiento de las grandes piedras con las que se molían los granos.

Hasta la edad media, no se volvió al aprovechamiento hidráulico. Fue en el siglo XII cuando se instalaron grandes ruedas que desarrollaban elevadas potencias, hasta casi 50 caballos de vapor (CV). Estas ruedas estaban construidas en madera. Fue el británico John Smeaton el primero en construirlas en fundición de hierro.

El impulso del agua sobre las norias (ruedas con cazos) también se empleaban para extraer agua de los ríos para regar cultivos o para abastecer ciudades.

Todavía hoy podemos ver grandes norias en Hama (Siria) en el río Orontes (todavía en servicio, después de varios siglos) o en Córdoba, sobre el río Guadalquivir.

En 1781, Cavendish obtuvo agua en la combustión del hidrógeno.

Lavoisier demostró que el agua era un compuesto de hidrógeno y oxígeno. La proporción sobre el peso total de una masa de agua es: 16 partes de oxígeno y 2,016 partes de hidrógeno.

Cuando llegó la industrialización se construyeron presas y canales con las que se aprovisionaba el agua que movía las ruedas, muchas de ellas para mover máquinas textiles, aserraderos y otras.

Las ruedas hidráulicas cayeron en desuso cuando apareció la máquina o motor de vapor, alimentado con carbón.

En los inicios del desarrollo de la energía eléctrica, el accionamiento hidráulico fue la fuente de energía utilizada para generar corriente eléctrica.

La primera central hidráulica se construyó en Northumberland (Reino Unido), en 1880. Poco después se construyó una gran central hidráulica en las cataratas del Niágara.

En los primeros años del siglo XX hubo una gran proliferación de centrales eléctricas que se construían en el curso de los ríos. Estos saltos eran de pequeño desnivel, el que proporcionaban las presas y que se utilizaban para alimentar pequeños núcleos de población e industrias.

9.2. El agua

El agua como se ha dicho y por todos es sabido, es un elemento básico para la vida sobre la Tierra.

El agua dulce es necesaria para muchas aplicaciones, como son las domésticas (beber, cocinar, lavar, limpiar, etc.), las industriales (lavar, limpiar, ser un componente más, etc.), los de servicios (hospitales, hoteles, piscinas, etc.), limpiezas de calles, regar parques, apagar incendios, regar los campos, limpieza y bebida de granjas, etc.

El agua del mar la utilizamos para extraer su sal, bañarnos en época de calor y ser la principal fuente de evaporización que luego nos proporciona la lluvia y la nieve.

También tienen otras aplicaciones como fuente de energía, la energía hidráulica, la energía mareomotriz y la energía undimotriz.

Figura 9.1. El río Ebro en su zona media.

9.3. El agua en el planeta Tierra

La superficie total de la Tierra es de: 510.065.284,702 km^2

El 71% de la superficie terrestre está recubierta de agua, de la que el 97% es agua salada y el 3%, agua dulce. El agua es básica para la vida sobre la Tierra.

El agua salada está en los océanos y en los mares, y el agua dulce en los ríos y los lagos.

El agua también está presente en la atmósfera y gracias a esta situación se producen las precipitaciones de agua en forma de lluvia o nieve, con la que se alimentan los cauces fluviales y los acuíferos y se riegan los campos.

La masa de agua almacenada en la atmósfera es muy importante y se comporta como un gas, en forma de pequeñas gotitas que forman las nubes o en cristalitos de hielo si es en forma de nieve.

La masa de agua almacenada en la atmósfera es muy importante, como lo demuestran las importantes cifras que se recogen a continuación:

- La atmósfera contiene unos 12.000 km^3 de agua.
- La mitad de la cantidad almacenada está entre 0 y 1.800 m de altitud.
- Cada año se evaporan alrededor de 500.000 km^3/año.

Figura 9.2. El mar. Gran masa de agua.

9.4. Características del agua

El agua es un líquido incoloro en pequeñas cantidades, azulado cuando está en grandes cantidades y es inodoro e insípido.

El agua hierve a 100 °C y se congela a 0 °C, a presión normal.

La fórmula química del agua es H_2O.

Figura 9.3. Pantano de Riaño (León).

9.4.1. Clases de aguas

a) Aguas duras

Las que llevan disueltas gran cantidad de sales de calcio y magnesio.

Cuecen mal las legumbres y dificultan el lavado de la ropa al dificultar que el jabón sea soluble en este tipo de agua.

b) Aguas blandas

Las que contienen en disolución muy pequeñas cantidades de sales de calcio y magnesio.

Tabla 9.1. Control de la dureza del agua.

Clasificación del agua	°F	mg de CO_3Ca/litro
· Agua muy blanda	3	30
· Agua blanda	4,5	45
· Agua neutra	10	100
· Agua dura	13	130
· Agua muy dura	17	170
· Agua extremadamente dura	25	250

9.4.2. Descalcificación

Proceso por el cual las aguas duras se convierten en aguas blandas.

El proceso de descalcificación se realiza mediante filtros especiales a base de permutitas o zeolitas. Al pasar el agua, las sales de calcio en disolución se permutan en sales de sodio.

9.4.3. Clasificación del agua respecto a su dureza

- Aguas blandas .... 7 °F
- Aguas medias .... 7 a 15 °F
- Aguas duras.......................... 15 a 25 °F
- Aguas muy duras................. más de 25 °F

Equivalencia: 1 grado francés (°F) equivale a 10 mg (0,01 g) de carbonato cálcico por litro de agua y que equivalen respecto a otros grados de dureza a:

1 °F = 0,7 Inglés
1 °F = 0,8 Americano
1 °F = 0,56 Alemán

El agua existe en la Tierra en tres estados: sólido (hielo, nieve), líquido y gas (vapor de agua) y está en los océanos, los ríos, las nubes. El agua que está en superficie se evapora a la atmósfera, formando las nubes y que luego se precipita en forma de lluvia, riega la tierra y se filtra, para aparecer en fuentes y acuíferos. Sin embargo, la cantidad total de agua en el planeta no cambia. La circulación y conservación de agua en la Tierra se llama ciclo hidrológico, o ciclo del agua.

9.4.4. Grado de acidez (pH)

pH – Potencial de hidrógeno.

De 0 a 7 de pH, el líquido es ácido.
Con 7 de pH, es neutro.
De 7 a 14, el líquido es básico.

9.4.5. Ciclo del agua

El agua realiza un ciclo que permite que se disponga de este líquido de una forma constante, como sucede con la:

- Evaporación del agua de los mares, lagos y ríos.
- Humedad en el aire y formación de las nubes.
- Variaciones en la situación de la atmósfera terrestre.
- Caída del agua en formas diversas (lluvia, nieve, granizo, rocío).

- Formación de fenómenos atmosféricos diversos (tormentas, tornados, nubes, claros, despejado, etc.).

Este ciclo del agua da lugar a que los bosques y los campos tengan el riego natural necesario para su subsistencia, y que los cursos fluviales tengan acuíferos que los aprovisionen.

El agua de los ríos es fuente de vida y de energía, que utilizamos industrialmente para producir energía eléctrica o mecánica, aprovechando la fuerza del empuje del agua en su camino hacia el mar.

9.5. Energía hidráulica

La energía hidráulica se viene utilizando desde la antigüedad. La primera información escrita que se tiene es de Vitrubio, en el siglo I a.d.C. que describe una rueda hidráulica compuesta por un eje vertical o normalmente horizontal con aspas o palas radiales que gira por el impulsos de una masa de agua, que puede ser la de un río o canal, estando las palas semienterradas en el agua que con el empuje, hace mover el eje.

Se tardaron muchos años en mejorar el sistema de aprovechamiento de la fuerza del agua y esto sucedió en el siglo XIX en plena industrialización y motorización de las industrias y de forma definitiva cuando hubo que generar electricidad a partir de la fuerza del agua (energía hidráulica).

James B. Francis mejoró en 1848 los diseños de Jean V. Poncelet que en 1820 había diseñado una turbina de flujo interno y de Bonoit Fourneyron que en 1826 había diseñado una turbina de flujo externo de alta eficiencia (80%).

Pelton en 1880 patentó una turbina que consistía en una rueda dotada de cucharas que reciben el chorro de agua y que se aplica a grandes saltos.

Víktor Kaplan en 1912 publicó un trabajo sobre una turbina con rotor en forma de hélice y sistema propio de orientación, que permitía aprovechar saltos con pequeños desniveles.

Cuando se trata de generar energía eléctrica y otras aplicaciones mecánicas, es importante asegurar la uniformidad de la rotación (rpm), en cualquier circunstancia de aprovisionamiento de agua a la turbina.

A continuación vemos las aplicaciones de las turbinas en función de las condiciones del salto de agua.

Se entiende por energía hidráulica aquella energía que se obtiene a partir de la fuerza del agua. Es una energía cinética renovable, que la encontramos en:

- El curso del agua de los ríos.
- El agua contenida en los pantanos.
- En el agua contenida de las presas y represas del curso de los ríos.
- En los canales.
- En las olas y las mareas.

La energía cinética del agua se aprovecha desde la antigüedad en diferentes aplicaciones, como son:

- Accionar las piedras de los molinos.
- Elevar agua (norias).

- Accionar máquinas diversas, especialmente textiles, a partir de mediados del siglo XIX.
- La generación de electricidad es una de sus principales aplicaciones.

En la actualidad, el 19% de la energía eléctrica generada en el mundo, tiene su origen en la energía hidráulica.

9.6. Ventajas e inconvenientes del empleo de agua para generar electricidad

Para asegurar regularidad en el suministro de agua a las turbinas a partir de las cuales se genera electricidad, el agua de los ríos se almacena en grandes pantanos. Esta forma de aprovechamiento de la energía cinética del agua en movimiento y desnivel, tiene sus ventajas e inconvenientes como se señala a continuación:

a) Ventajas

- Se trata de una materia que es renovable.
- No cuesta dinero su obtención, ni su almacenamiento y canalización.
- No hay consumo de agua en el proceso, se reintegra al curso del río.
- Se almacena y regula el curso de los ríos.
- Se puede aprovechar para otros usos (abastecimiento a poblaciones, industrias y regadíos).
- Su utilización no produce contaminación, ni genera calor ni emite gases.
- Regulariza el curso de los ríos. Se evitan inundaciones.
- Mayor aprovechamiento de las grandes avenidas de agua.
- Permite en sus aguas realizar actividades de recreo.
- Casi nula contaminación medioambiental.

b) Inconvenientes

- La construcción del pantano y su relleno de agua produce impacto medioambiental.
- Altera la biología de la zona (animales, aves y vegetales).
- Proporciona más humedad ambiental a la zona.
- Puede introducir microclima.
- Acumula sedimentos que quita al curso bajo del río.
- Altera la biología de la zona.
- Las aguas sufren estancamiento. Pierden propiedades.
- Gran inversión inicial para pagar terrenos, acondicionar la zona, hacer la presa, instalar la central, montar el parque eléctrico y la red eléctrica.
- Riesgo de catástrofes, por rotura de las presas.

Figura 9.4. Agua almacenada en un pantano.

9.7. Aplicaciones del agua como forma de energía

La energía del agua está en la fuerza de un caudal de agua bien sea por su empuje o acción sobre las paletas de una rueda o un chorro de agua que cae desde una altura determinada sobre una turbina.

9.7.1. En molinos

Todavía se sigue empleando el agua para mover las piedras de un molino y otros usos.

9.7.2. Pequeñas centrales hidroeléctricas

Utilizan pequeños saltos y caudales de agua para aprovechar toda posible energía y más en momentos de crisis de materias primas.

Recientemente se han puesto en servicio muchas minicentrales en presas que antes tuvieron otros usos o tenían equipos obsoletos. La energía eléctrica generada por este método empieza a tener un valor importante.

Las centrales hidroeléctricas y por tanto, el aprovechamiento del agua, fueron la primera forma de generar corriente eléctrica de forma industrial.

9.7.3. Centrales eléctricas

Varias formas de aprovechar el agua de los cauces fluviales y su energía potencial:

a) Centrales a filo de agua.

También denominadas *centrales de agua fluyente* o *de pasada*, utilizan parte del flujo de un río para generar energía eléctrica.

b) Centrales a pie de un pantano o embalse

Es el tipo clásico de central hidroeléctrica, que cumple dos funciones: regular el caudal del río o de la cuenca, y generar electricidad.

c) Centrales acopladas a uno o más embalses.

Cuando el curso del río tiene varios embalses consecutivos. Utilizan un embalse para reservar agua e ir graduando el agua que pasa por la turbina.

d) Centrales reversibles

Centrales hidráulicas que cuando la energía demandada es menor tienen la posibilidad de bombear el agua hacia el pantano con la misma energía que producen y no se consume en momentos valle (bajo consumo).

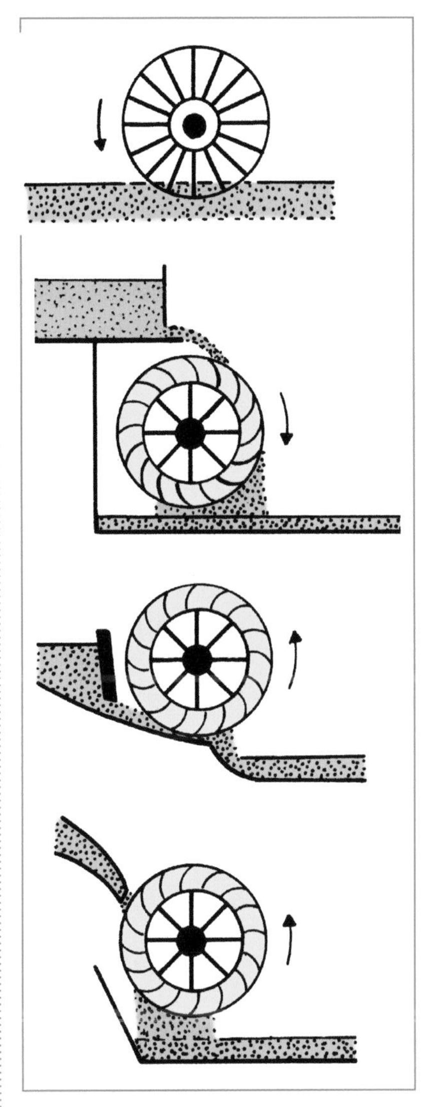

Figura 9.5. Ejemplos de ruedas de accionamiento de molinos.

9.7.4. Formas de suministrar energía eléctrica

La energía hay que suministrarla en función de la demanda. Hay momentos (puntas), en los que la demanda de energía es muy elevada y otros, en la que es muy baja.

Las centrales estarán equipadas para estos momentos poniendo en servicio más generadores o quitando, según proceda.

9.7.5. Grandes centrales hidroeléctricas

Todavía es importante esta forma de generar electricidad a partir de la fuerza del agua. En los años 20 del pasado siglo, la mayoría de la generación eléctrica se hacía a partir de la energía hidráulica.

Todavía hay países en los que la mayor parte de su energía eléctrica se obtiene a partir de la energía hidráulica, casos de Canadá, Noruega, Zaire, Brasil, Paraguay y otros.

Las principales centrales hidráulicas se encuentran Grand Coulee (EE.UU.) que genera en torno a 6.500 MW. La central de Itaipú situada en

Figura 9.6. Compuertas de una minicentral eléctrica.

Figura 9.8. Presa de agua con aprovechamiento de la energía potencial para generar electricidad, y usos varios (agricultura, uso doméstico e industrial).

la frontera entre Brasil y Paraguay que fue inaugurada en 1982 y tiene la mayor capacidad de generación mundial.

China también está potenciando la construcción de grandes presas para la generación de energía eléctrica.

Hay centrales hidroeléctricas que no tienen salto de agua por retención (presa), sino que son de caudal uniforme (agua fluente) como es el caso de las cataratas del río Niágara, aguas debajo de la catarata.

Aprovechar la energía hidráulica tiene muchas ventajas, como son:

- Es una forma de energía renovable.
- En el cauce de un mismo río, se puede reutilizar varias veces.
- La tenemos en nuestro país, por lo que no hay que importarla, aunque cada vez es más escasa (varía según las estaciones del año).
- No es contaminante. Es una energía limpia.
- Su utilización es rentable.

9.8. Descripción de una central hidroeléctrica

A continuación se representa de forma esquemática, una central que aplica la energía hidráulica, para generar electricidad.

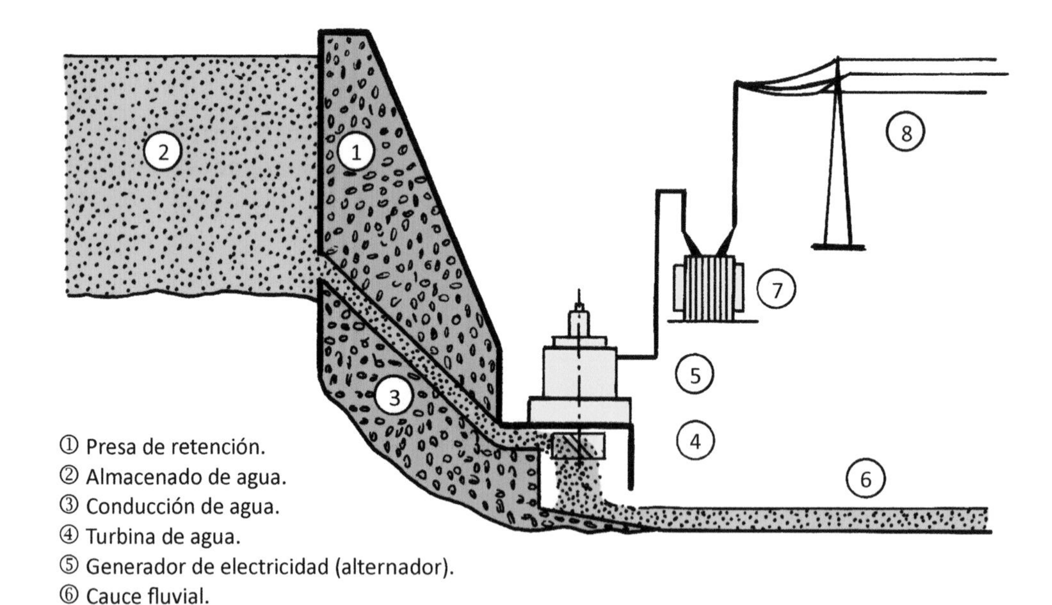

Figura 9.7. Esquema de una central hidroeléctrica.

En épocas secas, los pantanos y presas, además de generar electricidad, sirven para regular los cauces fluviales y garantizar los suministros básicos, como son el doméstico, industrial y agrícola.

9.9. Turbinas hidráulicas

Las turbinas transforman la energía de la fuerza del agua, en energía mecánica rotativa, que se emplea para accionar generadores de corriente alterna (alternadores).

a) Turbinas hidráulicas

Las turbinas hidráulicas están accionadas por la fuerza del agua que incide directamente sobre los álabes de la turbina para proporcionarle movimiento rotativo y mover, normalmente alternadores, con los que se genera energía eléctrica, al transformar la energía mecánica del agua.

Se trata de la primera forma de energía que se utilizó industrialmente para generar electricidad a partir de los saltos de agua.

La ventaja de la energía mecánica que proporciona el agua está en que la misma masa de agua puede ser reutilizada varias veces a lo largo del curso de un río.

b) Tipos de turbinas

La turbina es el dispositivo mediante el cual la fuerza del agua se transforma en energía mecánica y la energía mecánica en energía eléctrica por medio del alternador que mueve la turbina.

Existen tres modelos básicos de turbina hidráulica: Pelton, Francis y Kaplan (tabla 9.2).

Figura 9.9. Presa de Riaño (León).

Figura 9.10. Presa y pantano de Aguilar de Campoo (Palencia).

Tablas 9.2. Las principales turbinas utilizadas en saltos de agua.

Turbinas	Altura y caudal del salto de agua
a) Pelton	Para saltos grandes y caudales pequeños.
b) Francis	Para saltos medianos y mayores caudales.
c) Kaplan	Saltos pequeños y caudales muy grandes.

Tabla 9.3. Clasificación de los saltos de agua por su presión.

Presión	Altura (h) y caudal (Q) del salto de agua
Alta Presión	$h \geq 200$ m $Q \approx 20\ m^3/s$ por turbina. Turbinas Pelton y Francis.
Media Presión	h : entre 20 y 200 m $Q \approx 200\ m^3/s$ por turbina. Turbinas Francis y Kaplan. Pelton para grandes saltos.
Baja Presión	$h \leq 20$ m $Q > 300\ m^3/s$ por turbina. Turbinas Kaplan y Francis.

c) Representación de las turbinas hidráulicas

La figura 9.11 se representan los tres tipos de turbinas que se emplean en centrales hidroeléctricas, que tienen diferencias apreciables entres ellas.

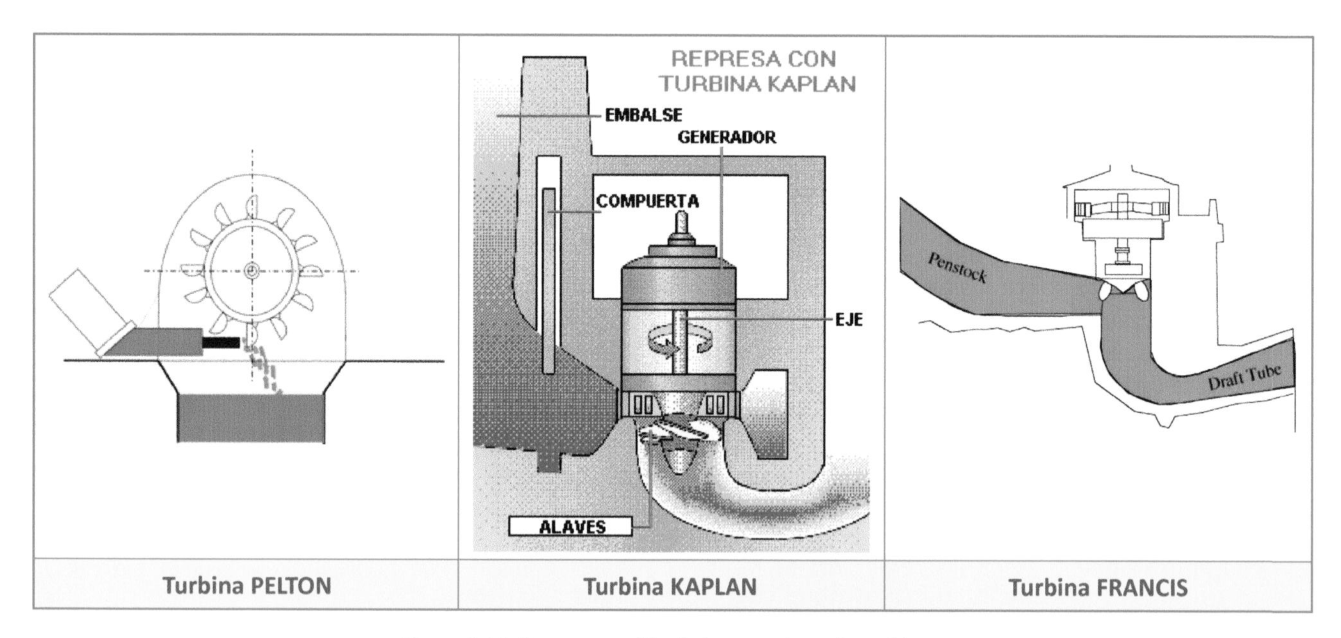

Figura 9.11. Representación de los tres tipos de turbinas.

9.10. Turbinas hidráulicas

Existen tres modelos básicos de turbina hidráulica: Pelton, Francis y Kaplan [1].
La turbina Pelton es una evolución del molino de agua, y utiliza agua impulsada como fluido de trabajo.

9.10.1. Potencia suministrada por una turbina hidráulica

La potencia de una turbina hidráulica depende de muchos factores y condiciones de suministro del fluido a la turbina, sin embargo, las fórmulas que aquí se presentan nos acercan a los valores que proporciona.

a) Potencia suministrada por una turbina hidráulica, en CV:

$$P = \frac{1000 \cdot Q \cdot H}{75} \cdot \eta_t \quad \text{(CV)}$$

P – Potencia, en CV (1CV=736W).
Q – Caudal, en metros cúbicos por segundo (m^3/s).
H – Altura del salto de agua, en metros (m).
η_t – Rendimiento de la turbina.
El rendimiento de las turbinas hidráulicas (η_t) oscila entre 0,8 y 0,95.

b) Potencia suministrada por una turbina hidráulica, en kW:

$$P = \frac{1000 \cdot Q \cdot H}{102} \cdot \eta_t \quad \text{(kW)}$$

[1] En minicentrales se utilizan las turbinas Mitchell-Banki.

c) Potencia suministrada por el alternador, en kW.

$$P = \frac{1000 \cdot Q \cdot H}{102} \cdot (\eta_t + \eta_a) \quad (\text{kW})$$

η_a – Rendimiento del alternador.
El rendimiento del alternador eléctrico oscila entre 0,92 y 0,98.

9.11. Potencia de una central hidráulica (P_u)

La potencia de una central hidráulica depende principalmente del desnivel del salto de agua (diferencia entre la parte superior del agua del pantano y la cota a que se encuentra la turbina).

a) Sin tener en cuenta la densidad del fluido y que se supone sea agua (1kg/dm³)

$$P_u = 9{,}81 \cdot \eta_t \cdot \eta_g \cdot Q \cdot H \text{ (kW)}$$

P_u – Potencia, en kW.
η_t – Rendimiento de la turbina (0,75 a 0,90).
η_g – Rendimiento del generador eléctrico (0,92 a 0,97).
Q – Caudal del agua que pasa por la turbina, en m^3/s.
H – Desnivel del salto entre nivel superior y el agua a nivel de la turbina, en m.

b) En función de la densidad del agua

$$P_u = 9{,}81 \cdot \rho \cdot \eta_t \cdot \eta_g \cdot \eta_m \cdot Q \cdot H \text{ (W)}$$

P_u – Potencia, en W.
ρ – Densidad del fluido, en kg/m^3.
η_t – Rendimiento de la turbina (0,75 a 0,90).
η_g – Rendimiento del generador eléctrico (0,92 a 0,97).
η_m – Rendimiento mecánico del acoplamiento turbina/alternador (0,95 a 0,99).
Q – Caudal del agua que pasa por la turbina, en m^3/s.
H – Desnivel del salto entre nivel superior y el agua a nivel de la turbina, en m.

c) Potencias a considerar

Potencia media: potencia calculada mediante la fórmula de arriba considerando el caudal medio disponible y el desnivel medio disponible.

Potencia instalada: potencia nominal de los grupos generadores instalados en la central.

9.12. Esquema de la parte eléctrica de la central

Las partes principales de la central hidráulica en lo que afecta al circuito eléctrico, son las siguientes:

- Generador eléctrico (alternador). La tensión eléctrica se genera entre 10 y 20 kV.

- Subestación eléctrica. Se eleva la tensión a alta tensión (AT) para efectuar en transporte de la energía eléctrica con las menores pérdidas de energía.
- Red eléctrica de transporte en alta tensión.

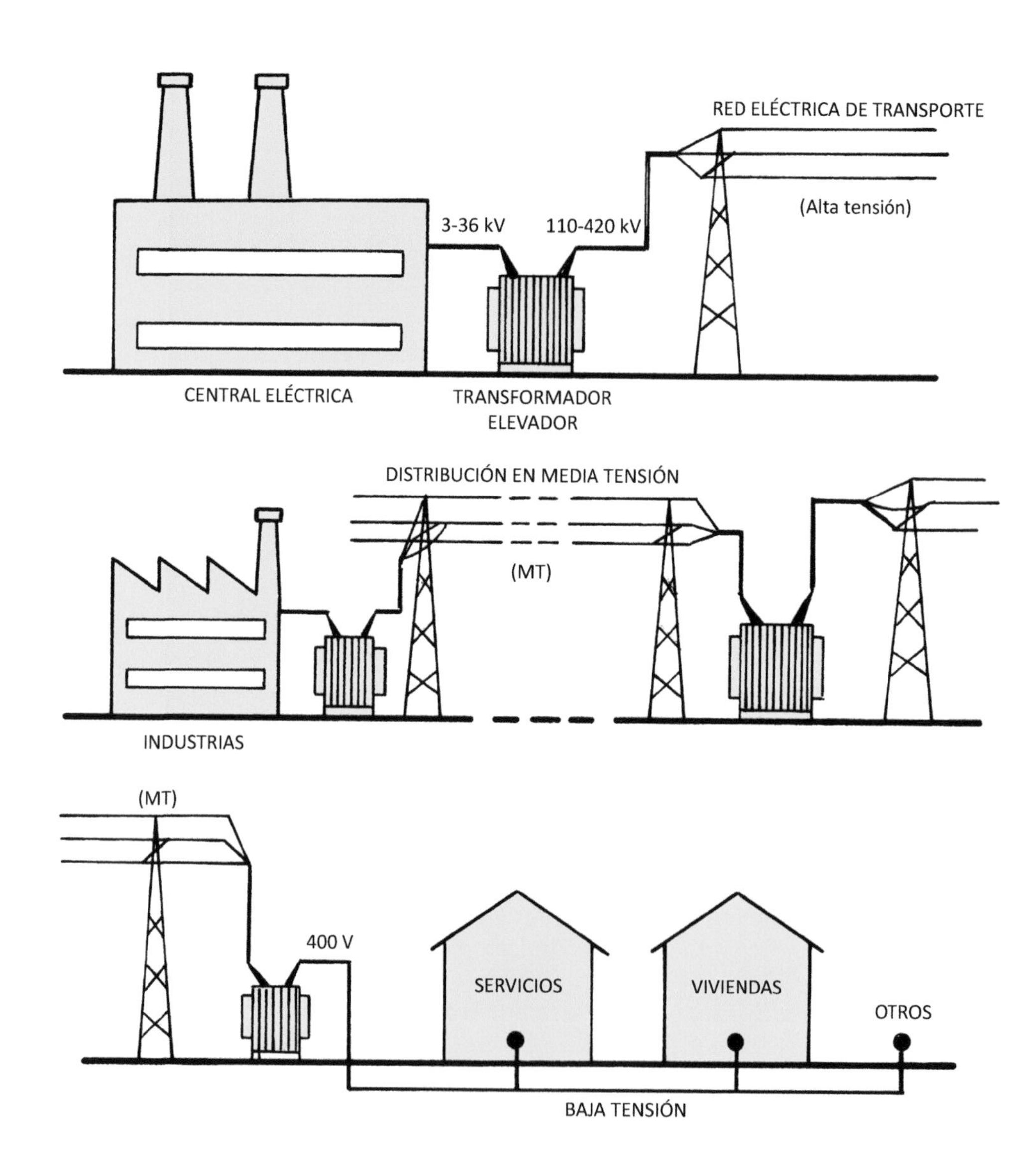

Figura 9.12. Esquema eléctrico de una central y su distribución.

9.13. La energía eléctrica de origen hidráulico en España

En España hay más de 150 grandes pantanos. También hay un número elevado de pequeñas minicentrales que aprovechan presas y represas que ya estaban construidas cuando se instaló la central.

La tabla 9.4 recoge las centrales más importantes en servicio.

Tabla 9.4. Principales centrales hidráulicas en España.

Centrales hidráulicas	Potencia instalada en MW	Río	Provincia
Aldeadávila	1.139,2 [(1)]	Duero	Salamanca
José Mª Oriol	915,2	Tajo	Cáceres
Cortes-La Muela	908,3 [(2)]	Júcar	Valencia
Villarino	810,0	Tormes	Salamanca
Saucelle I y II	570,0	Duero	Salamanca
Estany Gento-Sallent	451,0	Flemisell	Lérida
Cedillo	440,0	Tajo	Cáceres
Tajo de la Encantada	360,0	Guadalohorce	Málaga
Aguayo	339,2	Torino	Cantabria
Mequinenza	324,2	Ebro	Zaragoza
Puente Bibey	285,2	Bibey	Orense
San Esteban	265,2	Sil	Orense
Ribarroja	262,8	Ebro	Tarragona
Conso	228,0	Camba	Orense
Belasar	225,0	Miño	Lugo
Valdecañas	225,0	Tajo	Cáceres
Moralets	221,4	N. Ribajosa	Huesca
Guillena	210,0	Ribera de Huelva	Sevilla
[(1)] Aldeadávila II es una central mixta con bombeo de 421 MW. [(2)] Cortes-La Muela es toda de bombeo con 628,35 MW.			

9.14. La energía eléctrica en el mundo

Hay países que tienen grandes recursos hidráulicos y que dan lugar a que puedan construir grandes centrales hidráulicas, como es el caso de varios países de América del Sur y Canadá. China tiene la presa más grande del mundo

Tabla 9.5. Principales centrales hidráulicas en el mundo.

Central	País	Potencia instalada en MW
Tres Gargantas	China	18.460 [(1)]
Itaipu	Brasil/Paraguay	14.750
Raúl Leoni	Venezuela	10.055
Tucurui	Brasil	8.370
Grand Coulee	Estados Unidos	6.494
Krasnoyarsk	Rusia	6.000
Paulo Afonso	Brasil	3.935
Ilha Solteira	Brasil	3.240
[(1)] Sin concluir (se espera instalar una potencia hasta 22,5 GW.		

Tabla 9.6. Países en los que la energía hidráulica es su principal fuente de generación eléctrica.

País	Porcentaje de procedencia hidráulica de la energía eléctrica (%)
Noruega	99
Zaire	97
Brasil	96
Canadá	60
España	20

Presas más altas del mundo:

- La presa más alta del mundo es de Nurek (Tajakistán), con 300 m, y que será superada por otra de 335 m que se está construyendo en el mismo río.
- Presa Grande Dixence (Suiza), con 285 m.
- Presa Chicoacán (México), con 262 m.
- Presa Thai (India), con 261 m.

9.15. Posibilidades de aprovechamiento de la energía hidráulica

La energía hidráulica en España tiene aún algunas posibilidades de ampliar su aprovechamiento, sino se puede en grandes embalses porque para esto hay muchas dificultades, sí ampliar más el aprovechamiento por medio de minicentrales que tienen poco impacto medioambiental, y es posible generar una potencia importante.

La energía hidráulica debe ser aprovechada al máximo. El agua es energía y la que pasa por un punto y no se emplea, es energía perdida. En estos tiempos no estamos para perder energía.

Se deberían hacer estudios de viabilidad de todos los cauces fluviales con vistas a su aprovechamiento hidroeléctrico.

Existen países en los que la energía hidráulica tiene muchas posibilidades, tienen grandes ríos, zonas poco pobladas y posibilidades de construir grandes presas.

10 Biomasa

La biomasa es cualquier materia orgánica obtenida a partir de vegetales o de animales y residuos. La tecnología actual permite aprovechar y transformar muchos productos como energía.
La biomasa se obtiene en los bosques, en residuos forestales y la agricultura.
Después están los residuos sólidos urbanos, los residuos industriales, y los residuos de las explotaciones ganaderas
Hay productos que se cultivan únicamente con fines energéticos.
La biomasa es una fuente importante de producción de energía renovable y se pueden clasificar en:

- *Biomasa natural que es la que produce la naturaleza de forma espontánea.*
- *Biomasa residual seca es la que procede de actividades agrícolas forestales.*
- *Biomasa residual húmeda es la que procede de vertidos biodegradables, como son aguas residuales urbanas e industriales y residuos ganaderos.*
- *Biomasa procedente de cultivos energéticos forestales y agrícolas.*

Almacenado de paja en grandes pacas.

10.1. Pequeña reseña histórica

La biomasa son materias renovables de origen vegetal y animal.

La energía contenida en la biomasa está en la energía almacenada por sustancias orgánicas en los vegetales y animales. Los animales transforman esta energía, dando lugar a residuos que pueden utilizarse como recursos orgánicos.

Desde el principio de la humanidad, la biomasa ha sido una fuente de energía básica para el hombre.

Ya en la Edad de Piedra, los hombres se calentaban junto al fuego y asaban las piezas que cazaban.

El Tercer Mundo sigue quemando leña para calentarse y cocinar. En el primer mundo se utilizan la biomasa y residuos para generar calor (agua caliente, calefacción y vapor), gases y líquidos combustibles.

El Mundo Industrial ha reducido el consumo de la biomasa tradicional, para pasar a combustibles de origen fósil.

La gran cantidad de residuos que produce nuestro mundo avanzado lleva a que se tenga que hacer un tratamiento riguroso de estos productos con el fin de reducirlos y eliminarlos, y también, para recuperar una parte de los mismos como fuente de energía.

Figura 10.1. Los bosques proporcionan madera y combustibles.

10.2. ¿Qué es la biomasa?

El término biomasa abarca un campo muy amplio sobre cualquier tipo de materia que tenga origen inmediato con un proceso biológico. La biomasa comprende productos y materias que tienen su origen en los vegetales y los animales.

Desde el punto de vista de la ecología podemos distinguir tres formas de biomasa.

a) Biomasa primaria

Son productos que tienen su origen en la fotosíntesis. Biomasa vegetal, algas, plantas verdes y seres autótrofos.

Comprende toda la biomasa vegetal, incluidos residuos agrícolas (paja, restos de podas y forestales, leñas, etc.).

b) Biomasa secundaria

Está producida por seres heterótrofos que utilizan en su nutrición la biomasa primaria.

Ejemplo: carne y desechos de animales y las deyecciones de animales herbívoros.

c) Biomasa terciaria

Está producida por seres que se alimentan de biomasa secundaria.

Ejemplo: Carne de animales carnívoros que se alimentan de los herbívoros.

10.2.1. Clasificación de la biomasa:

a) Biomasa natural

La producen los ecosistemas silvestres.

b) Biomasa residual

Se extrae de residuos agrícolas, forestales, humanos, así como actividades agrícolas y ganaderas.

10.3. Tipos de biomasa

La biomasa la produce la naturaleza, bien de forma natural, o por cultivo. Se trata de una energía renovable y su aprovechamiento, aunque no siempre fácil, nos permite obtener energía alternativa que está a nuestra disposición, sin necesidad de importar. La biomasa la obtenemos de diferente forma, tal como se resume a continuación.

a) Biomasa natural

La produce de forma espontánea la naturaleza, como es por ejemplo un bosque del que se obtiene biomasa por podas y limpiezas, desechos de talas, desechos de aserraderos, etc.

b) Biomasa residual seca

Procede de recursos generados por actividades agrícolas y forestales.

También se obtiene biomasa de los procesos de la industria agroalimentaria (conserveras) y de la industria de la transformación de la madera.

c) Biomasa residual húmeda

Procede de vertidos biodegradables constituidos por aguas residuales e industriales, incluidos los residuos agrícolas.

d) Cultivos energéticos

Los cultivos energéticos pueden ser: agrícolas y forestales y están dirigidos a producir biomasa con fines alimentarios o energéticos.

La tabla 10.1 muestra el proceso, el producto obtenido y las aplicaciones del aprovechamiento de diferentes materias primas que tienen su origen o forman parte del proceso en el que la fuente principal es la biomasa.

Tabla 10.1. Resumen de materias primas y productos que se obtienen.

Materia prima	Proceso	Producto	Aplicaciones
· Aceites vegetales (limpios o usados) · Cultivos oleaginosos	Refino/transesterificación	Biodiesel	Aplicaciones mecánicas (motores diésel).
· Residuos vegetales · Cultivos alcoholícenos	Fermentación/destilación	Bioetanol	Aplicaciones mecánicas (motores de gasolina).
Residuos sólidos urbanos (RSU)	Descomposición anaeróbica	Biogás	Aplicaciones térmicas y eléctricas.
Residuos ganaderos	Digestión anaeróbica	Biogás	Aplicaciones térmicas y eléctricas.
· Residuos forestales · Residuos agrícolas · Cultivos	Gasificación	Gas pobre/gas de síntesis	Aplicaciones térmicas y eléctricas.
· Residuos forestales · Residuos agrícolas · Cultivos lignocelulósicos	Corte, secado, compactación, etc.	Pelets, cáscara de almendra, orujo, etc.	Aplicaciones térmicas y eléctricas.

Figura 10.2. Residuos forestales.

Figura 10.3. Almacenado de granos en silos para su tratamiento posterior.

Figura 10.4. Industria que transforma grasas para uso industrial y grasas vegetales alimentarias.

Figura 10.5. Recogida de residuos urbanos.

10.4. Aprovechamientos de residuos

Son muchos los residuos que genera el hombre en su actividad diaria, bien sea de forma directa o de forma indirecta. Si no los eliminamos de una forma reglada y ordenada, pueden dar lugar a muchos problemas. Por otro lado, los residuos que tienen relación con la biomasa pueden aprovecharse para generar otras energías.

Los residuos se pueden agrupar en cinco grandes grupos, que son:

a) Residuos forestales

Proceden de la naturaleza

Principales residuos: troncos, ramas, cortezas, virutas, serrín, restos de madera de aserraderos y carpinterías, hojarasca, raíces, restos de procesos de limpieza de montes, etc.

b) Residuos agrícolas

Proceden de las explotaciones agrícolas.

Principales residuos: paja de cereales, leña de la poda de frutales, viñedos, olivos, tallos de cultivos textiles y de oleaginosas, caña del maíz, restos leñosos de otros vegetales.

Tabla 10.2. Generación de residuos en diferentes cultivos.

Cultivos	Residuos	Cultivos	Residuos
Cultivos en grano	t/t	Cultivos frutales	t/ha
Trigo	1´20	Cítricos	2´00
Cebada	1´35	Frutales de pepita	3´50
Avena	1´35	Frutales de hueso	2´00
Maíz	2´00	Frutos secos	1´50
Arroz	1´50	Olivo	1´70
Sorgo	1´70	Vid	3´50

Tabla 10.3. Poder calorífico de algunos residuos procedentes de cultivos.

Residuo	PC Medio 10^7 kcal/t	Observaciones 10^7 kcal ≈ 1 tep
Cañote de maíz	0,365	El cañote y el zuro (corazón de la mazorca después de desgranada) se recogen juntos.
Zuro del maíz	0,388	
Cañote y cabezuela del maíz	0,29	
Ramón del olivar	0,43	Se produce a razón de unas 0,25 t/ha, pero el 40% son hojas que pueden secarse. El resto, denominado vareta, es lo que se suele aprovechar.
Sarmiento de la vid	0,28	Se genera a razón de 0,7 t/ha. El PCS puede llegar a 0,456 tep/t.
Residuos del tomate industrial	0,51	
Cañote del girasol	0,335	

c) Residuos ganaderos

Proceden de las explotaciones ganaderas.

Principales residuos: deyecciones de animales estabulados en explotaciones ganaderas.

d) Residuos industriales

Proceden de las industrias y principalmente de las industrias conserveras y donde se tratan productos agrícolas.

Principales residuos: provienen mayoritariamente de industrias de conservas vegetales, producción de aceites, vinos y frutos secos.

e) Residuos urbanos

Proceden de la actividad humana en viviendas, hoteles, hospitales, restaurantes, bares, cafeterías, etc.

Constituyen una biomasa residual de un volumen muy importante, que se concentra en puntos de recogida y que es imprescindible transportar a centros de almacenamiento. Su reciclado puede suponer un aprovechamiento importante de la energía que contienen.

Las aguas residuales proceden de la actividad humana, cuya depuración genera fangos que pueden contener cargas contaminantes que es necesario reducir. En este proceso, la fracción sólida contiene una apreciable cantidad de biomasa residual.

Figura 10.6. Recogida de desechos urbanos.

10.5. Poder calorífico de diferentes productos vegetales

La madera y muchos residuos agrícolas son una fuente energética que se puede aprovechar para generar calor y a partir de él, otras formas de energía. Las tablas recogen el poder energético de diferentes residuos vegetales.

Tabla 10.5. Poder calorífico superior de maderas y residuos agrícolas.

Combustible	Poder calorífico superior kJ/kg
Cáscara de almendras	36.800
Cáscara de nueces	32.000
Cáscara de arroz	15.300
Cáscara de pipa de girasol	17.500
Cáscara de trigo	15.800
Corteza de pino	20.400
Corcho	20.930
Orujillo de aceituna	17.900
Orujo de uva	19.126
Papel	17.500
Jara (8% humedad)	18.900
El poder calorífico de la madera verde disminuye según aumenta la humedad de la misma. En la tabla 10.6 se da el coeficiente por el que hay que multiplicar su poder calorífico para obtener el poder calorífico real.	

Tabla 10.4. Poder calorífico medio de maderas y residuos agrícolas.

Combustible	Poder calorífico medio kJ/kg
Bagazo húmedo	10.500
Bagazo seco	19.200
Cáscara de cacahuete	17.800
Cascarilla de arroz	13.800
Celulosa	16.500
Corteza escurrida	5.900
Cosetas de caña	4.600
Madera seca	19.000
Madera verde	14.400
Paja seca de trigo	12.500
Paja seca de cebada	13.400
Serrín húmedo	8.400
Viruta seca	13.400

Tabla 10.6. Coeficientes de la madera húmeda

Madera	Coeficiente	Madera	Coeficiente
Álamo negro	0,55	Haya	0,62
Castaño	0,48	Olivo	0,88
Chopo	0,30	Pino marítimo	0,58
Encina	0,68	Pino silvestre	0,49
Enebro	0,50	Roble	0,68

10.6. Unidades energéticas

En el aprovechamiento de la biomasa se utilizan diversas unidades de energía, que se emplean en función de la materia de que se trate.

Tabla 10.7. Unidades energéticas.

Unidad	Símbolo	Equivalencia
Julio	J	Unidad de energía y trabajo del SI y equivale a la fuerza de un newton (N) en un desplazamiento de un metro (m). 1 J = 1 W x s
Caloría	cal	Cantidad de energía necesaria para elevar la temperatura de un gramo de agua en un grado centígrado. Es unidad del Sistema Técnico.
Kilovatio-hora	kWh	Unidad práctica de energía. Equivale al consumo de un kW durante una hora. 1 kWh = 3.600.000 julios.
Tonelada equivalente de petróleo [1]	tep	Equivale a la energía que hay en una tonelada de petróleo, tomando el valor convencional de: 1 tep = 41.840.000.000 julios = 11.622 kWh.
Tonelada equivalente de carbón [1]	tec	Representa la energía liberada por la combustión de 1 tonelada de carbón (hulla).
Termia	th	Unidad de energía utilizada en el suministro de gas natural. 1 th = 1.000.000 calorías.
British Thermal Unit	Btu	Unidad del Sistema Inglés. 1 Btu equivale a la cantidad de calor necesario para aumentar la temperatura de una libra de agua, un grado Fahrenheit.
Kilocaloría/kilogramo	Kcal/kg	Aplicada a un combustible, indica el número de kilocalorías que obtendríamos en la combustión de 1 kg de ese combustible.

[1] Equivalencia: 1 tep = 1,428 tec

Tabla 10.8. Equivalencias entre diferentes unidades de energía.

1 J = 0,239 cal ≈ 0,24 cal
1 cal = 4,185 J
1 Wh = 3.600 J
1 kWh = 3,6 x 106
kcal/kg ⇒ kilo-calorías por kilogramo de materia.
kJ/kg ⇒ kilo-julios por kilogramo de materia.
1 kJ = 0,23892 kcal = 238,92 cal
1 kcal = 4,1855 x 103 J
1 tep = 41,34 x 109 J ⇒ tonelada equivalente de petróleo.
1 tec = 29,3 x 109 J ⇒ tonelada equivalente de carbón.
1 th = 106 cal

Tabla 10.9. Equivalencias entre unidades de energía

Multiplicar por: ⇒ Para pasar de: ⇓	Símb.	J	cal	kWh
Julio	J	1,000	0,239	$0,278 \times 10^{-6}$
Caloría	cal	4,184	1,000	$1,162 \times 10^{-6}$
Kilovatio-hora	kWh	$3,6 \times 10^{6}$	$0,860 \times 10^{6}$	1,000
Tonelada equivalente de petróleo	tep	$41,840 \times 10^{9}$	10^{10}	$11,622 \times 10^{3}$
Termia	th	$4,184 \times 10^{6}$	10^{6}	1,162
British Thermal Unit	Btu	$1,054 \times 10^{3}$	251,996	$0,293 \times 10^{-3}$

Ejemplo de aplicación de la tabla: para pasar de ***julios*** a ***calorías***, multiplicar 0,239.

Poder calorífico estimado de la biomasa:

- 1 kg de biomasa proporciona en torno a 3.500 kcal.
- 1 litro de gasolina proporciona 10.000 kcal.

Esta rentabilidad de la biomasa no se da en todos los casos.

10.7. Potencia calorífica de un combustible conocido su composición química

La potencia de un combustible puede calcularse en función de su composición química, de acuerdo con la siguiente fórmula:

$$Pc = 8.080 \cdot C + 34.462 \left(H - \frac{O}{8}\right) + 2.240 \cdot S \text{ (cal/kg)}$$

Pc – Potencia calorífica, en calorías por kg (cal/kg) de combustible.
C – % de carbono en el combustible.
H – % de hidrógeno en el combustible.
O – % de oxígeno en el combustible.
S – % de azufre en el combustible.

10.8. Transformación de la biomasa en energía

Podemos considerar dos métodos para transformar la biomasa en otra forma de energía, que son: métodos termoquímicos y métodos bioquímicos.

10.8.1. Métodos termoquímicos

a) Por combustión

Consiste en quemar biomasa con abundante oxígeno, para producir energía calorífica aprovechable en viviendas (cocinar, proporcionar agua caliente y calefacción), en la industria y para generar electricidad.

b) Por pirólisis

Un procedimiento que se utilizó mucho fue el del carbón vegetal, quemando leña en una combustión incompleta, con poco oxígeno a una temperatura entorno a los 500 °C.

Por este método, también se obtienen combustibles líquidos, semejantes a los hidrocarburos.

c) Por gasificación

Es una variante el método anterior y consiste en una combustión incompleta de la biomasa (leña) a elevada temperatura y poco oxígeno, según se le aplique aire u oxígeno puro se consiguen dos productos diferentes:

- Con el aire se obtiene gasógeno, también llamado gas pobre. Este gas se utilizó para mover automóviles y en la actualidad para generar vapor y electricidad.
- En un gasificador alimentado con gas, junto con oxígeno puro y vapor de agua, se obtiene un gas de síntesis que puede ser transformado en combustible líquido.

Figura 10.7. Propaganda de los años cuarenta de pasado siglo, anunciando un proveedor de gasógenos.

10.8.2. Métodos biológicos

Este método utiliza diversos tipos de microorganismos para degradar las moléculas a compuestos más simples y de alta densidad energética. Proceso adecuado para biomasa con alto contenido de humedad. Los métodos más utilizados son:

- Fermentación alcohólica para producir etanol.
- Digestión anaerobia para producir metano.

Se utiliza en explotaciones de ganadería intensiva, instalando digestores o fermentadores, en los que la celulosa procedente de excremento de animales se degrada, produciendo un gas que contiene cerca de 60% de metano.

10.9. Aplicaciones energéticas de la biomasa

La energía se puede recuperar de la biomasa por diferentes procedimientos, como son algunos de los que se citan a continuación.

10.9.1. Generación de energía térmica

Se quema biomasa (combustión directa) para producir calor (energía calorífica), con el que se efectúan las siguientes operaciones:

- Cocinar alimentos.
- Secar vegetales (alfalfa, maíz, granos, etc.).
- Proporcionar calefacción.
- Colaborar en procesos industriales.

10.9.2. Generación de energía eléctrica

Dependiendo del proceso de aprovechamiento que se haga de la biomasa y de la tecnología que se aplique, se podrá generar electricidad:

- Por combustión de la biomasa para producir vapor de agua con el que a través de una turbina de vapor, accionar un alternador y generar electricidad.
- A partir de la biomasa, obtener gas de síntesis por gasificación de recursos sólidos. Con el gas obtenido se alimenta una turbina de gas, que acciona un alternador que genera electricidad.
- Mediante un motor alternativo que consuma gas, accionar un generador eléctrico.

10.9.3. Energía mecánica

Se genera energía mecánica a partir de biocombustibles procedentes de la biomasa, con los que se alimentan motores de combustión tanto de gasolina como de gasoil. Se pueden utilizar biocombustibles mezclados en gasolinas y gasoil, o solos.

Figura 10.8. Bosques con grandes pinos que proporcionan madera.

10.10. La madera como fuente de energía

La madera tiene muchas posibilidades de aprovechamiento y una de ellas es como combustible. No hace muchísimos años, era el combustible utilizado en los hogares. Todavía hoy día se utiliza para calefacción en forma de pelets o cortado en trozos.

De los árboles se aprovecha como combustible, limpiezas de bosque, las ramas, las podas, las cortezas, los restos, el serrín, árboles que se arrancan por su vejez, etc.

Es una fuente de energía aprovechable, y de forma especial todas aquellas partes que no se utilizan como madera para fines concretos.

A la madera se le unen otros residuos que producen los árboles como es por ejemplo, la piña, los orujos, cáscaras, huesos, etc.

Otra variante de la madera es el carbón vegetal con diversas aplicaciones, como ya se ha indicado.

El poder calorífico de la madera seca es del orden de 19.000 kJ/kg.

10.11. Ventajas e inconvenientes del empleo de biomasa

Como toda transformación de energía, la recuperación de la biomasa tiene ventajas e inconvenientes como se recoge a continuación.

a) Ventajas

- Es una energía renovable.
- La vegetación absorbe en su desarrollo mucho más CO_2 del que desprende en la fase de recuperación (combustión), como biomasa. Proceso muy importante para evitar el efecto invernadero. De

aquí, la importancia de conservar las masas arbóreas e incluso aumentarlas para que hagan de absorbedoras de las grandes cantidades de CO_2 que diferentes procesos evacuan a la atmósfera.
- La biomasa no desprende en su combustión ni azufre ni hidrocarburos (o lo hacen en muy pequeña cantidad).
- Su recuperación o cultivo, genera puestos de trabajo.
- Las plantaciones energéticas pueden reducir la erosión de los suelos.
- Abre oportunidades respecto a ciertos cultivos.
- Aprovecha residuos que a veces son un problema si no se eliminan correctamente.
- Su aprovechamiento evita contaminación del medioambiente.
- Permite la obtención de productos biodegradables.
- La limpieza de bosques evita futuros incendios.

b) Inconvenientes
- No todos los productos biomasa son recuperables.
- La biomasa tiene una baja densidad energética.
- La biomasa tiene un rendimiento muy inferior a los combustibles fósiles.
- Para hacer rentable su aprovechamiento en instalaciones industriales se necesita tratar grandes masas de producto cuya recolección, transporte y almacenado dificulta el tratamiento.
- Si los combustibles fósiles están baratos, la rentabilidad de estos procesos de recuperación no suele ser competitiva. Si lo es, si se tienen en cuenta otras circunstancias estratégicas.
- Aún habiendo biomasa, en muchas ocasiones resulta muy difícil su recuperación, por estar muy dispersa la materia.
- La biomasa no se puede considerar una solución alternativa a las energías tradicionales, pero sí una ayuda.
- La producción de biomasa con fines energéticos puede agostar los terrenos.

11 Biocombustibles

La producción de biocombustibles colabora en la reducción de las importaciones de productos petrolíferos. La combustión de bioetanol puede reducir las emisiones de gases de efecto invernadero entre el 40 y el 80% al considerar que los cereales convertibles en alcohol o los cultivos oleaginosos con los que se produce biodiesel absorben una parte del CO_2 que después generan en la combustión.
El etanol y el biodiesel son combustibles renovables, que se pueden cultivar en muchos lugares, dando salida a las muchas toneladas de aceites usados (30 litros por hogar y año), que si se vierten con las aguas residuales, contaminan los ríos.

El girasol representa a las energías que se obtienen de la biomasa.

11.1. Pequeña reseña histórica

Los biocombustibles son derivados de la biomasa y por tanto, deberían formar parte del capítulo 10, pero dado la importancia de esta materia en lo que se refiere a su transformación como biocombustibles, parece lógico darle un tratamiento especial.

En cada una de las muchas crisis energéticas por las que hemos pasado en el último tercio del siglo XX, hasta hoy día, se ha ido relanzando la posibilidad de producir combustibles que tienen su origen en la biomasa por las siguientes razones:

- Lo podemos producir en el propio país.
- No hay que importarlo.
- Reduce el consumo de combustibles de origen fósil, y por tanto dependencia exterior.
- Es una alternativa real a los combustibles tradicionales.

Los mercados de los combustibles fósiles en los últimos años están sujetos a muchas variaciones de precio, dificultades de suministro y un encarecimiento muy importante, lo que lleva a pensar en alternativas, que aunque no sean definitivas por su volumen, sí supongan una ayuda a la estabilización del mercado.

Mirando hacia atrás, cuando Rudolf Diesel diseñó su motor de combustión, ya pensó en alimentarlo con aceites vegetales (el gasoil todavía no estaba generalizado), haciendo las primeras pruebas con aceite de cacahuete.

Los combustibles de origen fósil empezaron a ser más abundantes y baratos, lo que llevó a que todos los motores utilizaran gasolina o diésel (gasoil).

Cuando Henry Ford diseñó el famoso modelo T en 1908, pensaba alimentarlo con etanol.

A lo largo del pasado siglo XX se han realizado intentos de elaboración de biocombustibles, pero sin intensidad.

Cuando en 1973 se produjo la guerra judeo-árabe que significó una aguda crisis de suministro de petróleo, se volvió a impulsar la producción de biocombustibles.

El 1975, Brasil inició la fabricación de biocarburantes.

En 1985, la UE intentó introducir en Europa la fabricación de biocombustibles con el objetivo de reducir en un 25% el gasto de combustible de origen fósil, pero el intento quedó en eso. Después, el Consejo Europeo estableció en marzo de 2007 otro objetivo para el año 2020, para que el 10% del combustible utilizado en el transporte por carretera proceda de biocombustibles-biocarburantes.

Por las razones que se indican, no es fácil cumplir objetivos, porque sus ventajas se ven contrarrestadas muchas veces por sus desventajas.

11.2. ¿Qué son los biocombustibles?

Los biocombustibles son combustibles que se obtienen a partir de procesos diversos en los que la biomasa es la materia prima y cuyo objetivo principal es el de sustituir a los carburantes fósiles tradicionales y reducir la emisión de gases contaminantes de efecto invernadero.

Los principales biocombustibles son:

- Bioetanol (bioalcohol), como sustituto o acompañante de la gasolina.
- Biodiesel, como sustituto o acompañante del gasoil.

11.3. ¿Qué son los biocarburantes?

Los biocarburantes son carburantes que proceden de procesos biológicos, a partir de materias de origen vegetal (biomasa) y de desechos varios, y que presentan ventajas muy claras sobre los combustibles de origen fósil (carbón, gasolina, gasoil y gas natural.

Pueden ser una solución alternativa a los combustibles de origen fósil que tienen caducidad, además de ser menos contaminantes que los anteriores.

Se tienen puestas muchas esperanzas en estos productos, pero hay que buscar un equilibrio entre la materia que se destina a consumo humano y animal, y la que se destina a la fabricación de biocarburantes, con dos fines principales, como son el que no haya escasez de suministros y que los precios de las materias primas agrícolas no se eleven y se mantengan dentro de unos límites razonables.

El objeto de la fabricación de los biocarburantes es la de suplir una parte del consumo de los carburantes de procedencia fósil. Un ejemplo de la producción y utilización de biocarburantes (bioetanol) son Brasil y EE.UU. La Unión Europea produce biodiesel.

Los principales biocarburantes son: bioetanol y biodiesel, aunque hay muchos más.

11.4. Principales biocombustibles

Varios son los productos obtenidos de la biomasa y residuos que se aplican como biocarburantes, y entre los más importantes están:

La Unión Europea en una reciente legislación, considera como biocarburantes los siguientes productos:

- **Bioetanol:** etanol producido a partir de la biomasa o de la fracción biodegradable de los residuos.
- **Biodiesel:** éster metílico producido a partir de un aceite vegetal o animal de calidad similar al gasóleo.
- **Biogás:** combustible gaseoso producido a partir de la biomasa y/o a partir de la fracción biodegradable de los residuos.
- **Biometanol:** metanol producido a partir de la biomasa.
- **Biodimetiléter:** dimetiléter producido a partir de la biomasa.
- **BioETBE** (etil ter-butil éter): ETBE producido a partir del bioetanol.
- **BioMTBE** (metil ter-butil éter): combustible producido a partir del biometanol.
- **Biocarburantes sintéticos:** hidrocarburos sintéticos o sus mezclas, producidos a partir de la biomasa.
- **Biohidrógeno:** hidrógeno producido a partir de la biomasa y/o a partir de la fracción biodegradable de los residuos.
- **Aceite vegetal puro:** obtenido a partir de plantas oleaginosas mediante presión, extracción o procedimiento comparable, crudo o refinado, pero sin modificación química.

Nota:

Los biocombustibles están producidos para ser usados como biocarburantes.

11.5. Principales características de los biocarburantes

A continuación conoceremos la obtención, procedencia y aplicaciones de los principales biocarburantes obtenidos a partir de biomasa.

11.5.1. Bioetanol

Obtención. El bioetanol es un alcohol etílico deshidratado producido a partir de la fermentación de diversos productos de la biomasa que son ricos en azúcares, amiláceos y también lignocelulósicos.

- Azúcares: remolacha, caña de azúcar y melaza.
- Almidones: maíz, trigo, cebada y sorgo.
- Celulosas: maderas y residuos de poda.

Los principales cultivos son. Caña de azúcar, remolacha, maíz, sorgo, trigo, cebada y también la paja de cereales y residuos de vegetales.

Aplicaciones. Como sustitutivo de la gasolina o mezclada con ésta en diversas proporciones.

Observación. Su porcentaje en la gasolina no debe superar en climas fríos y templados el 5-10%, pudiendo llegar al 20% en zonas cálidas.

Para el empleo como combustible único, los motores deben tener características especiales para este combustible.

Brasil es el mayor productor y consumidor mundial de bioetanol como combustible. Produce más de 15 millones de m^3.

11.5.2. Biodiesel

El biodiesel es una mezcla de ésteres metílicos derivados de los ácidos grasos presentes en aceites vegetales que están obtenidos mayoritariamente por reacción de transesterificación de los mismos con metanol.

Obtención. El biodiesel es un éster metílico que se extrae de diferentes plantas oleaginosas y se obtiene mediante un proceso denominado transesterificación.

Procedencia de la materia.

- Aceites vegetales convencionales: de colza, girasol, palma, soja, coco.
- Aceites vegetales alternativos.
- Aceites de semillas modificadas genéticamente.
- Aceites de frituras.
- Aceites de otras fuentes.
- Grasas animales.

Aplicaciones. Como sustituto del gasoil en motores Diesel, para quemar en calderas de calefacción.

Las propiedades del bioetanol son muy parecidas a las del gasóleo.

Observación. Cuando se emplean mezclas con biodiesel superior al 5%, es preciso reemplazar los conductos de goma por los que pueda pasar la mezcla.

Pueden hacerse otras mezclas de acuerdo con los tipos de motores.

En la planta integral ACOR, los productos base para la obtención del biodiesel son el girasol y la colza (hasta 160.000 toneladas/año).

Tiene una capacidad inicial de producción de biodiesel de 100.000 toneladas/año, ampliables hasta 160.000 toneladas/año y en torno a 12.000 toneladas/año de glicerina.

Figura 11.1. Silos que almacenan semillas de girasol y colza en la planta ACOR de Olmedo (Valladolid) para la obtención de biodiesel.

Nota:

La planta integral ACOR de Olmedo para la fabricación de biogás que inició su actividad a finales del año 2009 tuvo una inversión de 60 millones de euros. Realiza todo el proceso completo de fabricación, desde el almacenado de la materia hasta la obtención del biodiesel. Parte de la materia prima la producen socios de la cooperativa ACOR.

Trata 90.000 toneladas/año de residuos de los productos tratados, que se convierten en harina para piensos.

a) Ventajas del biodiesel:

- Mezclado con gasoil mejora las condiciones del gasoil.
- No contiene azufre.
- Mejora la combustión del gasoil.
- Reduce el humo en el arranque en un 30%.
- Reduce las emanaciones de CO_2 y partículas e hidrocarburos aromáticos.
- Reduce la emisión de gases nocivos de la combustión.
- Es menos irritable para la piel humana.
- Es menos peligroso para la flora y la fauna marina en caso de vertido a ríos o mares, que los combustibles fósiles.
- Prolonga la vida útil de los motores.
- Menos peligroso en su almacenamiento y transporte.

b) Fabricar biodiesel:

Para producir 1.005 kg de biodiesel, son necesarios:

- 110 kg de metanol [1].
- 15 kg de catalizador.
- 1.000 kg de aceite.
- 4.290 litros de agua.

11.5.3. Biogás

Obtención. El biogás se obtiene por fracción biodegradable de los residuos a través de la fermentación anaeróbica de biomasa húmeda.

Procedencia de la materia. Es un producto que se obtiene por la descomposición anaerobia de compuestos orgánicos, por la acción de diversas bacterias. Es una mezcla de metano y CO_2.

Aplicaciones. El gas obtenido puede ser purificado y alcanzar una calidad igual o superior al gas natural.

Se emplea como biocarburante.

11.5.4. Biometanol

Obtención. El metanol CH3OH se produce a partir de gas de síntesis ($CO + H_2$) procedente de la biomasa.

Procedencia de la materia. A partir de residuos forestales y materiales lignocelulósicos.

Aplicaciones. Como combustible.

Se emplea en motores de ignición en sustitución de la gasolina, tiene elevado octanaje, aunque bajo contenido energético por volumen.

11.5.5. Biocarburantes sintéticos

Obtención. Hidrocarburos sintéticos o sus mezclas.

Procedencia de la materia. Biomasa.

Aplicaciones. Para uso como biocarburante.

---

1 Conocido como alcohol metílico, se utiliza como combustible y se obtiene industrialmente en procesos petroquímicos. También se emplea para incrementar el octanaje de las gasolinas.

11.5.6. Aceites vegetales

En ciertas aplicaciones (motores) es posible incorporar un determinado porcentaje de aceites vegetales a combustibles diesel o biodiesel.

11.5.7. Biodimetiléter

Obtención. A partir de metanol.
Procedencia de la materia. Biomasa.
Aplicaciones. Para uso como biocarburante.

11.5.8. Otros productos procedentes de la biomasa

- Etil ter-butil éter (BioETBE).
- Metil ter-butil éter (BioMTBE).
- Biohidrógeno.
- Otros.

11.6. Ventajas e inconvenientes de los biocombustibles

Como hemos indicado, en la generalización de este producto hay que ver las cosas que le son favorables y las que no lo son, lo que nos dará ocasión de ver ciertos problemas que tienen una importancia y que no son de fácil solución.

a) Ventajas

- Proporciona energía de fuentes reciclables y por tanto inagotables.
- Su obtención colabora en el suministro de combustibles.
- Reducen dependencia de combustibles de origen fósil.
- Los biocombustibles son menos contaminantes que los combustibles tradicionales.
- Ayudan a reducir el efecto invernadero.
- Su cultivo favorece a los agricultores.
- Pueden aprovecharse terrenos de poco valor agrícola que se hayan abandonado por su baja productividad.
- Más empleo de mano de obra en el campo y en la industria de transformación.

b) Inconvenientes

- La utilización de productos agrícolas para fabricar biocombustibles puede ser causa de la subida incontrolada de los precios y también, pueden dar lugar a su escasez.
- Su cultivo prioritario puede dar lugar a la pérdida de la biodiversidad y a ciertos recursos naturales de la zona.
- El cultivo intensivo de ciertas plantas libera gases de efecto invernadero (óxido nitroso).
- El proceso de obtención de los biocombustibles es caro y poco competitivo con los combustibles tradicionales.
- Necesitan ayudas de la administración para incentivar su producción.
- Los biocombustibles producidos a partir de palma aceitera, caña de azúcar y soja pueden conllevar impactos sociales y medioambientales en la zona.
- Se requieren grandes superficies dedicadas a estos cultivos.

- Los bioalcoholes pueden emitir mayor cantidad de óxido de carbono (CO_2), que la gasolina y el gasoil.

11.7. Situación actual de esta energía

A pesar de lo mucho que se ha hablado, escrito y promocionado sobre la energía de los biocombustibles, como una energía alternativa a los combustibles tradicionales, lo cierto es que su producción actual es casi testimonial, si exceptuamos algunos países entre los que se encuentran Brasil, Argentina y EE.UU.

En el año 2010, el bioetanol tenía en España una cuota de mercado del 4% en relación a la gasolina consumida. En el mismo año, el biodiesel alcanzó una cuota de mercado del 4,29% en relación al gasóleo consumido.

También en el año 2010, el 75% de las plantas españolas de biodiesel estaban paradas por la importación de biocombustibles desde otros países (Argentina y EE.UU.).

Se está teniendo una política muy errática en el tema de los biocombustibles y de las energías alternativas en general, lo que lleva a la inseguridad de los agricultores respecto a la continuidad de la rentabilidad de los productos y a que surjan inversores de plantas de transformación.

La Unión Europea también intenta su lanzamiento con objetivos claros, pero que se van diluyendo en el tiempo sin conseguir llegar a cotas de cierta importancia.

11.8. Incorporación de biocombustibles a los carburantes tradicionales

Hasta el mes de febrero de 2011 se incorporaba un 5% de biocombustibles a los combustibles de origen fósil (gasolina y gasoil). A partir de esta fecha, pasan a incorporar un 7% como consecuencia de la crisis en los países árabes del norte de África, y de forma especial, en Libia, quedando como sigue:

- Bioetanol en gasolina, 7%.
- Biodiesel en gasoil, 7%.

11.9. Potencia calorífica de un combustible, conocida su composición química

La potencia de un combustible depende de su composición química y su poder calorífico se calcula mediante la fórmula siguiente:

$$Pc = 8.080 \cdot C + 34.462\ (H - \frac{O}{8}) + 2.240 \cdot S\ \text{(cal/kg)}$$

Pc – Potencia calorífica, en calorías por kg (cal/kg) de combustible.
C – % de carbono en el combustible.
H – % de hidrógeno en el combustible.
O – % de oxígeno en el combustible.
S – % de azufre en el combustible.

11.10. Petróleo a partir de algas

En los últimos años se están realizando experimentos para la obtención de petróleo a partir de algas. En España hay un centro experimental que está situado en la provincia de Alicante. Otros países también están desarrollando este proceso o similares.

El biopetróleo producido a partir de microalgas vendría a sustituir al biodiésel que es un biocombustible, que como se ha indicado, es una materia orgánica utilizable como fuente de energía y que se obtiene a partir de lípidos naturales como aceites vegetales o grasas animales, con o sin uso previo, mediante procesos industriales y que se aplican en la preparación de sustitutos totales o parciales de productos obtenidos del petróleo. Como materia base para su obtención utilizan productos agrícolas, algunos de primera necesidad, que dan lugar a que se tengan dificultades en el suministro regular, dependiendo de las cosechas y de la especulación.

El empleo del petróleo proveniente de algas reduciría el consumo de productos agrícolas para fabricar biocombustibles, que se derivan del consumo animal o humano, dando lugar a su encarecimiento y a que pueda darse escasez de los mismos.

Respecto al petróleo obtenido a partir de algas, también se puede refinar y convertir en combustible para vehículos de transporte, como gasolina, diésel, etanol y biodiesel, que son totalmente compatibles con los motores existentes. Por otro lado, el biodiesel supone un ahorro de entre un 25% a un 80% de las emisiones de CO_2 producidas por los combustibles derivados del petróleo, constituyendo así un elemento importante para disminuir los gases invernadero producidos por el transporte.

Las algas necesitan tres componentes esenciales para su desarrollo: luz, CO_2[(2)] y agua. Gracias a que algunas especies de algas contienen un alto contenido en grasas resultan ideales para la producción de biodiesel.

Se cultivan en balsas, tubos o canales de escasa profundidad para permitir una mayor entrada de la luz. En su interior se mantiene un flujo y temperaturas constantes, y se inyecta CO_2 y nutrientes. Una vez desarrolladas en cuestión de horas, se extraen de su medio de crecimiento mediante un adecuado proceso de separación y se obtiene el aceite sin necesidad de secarlas de antemano. Las algas y el agua se vuelven a reutilizar.

Existen algas capaces de producir 130.000 litros de biodiesel por hectárea, mientras que si se cultivase la misma superficie con girasol, solo se obtendrían 500 litros. Los números hablan por sí mismos.

Resumiendo:

El petróleo obtenido de algas aportaría las siguientes ventajes:

- Reduciría el consumo de productos derivados del petróleo.
- Se reduciría sensiblemente las emisiones de CO_2.
- En el proceso de elaboración del petróleo de algas, se consume CO_2.

---

2 Una de las ventajas de este producto es que tiene la posibilidad de consumir CO_2 en su proceso de transformación, y además, que como combustible es mucho menos contaminante que los derivados del petróleo como combustibles.

- Se mejoraría de una forma apreciable el llamado "efecto invernadero" debido al transporte.
- Se reduciría la dependencia energética respecto a terceros países.
- Se abarataría el coste de la energía.
- Estaríamos hablando de una energía alternativa.

Hay que darnos un tiempo de espera, para ver si surge una alternativa real a las energías que hoy día utilizamos, algunas de ellas muy contaminantes y cada vez más escasas, por que no se renuevan.

12 Otras energías alternativas

Las energías que se obtienen del calor de la tierra y de la fuerza del mar tienen su importancia, habida cuenta, que toda energía que podamos obtener en nuestro propio territorio, tienen una importancia considerable. En el momento actual, estas energías son poco significativas, pero deben ser consideradas.
En este capítulo, se repasan las energías geotérmica, mareomotriz, undimotriz y de gradiente térmico. Todas estas energías requieren unas condiciones especiales del terreno (orografía) y del mar, para que pueda recuperarse la energía que almacenan de una forma rentable.

El mar.

12.1. Energía geotérmica

Se trata de una energía térmica que se encuentra bajo la superficie terrestre y que podemos aprovechar para diferentes fines.

La energía geotérmica es posible que tenga su origen en la descomposición de los isótopos radioactivos presentes en las zonas interiores de la Tierra, que al degradarse liberan gran cantidad de energía. Esta energía térmica provoca la fusión de rocas y el calentamiento de las aguas, etc.

12.1.1. Pequeña reseña histórica

El primer aprovechamiento geotérmico a nivel industrial para generar energía eléctrica se produjo en Larderello (Italia) el año 1904 y después de un siglo todavía sigue en servicio.

Islandia, por sus condiciones particulares, realiza un aprovechamiento especial de esta forma de energía que supone el 60% del total de la energía que consume.

Hay instalaciones puntuales para el aprovechamiento de esta energía. Son pocas, ya que resulta difícil encontrar zonas en que se den las condiciones idóneas para su aprovechamiento rentable, que además supone inversiones de dinero muy elevadas. Desde la antigüedad se vienen utilizando las aguas termales con fines medicinales, baños y termas. Las aguas afloran calientes al exterior de la tierra, después de que hayan pasado por zonas interiores con elevada temperatura.

12.2.2. La Tierra, fuente de calor

Nuestro planeta tiene mucho calor en su interior. El planeta Tierra se formó hace 4.600 millones de años y en su fase de formación se concentró mucho calor, que se encuentra en su interior. La Tierra inició la liberación de calor en el momento que completó su formación. Con el paso del tiempo, la temperatura en la superficie terrestre permitió que hubiera vida. El calor disipado se envía al Universo.

La densidad de la Tierra es de 5.515 kg/m^3.

12.2.3. Capas terrestres

Atendiendo a la composición y temperatura de la Tierra, se divide en: corteza, manto superior, manto inferior, núcleo exterior y núcleo interior. La tabla 12.1 recoge las temperaturas y espesor de las diferentes capas.

Tabla 12.1. Características de las diferentes capas terrestres.

Capas terrestres	Características
Corteza	· Capa exterior e irregular del planeta Tierra. · Grosor: entre 7 y 70 km. Materiales predominantes: silicio, magnesio, aluminio, etc.
Manto superior	· Grosor: entre 7 y 400 km
Manto inferior	· Grosor: entre 400 y 2.900 km · Temperatura: 1.000 a 3.000 °C
Núcleo exterior	· Grosor: entre 2.900 y 5.100 km · Temperatura: 4.000 °C
Núcleo interior	· Grosor: entre 5.100 y 6.371 km · Temperatura: 5.000 °C

12.2.4. Gradiente térmico

A medida que nos alejamos de la superficie terrestre hacia su interior, va subiendo la temperatura de acuerdo con los datos siguientes:

Se entiende por gradiente geotérmico al número de metros que es necesario "bajar" para que la temperatura aumente un grado.

- El gradiente geotérmico medio, para la corteza terrestre es de 1 °C por cada 33 m.
- El gradiente geotérmico mínimo es de 1 °C por cada 100m.
- El gradiente geotérmico máximo es de 1 °C por cada 11 m.

Por lo que apreciamos, el gradiente geotérmico es irregular y depende de la zona.

12.2.5. Clasificación de las fuentes de energía geotérmica

Cuatro niveles de obtención de la energía geotérmica atendiendo a su temperatura:

a) Energía geotérmica de alta temperatura

La temperatura está comprendida entre 150 y 400 °C.
Se obtiene en zonas activas de la corteza terrestre.
Se utiliza preferentemente para generar electricidad.

b) Energía geotérmica de media temperatura

La temperatura está comprendida entre 70 y 150 °C.
Se utiliza como fluido de calefacción.

c) Energía geotérmica de baja temperatura

La temperatura está comprendida entre 50 y 70 °C.
Se utiliza para usos domésticos, urbanos y agrícolas.

d) Energía geotérmica de muy baja temperatura

La temperatura está comprendida entre 20 y 50 °C.
Se utiliza para usos domésticos, urbanos y agrícolas.

12.2.6. Ventajas e inconvenientes de la energía geotérmica

a) Ventajas

- Es una fuente de energía que está ahí, aunque hay que considerarla con fecha de caducidad, aunque sea a muy largo plazo (millones de años)
- Se la puede considerar como una energía renovable.
- Su aprovechamiento supone una economía real en la factura energética.
- Los recursos geotérmicos son abundantes, aunque no fáciles.
- Reduce la emisión de gases de efecto invernadero.
- Su recuperación tiene un impacto medioambiental menor que el de otras energías (hidráulica, térmica, nuclear, eólica, fotovoltaica).
- No genera ruidos.

b) Inconvenientes

- Es una energía que tiene fin, aunque sea a muy largo plazo.
- Las instalaciones de recuperación de esta energía son complicadas y costosas.
- Puede tener un impacto medioambiental por el volumen de las instalaciones, especialmente en tuberías.
- No tiene posibilidad de extracción en todos los lugares. Las zonas deben tener unas condiciones especiales del terreno.
- Puede dar lugar a contaminación de aguas por arrastre de arsénico y amoníaco. También puede dar lugar a la emisión de ácido sulfhídrico que se detecta por su olor característico a huevos podridos, y que en grandes cantidades no se percibe y puede ser mortal.
- La energía térmica recuperada no tiene posibilidad de ser transportada a distancia. Debe aplicarse en la zona.
- La contaminación en este caso es térmica.
- Problemas de corrosión en los materiales de sondeo.
- No es una fuente de energía importante a nivel general, lo es puntual donde se dan las condiciones favorables, como es el caso de Islandia.

12.2.7. Aprovechamiento de la energía geotérmica

- Generar electricidad.
- Generar agua caliente para calefacción.
- Generar agua caliente sanitaria (ACS) para usos domésticos, servicios e industria.

El valor total de la energía geotérmica recuperada para generar electricidad no llega a los 10.000 MW de los que EE.UU. genera casi 3.000 MW.

Las aguas obtenidas entre 10 y 130 °C se aprovechan en:

- Viviendas, servicios e industria para suministrar agua caliente sanitaria.
- En balnearios (aguas termales).
- Para cultivos de invernadero, especialmente en épocas frías del año.
- Para diversos usos industriales.
- Para calefacción de viviendas en ciudades.
- Para calentar el agua en piscifactorías. Se acorta el tiempo de cría de pescados y crustáceos.

12.2.8. Ejemplo de instalación para generar energía eléctrica.

La figura 12.1 muestra el principio en que se basa una generación eléctrica que aprovecha la energía geotérmica.

Este tipo de instalaciones se instalan en la tierra hasta la profundidad calculada, dos tuberías a la distancia que convenga. Por una tubería se inyecta agua a una zona con suficiente calor para que se convierta en vapor de agua a la mayor presión posible, para recogerlo por la segunda tubería que lo conduce al exterior hacia la turbina de vapor transformándolo en energía mecánica rotativa con la que se mueve un alternador que genera corriente eléctrica.

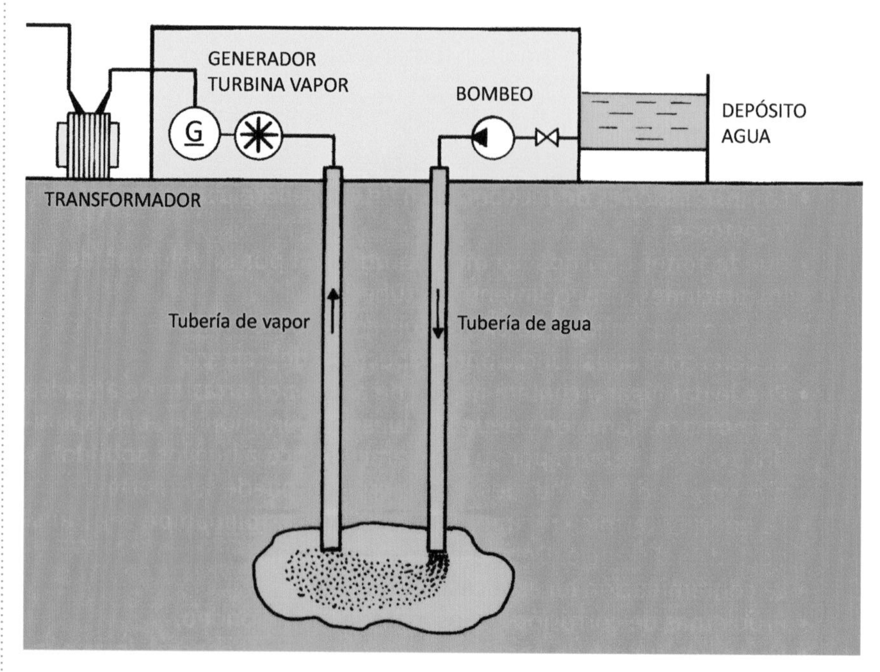

Figura 12.1. Esquema de una instalación para generar electricidad aprovechando energía geotérmica.

12.2. Energía mareomotriz

La energía mareomotriz es de reciente recuperación, dado que se necesitan grandes obras civiles para realizar las presas de retención y las instalaciones con turbinas y alternadores eléctricos. Además, no siempre es fácil encontrar el emplazamiento adecuado que permita edificar una instalación con rendimiento adecuado. Normalmente suelen ser las desembocaduras de los ríos los que mejores condiciones presentan, siempre que haya buenas mareas.

12.2.1. Las mareas

Las mareas se deben a la atracción gravitatoria de la Luna y en menor proporción del Sol, sobre las grandes masas de agua de los mares.

La energía mareomotriz aprovecha las mareas para recuperar energía del movimiento de grandes masas de agua, en las que se producen desniveles en la superficie del mar. Las mareas se dan en las costas y no son iguales en todos los mares y situaciones geográficas. En altamar, la amplitud de las mareas no llega a un metro. La parte de mar que está en el ecuador terrestre, tienen mareas muy débiles.

12.2.2. Fases de las mareas

Las mareas se ajustan a un horario que se va moviendo como consecuencia de la diferencia entre el día de la Tierra (24 horas) y el día en la Luna (23 horas, 50 minutos y 28 segundos). En cada ciclo o día lunar se producen dos mareas altas y dos mareas bajas.

- Marea alta o pleamar. El agua alcanza su máxima altura.
- Marea baja o bajamar. El agua alcanza el nivel más bajo.

El tiempo entre pleamar y bajamar es de casi 6 horas. Completando un ciclo en 23h, 50 min y 28 s.

- Flujo. Agua que asciende en pleamar.
- Reflujo. Agua que desciende en bajamar.

12.2.3. Ventajas e inconvenientes de la energía mareomotriz

a) Ventajas

- Es una energía renovable.
- La energía base es gratuita.
- Procedimiento silencioso.
- Tiene un proceso no contaminante.
- Tiene bastantes posibilidades de recuperar esta energía para transformarla en electricidad.

b) Inconvenientes

- No es posible su recuperación en todos los mares.
- Desembolso inicial muy importante.
- Localizaciones puntuales en las que se dan condiciones favorables.
- La posibilidad de recuperación depende de la amplitud de las mareas.
- Puede haber impacto visual y estructural sobre el paisaje costero.
- Efecto negativo en la flora y la fauna marina.
- No en todos los lugares es posible el almacenamiento de agua. Se suele utilizar las desembocaduras de los ríos.

12.2.4. Aplicaciones de la energía mareomotriz

La aplicación principal de la energía mareomotriz está en la generación de energía eléctrica, con centrales cuyos alternadores están accionados por la fuerza del agua a través de turbinas, que transforman la energía del agua en energía mecánica rotativa.

Estas centrales no son de funcionamiento continuo pero aprovechan el flujo y el reflujo para generar electricidad.

En 1967 en el estuario del río Rance, la Compañía Francesa de Electricidad (EDF) construyó una central eléctrica que aprovechaba la fuerza de las mareas. En esta zona las mareas son muy intensas, hasta casi 12 m.

Las mareas pueden llegar a alcanzar los 8 y 13 m de desnivel. En estos lugares, la rentabilidad está asegurada si la zona es adecuada.

12.2.5. Otras formas de recuperar la energía del mar

Encontramos energía en diferentes accidentes del mar, como son:

- Las mareas (mareomotriz)
- Las olas (undimotriz).
- Las corrientes marinas.
- La diferencia de temperatura entre masas de agua a diferentes profundidades.

12.2.6. Ejemplo de instalación mareomotriz

La figura 12.2 muestra el esquema de una instalación para el aprovechamiento de energía mareomotriz y producir electricidad.

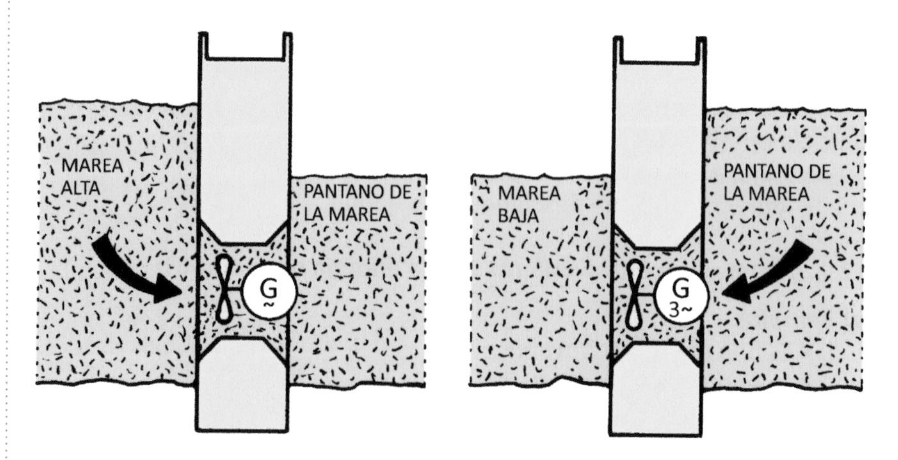

Figura 12.2. Esquema de principio de funcionamiento de una central eléctrica mareomotriz.

12.3. Energía undimotriz

Se entiende por energía undimotriz a la energía producida por el movimiento y la fuerza de las olas. Las olas se desplazan a grandes distancias sin apenas pérdida de energía. Esta energía se aprovecha preferentemente en la generación de electricidad, aplicando diversos sistemas, como:

- Columna oscilante de agua.
- Sistemas totalizadores.
- Sistemas basculantes.

- Sistemas hidráulicos.
- Sistemas de bombeo.

Muchos países tienen costas y las podrían aprovechar para generar electricidad, sin embargo, no es fácil obtener un sistema eficaz y relativamente fácil de instalar.

12.4. Energía de gradiente térmico oceánico

El agua del más se encuentra a diferentes temperaturas, más elevada en superficie y relativamente fría a grandes profundidades. El mar se calienta por la acción del calor del Sol.

Se entiende por gradiente térmico a la diferencia en grados centígrado (°C) que hay entre la temperatura en superficie y a una cota de profundidad determinada.

12.4.1. Temperatura del agua en función de su profundidad

La temperatura del agua en el mar varía en función a su profundidad, tal como de indica a continuación:

a) Aguas superficiales

Entre 0 y 200 m de profundidad.

La temperatura del agua del mar es variable en función del lugar geográfico en que se encuentre. La temperatura superficial del agua es mayor en zonas del ecuador y subtrópico.

b) Aguas intermedias

Entre 200 y 400 m de profundidad.

Las aguas intermedias actúan como si de una barrera térmica se tratara entre las aguas superficiales y profundas.

c) Aguas profundas

A partir de 400 m de profundidad

- A 1.000 m de profundidad, el agua puede estar a 4 °C de temperatura.
- A 5.000 m de profundidad, el agua puede estar a 2 °C de temperatura.

12.4.2. Aprovechamiento del gradiente térmico oceánico

El funcionamiento de una instalación para la recuperación de la energía térmica del mar (gradiente térmico) es necesario un intercambiador de calor que utilizando el calor del agua del mar (mínimo 20 °C), transformar en vapor un líquido de bajo punto de ebullición, con el que mover una turbina y esta un alternador que genera corriente eléctrica. El vapor una vez que ha pasado por la turbina hay que enfriarlo, lo que se consigue con el agua fría tomada de la profundidad del mar. Una vez licuado se vuelve a iniciar el ciclo, muy similar al de un frigorífico.

También es posible el aprovechamiento del agua a gran temperatura que emerge de los fondos marinos (fuentes hidrotermales) cuya temperatura puede llegar a los 400 °C.

Aunque en teoría es posible esta recuperación y posterior transformación en energía eléctrica, en la práctica no resulta ni tan fácil, ni tan rentable.

12.5. Problemas en el aprovechamiento de estas energías

Aunque se trata de energías que están a nuestro alcance, tienen el problema de que resulta muy difícil su aprovechamiento, tanto por las características de las instalaciones (grandes obras civiles) como por las dificultades que impone el medio (grandes masas de agua, temporales, golpes de mar, etc.).

Las inversiones de dinero para financiar estas instalaciones son muy elevadas y la energía eléctrica generada en algunos casos no suele compensar la inversión, teniendo el problema añadido de que el mantenimiento de estas instalaciones es dificultoso y caro.

En este momento, estas fuentes de energía no tienen un peso importante en el balance global de las energías renovables, por lo que están consideradas como unas fuentes de energía en espera de que puedan ser aprovechadas en situaciones más favorables que las actuales..